W9-BSV-089

Engineering and the Environment

Engineering and the Environment

Martin P. Wanielista
Yousef A. Yousef
James S. Taylor
C. David Cooper

UNIVERSITY OF CENTRAL FLORIDA

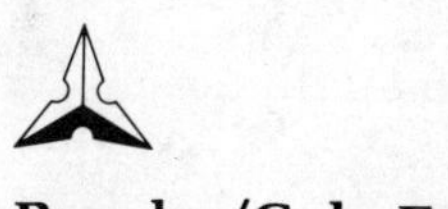

Brooks/Cole Engineering Division
MONTEREY, CALIFORNIA 93940

Brooks/Cole Engineering Division
A Division of Wadsworth, Inc.

Printed in the United States of America

10 9 8 7 6 5 4 3 2 1

Library of Congress Cataloging in Publication Data

Engineering and the environment.

Includes bibliographical references and index.
1. Environmental engineering. I. Wanielista, Martin P.
TA170.E54 1984 628 83-24075
ISBN 0-534-03043-2

ISBN 0-534-03043-2

Sponsoring Editor: Ray Kingman
Production Services Coordinator: Bill Murdock
Production: Ron Newcomer & Associates, San Francisco
Manuscript Editor: Andrew Alden
Interior Design: Jamie Brooks
Cover Design & Illustration: Sharon Smith
Illustrations: Irene Imfeld
Composition: Syntax International, Singapore
Printing & Binding: Halliday Lithograph

About The Authors

Martin P. Wanielista, Professor of Engineering and Chairman of the Civil Engineering and Environmental Sciences Department at the University of Central Florida, is a graduate of the University of Detroit, Manhattan College, and Cornell University. At the University of Central Florida, Dr. Wanielista teaches hydrology, hydraulics and environmental engineering. He has published three other textbooks, and his work has resulted in over 100 publications. His major research interests are in stormwater management and application of hydraulics and hydrology principles to general environmental engineering problems. He is a registered Professional Engineer in the state of Florida. For continuing service in education and research, he has received 18 awards.

Yousef A. Yousef, Professor of Engineering in the Department of Civil Engineering and Environmental Sciences at the University of Central Florida, is a graduate of the University of Texas at Austin. At the University of Central Florida, Dr. Yousef teaches physical, chemical, and biological treatment systems in environmental engineering. His major research interests are water quality and the fate of pollutants in natural systems. He has published many articles and edited several proceedings for national conferences. He is a registered Professional Engineer in the states of Florida and Texas.

James S. Taylor, Associate Professor in the Department of Civil Engineering and Environmental Sciences at the University of Central Florida, teaches courses in water chemistry and potable water treatment and has done research in trace contaminant removal from potable water and stormwater. He is a registered Professional Engineer and a member of several professional groups, having been chairman of the Florida Section of the American Water Works Association. He has published many articles in his research areas, contributed to textbooks on water treatment, and is currently engaged in research involving trace organic removal from drinking water.

C. David Cooper is an Associate Professor in the Civil Engineering and Environmental Sciences Department at the University of Central Florida. His teaching interests lie in the general areas of air pollution control and industrial waste treatment. He has degrees in Chemical Engineering from Clemson University and Rice University. Dr. Cooper entered the academic world in 1976, after six years with a major oil company in one of their large refineries. He is a registered Professional Engineer in the states of Florida and Texas. He has authored several papers and has worked as a consultant for other consultants and for process industries.

Preface

The nature of engineering has changed significantly in the past two decades. Today, more than ever, engineering is an interdisciplinary effort. It is important for all engineers to have some knowledge of all disciplines. A relatively recent branch of engineering, knowledge of which is needed and desired by many people today, is that of environmental engineering. Rarely can one pick up the newspaper without reading of some environmental impact, such as a hazardous waste dump site that has been uncovered, a sewage treatment need, or an accidental release of a toxic chemical. Another important interdisciplinary field of knowledge is that of energy. Everyone in the United States is greatly affected by energy usage from both an economic and an environmental point of view. This book, therefore, has been developed to address these interdisciplinary needs in a quantitative manner suitable for a sophomore or junior level engineering course.

This book has two main objectives. The first is to introduce concepts of materials and energy balances as applied to environmental unit processes. We want to drill the principles of making materials and energy balances into the student's way of thinking. This approach, which traditionally has remained in the realm of chemical engineers, has proved most useful to all engineers in clarifying their thinking concerning the analysis of any process. The second major objective is to introduce some basic concepts of environmental engineering to all engineering students. All branches of engineering sooner or later meet and affect the environment, and if knowledge is present about what some of the possible consequences are, then perhaps a better design can be produced.

This book fills a gap in engineering textbooks at this time. In the early 1970s, several informative environmental texts were published for students of all majors. However, because of the need to pass on a wide variety of information concerning a broad area to a broad audience, it was difficult to be quantitative. In addition to being informative in this book, we have tried to be quantitative. We believe that this approach is rewarded as students gain competence in analyzing the flow of materials and energy and are exposed to environmental issues from a quantitative viewpoint instead of a purely rhetorical one.

This book is organized as follows. First, the basic concepts are presented in two chapters. Next, a detailed chapter is presented on materials and energy balances. Then come two comprehensive chapters on chemical and biological systems, oriented to applying the principles of materials balance to the solution of engineering problems. The remaining eight chapters apply the concepts of the first five chapters. Water procurement, treatment, and distribution are next, followed by wastewater treatment. Discussion of energy supplies and power generation features a survey of fossil fuels; then comes a chapter on air pollution. The book next presents chapters on solid and hazardous wastes in which the concepts of sanitary landfill are introduced, the definition of hazardous waste, and the occurrence of hazardous waste in the United States is reviewed and some treatment alternatives are presented. The book ends with a discussion of environmental impact with special notes on the social, economic, and noise environments.

The scope of environmental engineering is so broad that in a course of this nature one cannot possibly hope to go into complete detail in any one textbook. This is particularly true of a textbook designed for the entry level student. In this particular book, we hope that we have struck a proper balance between detailed quantitative information and the more general introductory knowledge that is necessary for an understanding of the field.

In order to best accomplish the quantitative aspects of this course, there are certain prerequisites. We have assumed that the student before taking this course will have had courses in college chemistry, a first and second course in college physics, and at least two courses in calculus.

The text materials were reviewed by engineering students at the University of Central Florida. We also deeply appreciate the technical reviews provided by Dr. Thork Hvitved-Jacobsen of the University of Aalborg, Denmark, Dr. John Dietz of the University of Central Florida, E. Joe Middlebrooks of Clemson University, Richard B. Kapuscinski, The University of Michigan, Dr. Dragoflav Misic, California Polytechnic State University, K. Keshavan, Worcester Polytechnic Institute, and Chet A. Rock, University of Maine at Orono.

Typing of the final manuscript was the responsibility of Ms. Sharon Darling, Ms. Betsy Swayne, and Ms. Betty Wanielista. Their patience and experience were exemplary. The initial idea for a text of this nature came from Dr. Waldron "Mac" McLellon, Professor Emeritus at the University of Central Florida. His encouragement and his dedication to environmental protection are noteworthy.

Contents

3 MATERIALS AND ENERGY BALANCES 29

4 CHEMICAL SYSTEMS 69

5 BIOLOGICAL PROCESSES 113

6 WATER MANAGEMENT 143

7 WATER TREATMENT 157

8 WASTEWATER TREATMENT 203

Engineering and the Environment

1

Introduction

Should we account for all our actions? We should certainly account for our discharges of pollutants into the environment. But how can we account for all potential pollutants if we know not where they go or, of equal importance, what happens when they get there? A knowledgeable use of the principles of materials and energy balances can help answer these questions. Materials and energy balances are those basic areas of engineering science in which we utilize fragmentary data to develop more complete information, using basic principles of chemistry, physics, mathematics, and economics to help solve many of the interactive problems in a society.

Consider the simple act of driving an automobile. What interactions and what impacts result from the car? Air pollution, water pollution, energy utilization, noise, social redirection, and economic benefits will result, to name a few. Most people can define the effects qualitatively, but engineers also do it quantitatively. Furthermore, the engineer can plan, design, and operate those facilities necessary to reduce or eliminate unwanted impacts. Thus, the engineer must be responsive to society and be technically competent. Imagine a town meeting on industrial waste treatment in which the engineer is being questioned by the town council concerning the new treatment plant. At one point, a council member asks the engineer, "How many pounds of pH are being discharged?" Knowing that pH is a quantitative measure of the concentration of hydronium ions in an aqueous solution, the engineer can explain diplomatically that a better question would be "How many pounds of a chemical would be needed to neutralize the discharge of a pH that is acceptable in the environment?"

The purpose of this book is to provide introductory concepts of material and energy balances as applied to environmental processes. This approach should clarify thinking concerning the analysis of any process. Also, basic concepts of environmental engineering are developed that all engineers can use. All works of engineering sooner or later affect the environment, and proper plans, designs, and operations can minimize the impacts. Perhaps better engineered works can be devised to serve society's needs.

There are many materials that can modify human behavior such that expected functions are affected. In addition, there are materials that create unfavorable alterations of any environment, the environment being the aggregate of social and cultural conditions which influence the life of any organism. Any alteration which is unacceptable to society is considered a pollutant. Society's tolerance to a stress varies considerably given the particular environment and the particular pollutant. The determination of an unfavorable effect is not only a technical question but a social and cultural one. An alteration may be unfavorable to some who consider the risk too high, but to others the risk may be acceptable. The engineer is frequently charged with the responsibility of defining pollution and thereby protecting the public health of a community. To do this, the engineer relies on knowledge of the basic sciences such as mathematics, physics, and chemistry. Using this basic knowledge and the formal training of an engineering education, he or she can develop a model in quantitative terms that can help describe or predict tolerance conditions or pollution levels for various activities. At the start of this study, it is helpful to summarize some of the more frequently used introductory terms used in pollution studies and control. These are shown in Table 1.1. Some of the more frequently used nomenclature is shown in Appendix A.

Threshold or tolerance levels for pollutants vary with the environment and the pollutant. Would a teaspoon of arsenic in a coffee cup of water be acceptable to drink? Certainly not. But it would probably be acceptable in a large body of surface water, like Lake Erie. The tolerance levels differ depending on the use of that water. These levels are frequently expressed as parts per million (ppm) or even parts per billion (ppb). If the teaspoon of arsenic weighs one-tenth of an ounce and a bath tub holds 30 gallons, then the bath would have roughly 25 ppm of arsenic in it—not tolerable for bathing purposes.

TABLE 1.1 Definitions of some environmental engineering terms

Environment: The aggregate of social, physical, and cultural conditions that influence the life of an individual or a community: the surroundings.

Pollution: That which modifies the environment such that its use is affected.

Environmental Engineer: An engineer who applies engineering principles to the problems arising in the interaction of humans and their environment so as to ensure that ecological balances are preserved.

Technology: Any tool or technique, any product or process, any physical equipment or method of doing or making by which human capability is extended

Ecology: The study of relationships between organisms and their environments.

Ecosystem: The organisms of a locality, together with their functionally related environment, considered as a unit.

System: A group of many different connected parts that interact in a complicated manner.

Model: A description or analogy that is used to help visualize something that cannot be directly observed or to predict a future event; a statement in quantitative terms of those aspects of the situation that are relevant.

Biosphere: The earth's surrounding envelope of soil, water, and air, where life can exist.

Public health: Protection and promotion of the well-being of individuals and a community.

Environmental science: Encompasses the earth sciences, atmospheric sciences, and hydrographic sciences. Deals with physical matter, energy sources and pollution, and energy transformations.

1-1 Case studies

From history we can learn that many societies have developed products that tended to make life more rewarding or at least easier relative to the immediate time and environment. However, many technological advances are not without consequences. There are proven effects from some. As an example, the use of lead in utensils and plates by high Roman society is thought to have been the cause of physical symptoms and illness in the upper classes. Frequently, these effects occur because of a lack of knowledge of the consequences; however, in some cases it may be felt that the benefits outweigh the risks or the consequences. Here are some modern examples.

Hazardous waste

Between 1947 and 1952 near Niagara Falls in New York State, the Hooker Chemical Company placed used chemicals in steel drums and buried them in the long-disused Love Canal. It was believed that these chemicals would remain harmless. The land was then sold in 1953 to the school board for one dollar, but the board was warned by Hooker not to construct buildings on the site. However, about 1000 families built homes on or near the site. In 1977, most topsoil had been removed and the 200-plus different chemical compounds began leaking from the dumpsite. Epidemiological studies indicated apparent high incidence rates of birth defects, miscarriages, respiratory disease, kidney disorders, and various cancers. Much money and time was spent on relocating families from the site and lawsuits of over $2 billion were advanced. The Love Canal episode is an example of the need to account for the whereabouts of potentially hazardous chemicals and the need to study impacts. In reality, out of sight is not out of mind.

Greenhouse effect

As the demand for energy increases, the by-products of the processes for generating energy become of more concern. When coal, oil, natural gas, wood, or any carbon-containing fuel is burned, carbon dioxide (CO_2), water (H_2O), and heat are released to the atmosphere. Increases of CO_2 in the atmosphere are possibly correlated with increases in the average global temperature because heat that normally would return to space as infrared radiation is retained by CO_2 in the air through what has been called the greenhouse effect. As engineers and scientists learn more about the greenhouse effect, adjustments in energy generation may have to be made, or environmental pollution control activities may have to increase.

Locally, carbon monoxide also is released, and if the concentration to which people and animals are exposed becomes too high, significant harm can result. Engineers are responsible for making measurements of pollutants, identifying problems, and subsequently bringing about solutions. Knowledge of the consequences of a technological improvement is important. Frequently the consequences are both short-term and long-term in nature.

Acid rain

Effluents that contain sulfur dioxide from coal-fired plants and nitrogen oxides from industrial plants and automobiles can be converted to sulfuric acid and nitric acid in the atmosphere. These acids are then removed from the atmosphere in the form of acid precipitation. In the northeastern United States and adjacent Canada, hundreds of lakes have been acidified to the point that they contain no fish life. A materials-balance study shows that other sources of acid precursors exist, including runoff from nitrogen enhanced lands. Whatever the major source of the problem, it exists primarily because of our technological advances and ignorance of the consequences. The engineering profession can now design, construct, and operate nitrogen and sulfur effluent control systems for most industries. Is society ready to pay for them?

Fish kills

As well as from acid rain, fish also die because of oxygen starvation, temperature changes, toxic elements, and limited food supplies. Effluents from municipal and industrial treatment plants can contain toxins, nutrients, and oxygen-demanding materials. One toxin is polychlorinated biphenyls (PCBs). This class of chemicals was widely used as additives in electrical transformer fluids, inks, paints, rubber, coolants, duplicating paper, and other products. They are very resistant to biological degradation but are soluble in fats. In 1968, about 1100 Japanese came down with skin disease and liver damage after eating rice accidentally contaminated with PCBs. Methods to destroy PCBs are expensive: two known methods are incineration at high temperatures and exposure to ozone and ultraviolet light.

Oxygen-demanding materials are degraded by lower life forms which utilize much of the oxygen in the water, leaving very little for fish life. The ecology of the water will define the impacts. Nutrient discharges, for example, can stimulate plant growth, some of which utilize oxygen at night, thus causing fish kills in early morning hours. In addition, different species of animals may compete for food usually eaten by fish. The "death" of Lake Erie was characterized by a deterioration in water quality that included a change in species of fish. Now, there exists a greater biomass of less sporting fish. However, the lake has shown signs of recovery in some areas since controls were instituted for wastewater plants and stormwater discharges.

Heated water from industrial cooling and power plants can cause plant and animal damage. As we shall learn later in this book, waste heat is inevitable. Some heat may increase productivity; however, the threshold of harm usually must be determined for each situation.

Thus, the effluents of our society structure can cause the destruction of fish or other aquatic life. Not only municipal and industrial point-source discharges (wastewater treatment plants) but nonpoint discharges, such as river channel modifications, stormwater runoff, groundwater infiltration, and fallout must be studied to determine the cause of the problem. Again, as engineers, we must be able to alert people to the consequences of their actions and ways they can be mitigated or lessened. Although technical solutions are available, behavioral, legal, or fiscal measures may augment them.

1-2 Units of measurement

Modern measurements and accuracy of description would be impossible without a consistent precise definition of primary and secondary quantities. The primary quantities are defined as mass, length, time, and temperature. A secondary quantity is one defined in terms of primary quantities, such as volume in cubic meters (m^3), density in kilograms per cubic meter (kg/m^3), and energy expressed as joules, which are kilogram-meters squared per second squared ($kg\text{-}m^2/s^2$). The U.S. Customary System and the International System of Units (SI) are both used. It is important for engineers to think and act with a knowledge of both systems of units. Thus, a mixture of SI and customary units is found in the text with some conversion factors included. An extensive list of conversion factors is found in Appendix B. The primary units of the customary system are the yard for length, pound-mass for mass, second for time, and degrees Fahrenheit for temperature. The primary units of the International System are shown with other common measurements in Table 1.2.

Concentrations of dissolved chemicals and toxic substances are expressed in milligrams per liter (mg/L). If concentrations are generally less than 1 mg/L, it may be more convenient to express results in micrograms per liter (μg/L); always use μg/L when concentrations are less than 0.1 mg/L. Concentrations greater than 10,000 mg/L can be expressed in percentages, 1% being equal to 10,000 mg/L when specific gravity is 1.00. In solid samples and liquid wastes of high specific gravity, a correction is made as follows:

$$\text{ppm by weight} = \frac{\text{mg/L}}{\text{sp. gr.}} \tag{1.1}$$

$$\text{percent by weight} = \frac{\text{mg/L}}{10{,}000 \times \text{sp. gr.}} \tag{1.2}$$

In such cases, if the result is given as mg/L, state specific gravity. The unit grains per gallon (gr/gal) is equal to 17.1 mg/L and may be encountered in water hardness measurements.

In the case of air standards the unit ppm is referred to as 1 milliter per cubic meter (mL/m^3), or ppm by volume. Air standards

TABLE 1.2 Some units of the international system of measurement.

Quantity	*Unit*	*Symbol*	*Equivalence*
Length	centimeter	cm	0.01 m (meter)
Mass	kilogram	kg	1000 g (gram)
Time	second	s	
Temperature	kelvin	K	0°C = 273°K (approx.)
Moles	gram-mole	g-mol	1 lb-mol = 454 g-mol
Area	square meter	m^2	
Volume	liter	L	1000 cm^3 or 0.001 m^3
Force	newton	N	1 kg-m/s^2
Energy	joule	J	1 N-m = 1 kg-m^2/s^2
Power	watt	W	
Concentration	milligram/liter	mg/L	1000 mg = 1 g and 1000 g = 1 kg

expressed in terms of ppm could be converted to μg/L or mg/m^3 using the appropriate conversion factor for various gases at different temperature and pressure derived from the ideal gas law.

In the atmosphere at a temperature of 298°K (25°C) and pressure of 1 atmosphere, the relationship between parts per million and micrograms per cubic meter is

$$\mu g/m^3 = \frac{\text{ppm} \times \text{molecular weight} \times 10^3}{24.5} \tag{1.3}$$

The denominator is the volume per mole of the substance. Thus, it has the value 22.4 at 273.2°K (0°C) and 1 atm, as 1 g-mole of an ideal gas has a volume of 22.41 at 0°C and 1 atm.

Example Problem 1.1

Compute the work (energy) when a steady force of 6 pounds is needed to move a mass a distance of 200 feet.

Solution Energy = (6 lb)(200 ft) = 1200 ft-lb

= (1200 ft-lb)(0.3048 m/ft)(4.448 N/lb)(J/N-m)

= 1628 J

Example Problem 1.2

Compute the molecular weight of H_2SO_4 in pounds per lb-mole.

Solution M = 98.08 g/g-mole (1 lb/454 g)(454 g-mole/lb-mole)

= 98.08 lb/lb-mole

Example Problem 1.3

How many pound-moles and moles of CO_2 are in 10 lb of CO_2? How many pounds of O_2?

Solution

$10 \text{ lb } CO_2(1 \text{ lb-mole}/44 \text{ lb } CO_2) = 0.227 \text{ lb-moles of } CO_2$

$0.227 \text{ lb-moles } CO_2(454 \text{ moles/lb-mole}) = 103 \text{ moles } CO_2$

$10 \text{ lb } CO_2(32 \text{ lb } O_2/44 \text{ lb } CO_2) = 7.3 \text{ lb } O_2$

Example Problem 1.4

If a person discharges 0.17 pounds of oxygen-demanding materials in 100 gallons of water per day to a receiving water body, what is the contribution in terms of milligrams per day and milligrams per liter?

Solution

$$\text{Milligrams/day} = 0.17 \text{ lb/day}(1 \text{ kg}/2.2 \text{ lb})(10^6 \text{ mg/kg}) = 77\,272 \text{ mg/day}$$

$$\text{Milligrams/liter} = 77\,272 \text{ mg/day } (1 \text{ day}/100 \text{ gal})(1 \text{ gal}/3.79 \text{ L}) \cong 200 \text{ mg/L}$$

Example Problem 1.5

Express 747 micrograms per cubic meter in terms of milligrams per liter.

Solution

$(747\ \mu g/m^3)(1 \text{ mg}/10^3\ \mu g)(1 \text{ m}^3/10^3 \text{ L}) = 7.47 \times 10^{-4} \text{ mg/L}$

In other words: $1 \text{ mg/L} = 10^6\ \mu g/m^3$

Example Problem 1.6

The average 8-hr concentration of carbon monoxide near a highway is 20 mg/m^3 at 298°K and 1 atmosphere. What is the concentration in parts per million?

Solution

Using equation (1.3):

$$(20 \text{ mg/m}^3)(10^3\ \mu g/mg) = \frac{(\text{ppm})(28)10^3}{24.5}$$

Solving, we get

$$\frac{20(24.5)}{28} = 17.5 \text{ ppm}$$

Example Problem 1.7

One gram of water occupies approximately 1 mL of volume (1 kg of water occupies about 1 L), or 1 g-mole of water has a volume of 0.018 L. Thus, show using an equation analogous to (1.3) that 1 mg/L equals 1 ppm when using water.

Solution First, calculate $\mu g/m^3$, given 1.0 ppm.

$$\mu g/m^3 = 1.0 \text{ ppm}(18 \times 10^3/0.018) = 1.0 \times 10^6 \ \mu g/m^3$$

Next, convert to mg/L.

$$1.0 \times 10^6 \ \mu g/m^3 \ (1 \text{ mg}/10^3 \ \mu g)(1 \text{ m}^3/10^3 \text{ L}) = 1.0 \text{ mg/L}$$

Therefore, 1.0 ppm = 1.0 mg/L for water.

1-3 Complexity

The complexity of a system is related to the size and technical understanding of the fragments and interactions in the system. The system can be either natural or artificial. Natural systems are perhaps the most complex because we understand little about them, and what is understood is usually very complex in terms of mathematical representation. By their nature artificial systems must be well understood, but frequently they are also mathematically complex. Thus, complexity is defined by both a qualitative understanding and a mathematical abstraction. A system is considered complex if we do not understand how or why it works, and a system is considered complex if the mathematical abstraction is difficult.

Consider the generation of electricity using coal with high sulfur content. One knows that sulfur dioxide and other compounds may be discharged to the environment in quantities that can be estimated fairly accurately. However, the inputs must be defined, including oxygen requirements for combustion. Nevertheless the situation is not complex and is well understood. But our concerns may not stop at the air discharge. Are there any water or solid waste discharges? Also, where do the sulfur compounds go and what impacts do they have? If the boundary of the problem were expanded, the complexity would grow. Does acid rainfall occur? What effects will the acid have on nearby structures, lakes, or far-away places? After the system boundary has been defined, the complexities of the problem can be better defined.

1-4 Future outlook

As the complexities of the world in which we function become more enmeshed in our social structure and the field of engineering becomes more responsive to the interdisciplinary nature of engineering problems, the engineer will be called upon to know much more about the systems involved and the impacts on each. Technology changes in energy production and environmental protection are areas where engineers have a major role.

There are many problems affecting society that the engineer must work to define and solve: energy generation, hazardous waste disposal, suspect carcinogens in drinking water, unintentional additives to food, stormwater pollution, and sewage treatment are but a few. The fundamentals presented in this chapter will enable us to approach understanding the many complex interactions of materials and energy that are so much a part of environmental engineering.

1-5 Problems

1.1 Define in your own words: pollution; public health; concentration.

1.2 Consider the operation of an automobile. What materials are consumed and what pollutants are produced? Be general and not quantitative.

1.3 If 200 kg of sulfur dioxide is measured in a power plant stack discharge each day, how many kilograms and pounds of sulfur are discharged to the environment each week?

1.4 If carbon monoxide near a highway is 30 mg/m^3 at 298°K and 1 atm, what is the concentration in parts per million (ppm)?

1.5 The average daily concentration of nitrogen dioxide in air is observed as 600 μg/m^3 at 25°C and 1 atm in an industrial building. What is the concentration in ppm?

1.6 Develop the conversion factors for

$$\frac{\text{acre-inch}}{\text{hour}} \text{ to } \frac{\text{cubic feet}}{\text{second}} \text{ to } \frac{\text{cubic meters}}{\text{second}}$$

$$\frac{\text{pounds}}{\text{gallon}} \text{ to } \frac{\text{milligrams}}{\text{liter}}$$

foot-pounds to joules

megawatts to BTU/day

1.7 Use as an example from your field of engineering a technological advance that has a potential for environmental impact. Explain the impact in terms of pollution and then discuss pollution control methods.

2

Ecosystems and Engineering

All life on earth exists within a relatively thin shell of air, water, and soil surrounding the earth. This shell is called the biosphere; in it all interactive parts must function to preserve the overall system. It is approximately 14 kilometers thick from the ocean floor to the lower atmosphere. The earth and its parts are a cybernetic system, or one which maintains its controls and adapts to various environmental conditions by responding to information fed back into the system. Engineers must have some knowledge of specific aspects of this biosphere so as not to destroy small parts of it—or endanger the whole.

Ecology is the study of living organisms within their environment. Living organisms use the feedback mechanism to adjust to their environment. The nonliving environment and its living inhabitants are together called an ecosystem. An ecosystem can be as small as a drop of water or as large as our planet. Its boundary is defined by which inputs or outputs can be measured. Boundaries are somewhat arbitrary and selected for the convenience of study: aquatic ecosystems include lakes, estuaries, oceans, and swamps; terrestrial ecosystems include grasslands, forests, and urban areas. All ecosystems are connected to make the biosphere.

Energy flows into the biosphere from the sun; some is used and some is wasted back to space. Materials on the earth are essentially constant (except for nuclear transformations, fall of meteorites, and spaceship launches, which are negligible); thus, materials are recycled throughout the biosphere. This energy pass-through and recycling of materials is necessary to maintain life. A careful accounting for materials and energy flow is aided if one constructs a balance for the system.

A materials balance is a careful accounting of fragmentary information on material flow. It is in reality a restatement of the law of conservation of mass, which is that matter is neither created nor destroyed. An energy balance is similar to a materials balance, but is a careful accounting of energy flow. We will discuss materials and energy balances in detail in Chapter 3.

2-1 Properties of ecosystems

For all practical purposes, the total amount of matter in our world is fixed, and the chemicals necessary for life functions must be cycled and recycled. A typical example is a food chain, if it includes the decomposition pathways. Also, energy is transferred via the food chain unidirectionally from the sun to plants to animals. We can trace the life of humans back to the energy supplied by plants. Thus, if plants are contaminated, this poison may be transferred to animals and eventually to humans. When energy and matter are transferred up the food chain, energy is lost or wasted; thus, the shorter the food chain, the less waste.

Natural and artificial ecosystems are complex because they have many diversified responses to stimuli of both internal and external nature. If the complete spectrum of stimulus and response is known, then this complexity could be understood. Interest in defining the complete system has arisen only within the past three or four decades.

At a fundamental level, an ecosystem can be divided into living and nonliving components, which function together for their existence. The living organisms can be further divided into producers, consumers, and decomposers. Plants are examples of producers, herbivores (such as cattle or rodents) are examples of consumers that feed directly on plants, and decomposers include fungi and bacteria. Also, animals that feed on detritus are important to materials recycling. The nonliving components of the system are further classified as chemicals and are defined by energy and material flows. These classifications are shown in Figure 2.1.

Chemical elements, of course, make up both living and nonliving forms. There are about 40 elements essential for life; the main elements are listed in Table 2.1. The first three elements, carbon, hydrogen, and

Table 2.1 Elements necessary for life and some of their functions.

Category	*Element*	*Symbol*	*Some known functions*
Major constituents	Hydrogen	H	Universally required for organic compounds of cells
	Carbon	C	
	Oxygen	O	
Macronutrients	Nitrogen	N	Essential constituents of proteins and amino acids
	Sodium	Na	Important counter-ion involved in nerve action potentials
	Magnesium	Mg	Cofactor of many enzymes, such as chlorophyll
	Phosphorus	P	Universally involved in energy transfer reactions and in nucleic acids
	Sulfur	S	Found in proteins and other important substances
	Chlorine	Cl	One of the major anions
	Potassium	K	Important counter-ion involved in nerve conduction, muscle contraction
	Calcium	Ca	Cofactor in enzymes, important constituent of membranes, and regulator of membrane activity
Some micronutrients (trace elements	Boron	B	Important in plants, probably as cofactor of enzymes
	Silicon	Si	Found abundantly in many lower forms such as diatoms
	Manganese	Mn	Cofactor of many enzymes
	Iron	Fe	Cofactor of many oxidative enzymes, such as hemoglobin
	Copper	Cu	Cofactor of many oxidative enzymes
	Zinc	Zn	Cofactor of many enzymes, such as insulin

Adapted from Anderson (1981).

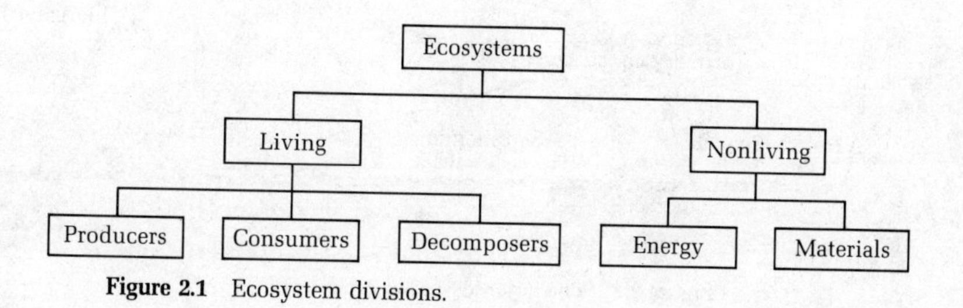

Figure 2.1 Ecosystem divisions.

oxygen, form most of the material in the biosphere. Over 75 percent of living things is water. Macronutrients include nitrogen, the fourth most abundant element in the biosphere. Animals and many plants are dependent upon nitrogen-fixing organisms for a supply of nitrogen. The final group of micronutrients are necessary in small quantities. Other elements do not have a metabolic role but may instead have an inhibitory role in living organisms. Typical inhibitory elements are lead and mercury.

Let us consider a lake as an example of an ecosystem. In this ecosystem, the wastes of some of the members becomes food for others, thus the system recycles food materials. At the same time, energy is transferred in one direction up the food chain. The food cycle of this lake is schematically shown in Figure 2.2 where it is called the aquatic animal cycle.

Within the aquatic animal cycle, materials are recycled to maintain a balanced system. The more external inputs there are, the more difficult it is to balance. For instance, urban lakes generally have more external inputs than rural lakes. The urban environment (Figure 2.3) has more people, more impervious areas producing run off, and possibly more atmospheric pollution ending in fallout on the lake or within the watershed (the area supplying water to the lake). Rural areas (Figure 2.4) generally have less runoff water and have more buffered areas of vegetation and fewer impervious areas. If external stimuli are added in the form of additional nutrients, heat, or organic matter, the

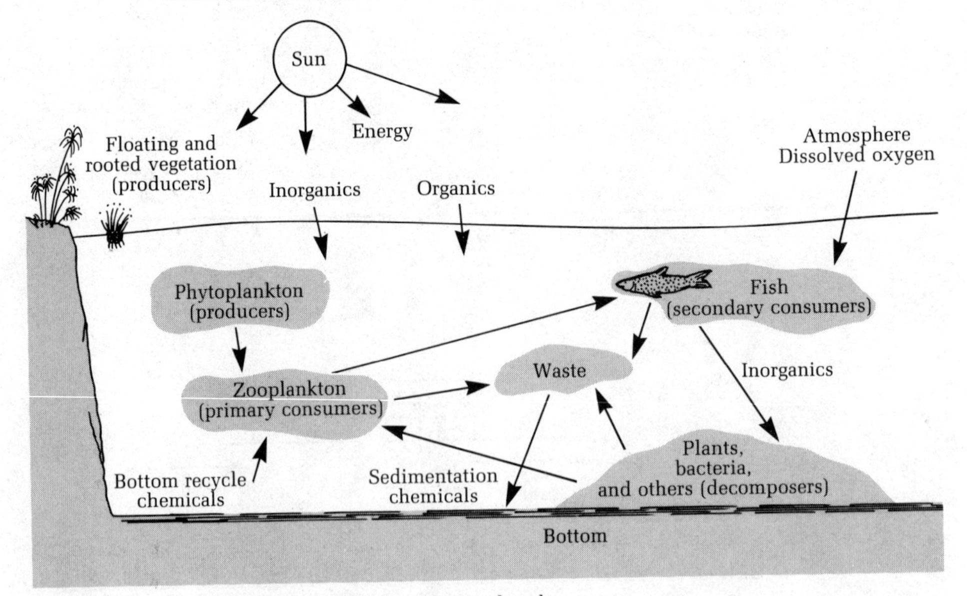

Figure 2.2 The aquatic animal cycle.

Figure 2.3 An urban lake illustrating watershed activities, such as those found on streets and park lands.

Figure 2.4 A rural lake illustrating vegetation on the shore line, which reduces runoff into the lake.

system must adjust. The bacteria will consume organics and oxygen; thus the dissolved oxygen in the water will be reduced and become less available for fish and the aerobic bacteria. Algae and phytoplankton may also multiply faster because of a rise in nutrients or temperature. Algae produce oxygen during sunlight hours. Thus, when a water body is stressed from external sources, the net amount of oxygen dissolved in the water may be in excess during the day but may be deficient at night.

Special conditions of the physical, chemical, and biological environments must be maintained. The aerobic bacteria exist because there is dissolved oxygen. Normal waste materials are quickly consumed by the aerobic bacteria, but under anoxic (little oxygen) or anaerobic (no oxygen) conditions, waste is degraded at a slower rate. As a result the system will be changed, but with a reintroduction of the original conditions, natural systems can be reestablished. An engineer learns to protect these systems and designs artificial ones.

Intensification is a process that may take place in the food chain. Autotrophs (plants) represent the first trophic (energy-using) level of a system. Next come the heterotrophs (animals). Some animals eat plants and are called primary consumers; some eat the animals that eat plants and are called secondary consumers. Tertiary consumers eat

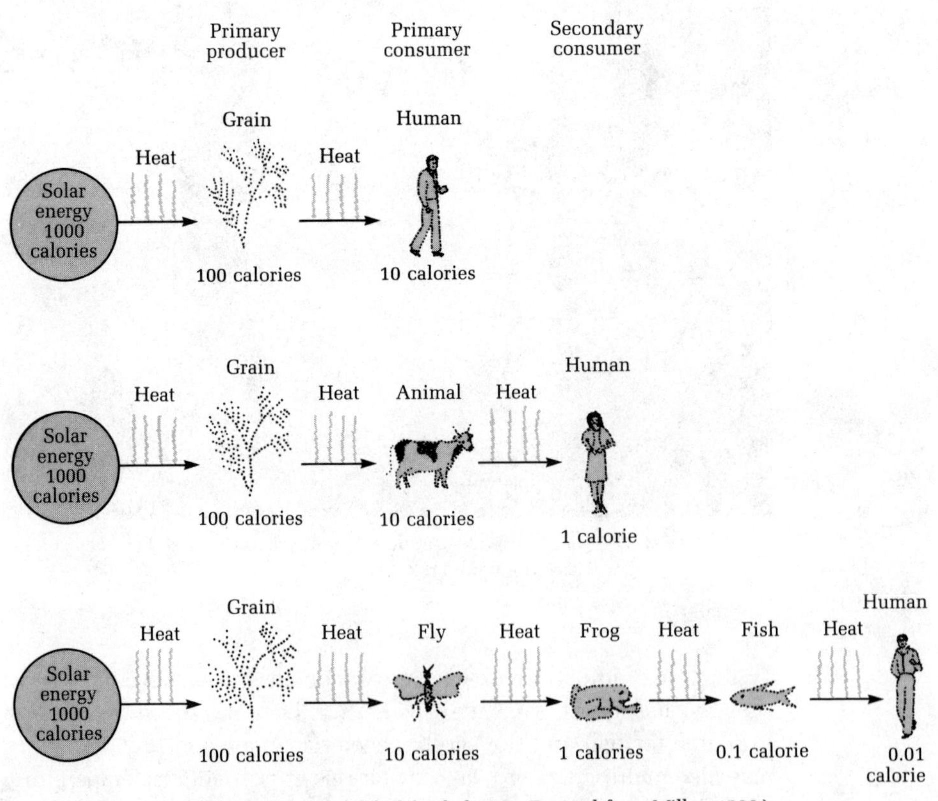

Figure 2.5 Simplified food chains. (Derived from Miller, 1982.)

the animals that eat the animals that eat plants. Some consumers eat both plants and animals. The connection between the primary producer and ultimate consumer is known as the food chain. Some simplified examples of food chains are shown in Figure 2.5. These linear food chains rarely appear in nature; for instance, many different plants provide food for many herbivores. A simplified terrestrial food web, or network of food chains, is shown in Figure 2.6. Minute quantities of toxic material may be present for the primary producers. If the substance acts as a substitute for a needed element (radioactive strontium for calcium as one example) or is one that cannot be discharged (such as heavy metals), the material may be passed on to higher forms of life and intensify in effect. An example of this is shown in Table 2.2. Materials balances for toxic materials are necessary to help define potential problems to society. Frequently an engineer can measure with accuracy one or more parts of an ecosystem, then make estimates for other parts using materials and energy balances.

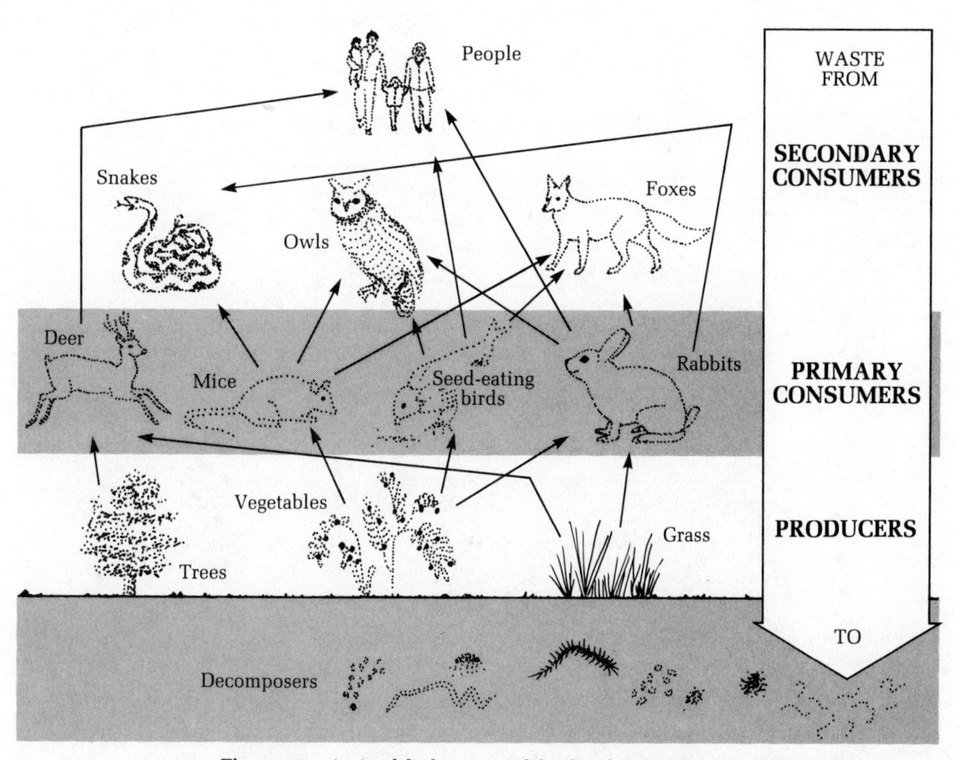

Figure 2.6 A simplified terrestrial food web. (Derived from Miller, 1982.)

TABLE 2.2 Intensification of DDT.

Medium	*Parts per million of residue*
Water	0.00005
Plankton	0.04
Silverside minnow	0.23
Pickerel (predatory fish)	1.33
Heron (feeds on small animals)	3.57
Herring gull (scavenger)	6.00
Merganser (fish-eating duck)	22.80

From Odum (1971)

Example Problem 2.1

Lead in dissolved and particulate form is discharged in stormwater runoff to an aquatic system. The particulate form settles to the bottom and is not readily available to consumers, but the dissolved form is available. Atmospheric fallout is negligible, and the mass of lead released from the bottom mud to organisms is 10% per year of the yearly particulate load. What is the yearly accumulation of lead in the biomass if the yearly runoff is 10^8 liters of water with an average lead concentration of 0.25 mg/L of which 20% is dissolved? There is no discharge from the lake, and no fishing allowed.

Solution

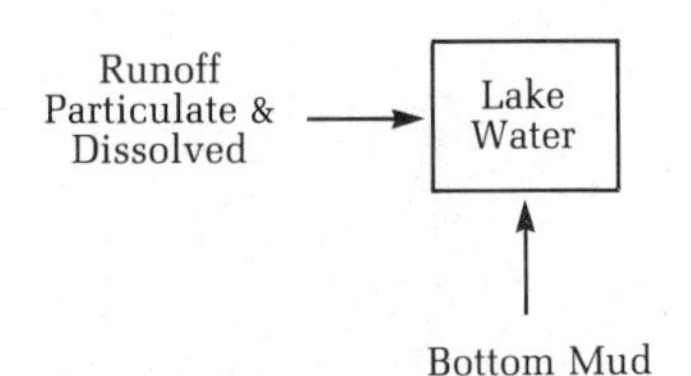

From the materials balance relation, we can write the equation

$$\text{Input} - \text{Output} = \text{Accumulation}$$

$$\text{Dissolved Lead} + \text{Recycled Particulate Lead} - \text{Output} = \text{Acc}$$

$$10^8 \text{L/yr}\,(0.20)(0.25\ \text{mg/L}) + 10^8(1 - 0.20)(0.10)(0.25) - 0 = \text{Acc}$$

$$10^8(0.25)(0.20 + 0.80(0.10)) - 0 = \text{Acc}$$

$$\text{Acc} = 0.25(10^8)(0.28) = 7 \times 10^6\ \text{mg/yr}$$

Mercury is another material that can be dangerous to organisms. Proper accounting of mercury in systems will minimize the threat of intensification or effects of spills in industrial accidents. Inorganic mercury was once believed to be nontoxic if discharged to the bottom of water bodies. However, mercury combines with methane formed in anaerobic zones to form methyl mercury, which is available to certain animals forms. In Japan, people near Minamata Bay in the late 1950s ate crabs which contained as much as 24 ppm of mercury. The kidneys of those who contracted nervous disorders and died contained as much as 144 ppm mercury.

In addition to these health problems, mercury is relatively expensive, thus the accounting of mercury within an industry is important from an economic viewpoint.

2-2 Hydrologic cycle

Water is a renewable resource in the sense that it is continually being recycled through a complex distribution system. Water is not destroyed; however, it can be made useless through careless management or polluted to the extent that it becomes unfit for its intended use. Most of the earth's water (about 97%) is found in the oceans and is too salty for most uses. However, water is distributed in the atmosphere, lakes, rivers, ice bodies, soil, and ground, and much of it can be made useful.

Water evaporates from the oceans and other water bodies into the atmosphere. Also, transpiration from plants adds water to the atmosphere. The water vapor eventually condenses and returns to earth in the form of precipitation. The materials balance for water is called the hydrologic cycle and is shown in Figure 2.7. An equation for estimating rainfall excess (that available for runoff) for a very simplified system is

$$R = P - I - I_A \tag{2.1}$$

where

R = rainfall excess

P = precipitation

I = infiltration

I_A = initial abstraction or storage

The equation can be made more complex by including terms for snow melt, spring flow, evaporation, transpiration, groundwater, and streamflow. The system also can be made more complex by including the time variability of each of the items shown in Figure 2.7.

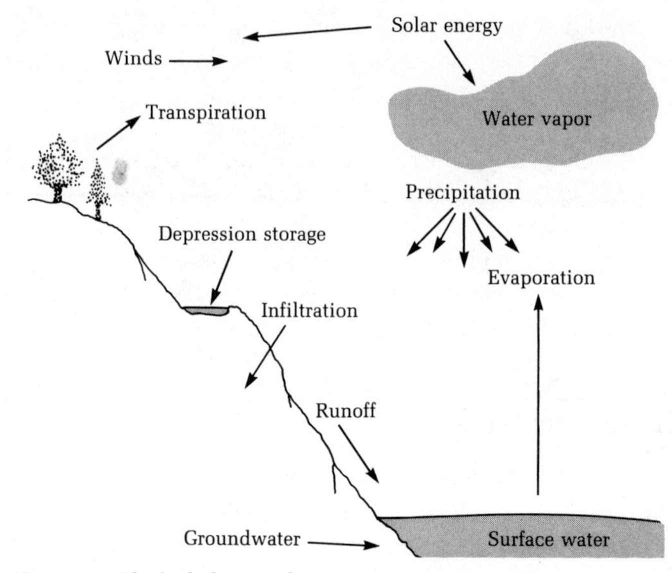

Figure 2.7 The hydrologic cycle.

The two primary sources of drinking water are surface water and groundwater. About 95% of the world's fresh water is found in the ground. Some groundwater is not suitable for direct human consumption because of pollution, mineral content, salt water intrusion, or other reasons. Other usable supplies of groundwater suffer from overpumping and thus rapid depletion. At present, pollution of groundwater is essentially irreversible because of economic considerations. Thus, pumping and pollution must be given extensive consideration along with the forces of the hydrologic cycle in order to maintain our fresh water. The quality and quantity of water are both important for human use.

Example Problem 2.2

One has to estimate the volume increase in a lake after a rainfall-runoff event. Six centimeters of rain fell over a watershed area of 2 km^2. If 40% of the rainwater infiltrates to groundwater, which does not reach the lake until a much later date, and 10% is stored in depression areas, what is the lake volume increase in liters? Direct precipitation on the lake amounts to 0.5×10^6 liters. No water leaves the lake, either by surface or ground discharge.

Solution From the materials balance,

$$\text{Input} - \text{Output} = \text{Accumulation}$$

$$\text{Runoff} - \text{None} = \text{Accumulation (neglecting direct precipitation for now)}$$

From equation (2.1),

$$\text{Precipitation} - \text{Infiltration} - \text{Storage} = \text{Accumulation}$$

$$6 \text{ cm} - 6 \text{ cm}(0.4) - 6 \text{ cm}(0.1) = 3 \text{ cm}$$

$$\text{Accumulation} = 3 \text{ cm}\left(\frac{1 \text{ m}}{100 \text{ cm}}\right)\left(2 \text{ km}^2\right)\left(\frac{10^6 \text{ m}^2}{\text{km}^2}\right)\left(\frac{10^3 \text{ L}}{\text{m}^3}\right)$$

$$\text{Accumulation} = 60 \times 10^6 \text{ L}$$

$$\begin{aligned}\text{Total Input} &= \text{Accumulation} + \text{Direct Precipitation}\\ &= 60.5 \times 10^6 \text{ L.}\end{aligned}$$

2-3 Energy

In the biosphere, biological activity is dependent on the recyling of chemicals and a supply of energy. This is evident from the elementary discussion of food chains. The abundance of energy at any time is dependent on the rotational position of the earth (that is, the time of day), the hydrologic cycle, and the abundance of nutrient minerals in the earth's crust.

Incoming solar radiation varies with the time of the year and day. About 25–30% of solar energy never reaches earth and is reflected back into space by clouds, gases, and dust. Another 25% is absorbed by the clouds and the atmosphere. Of the remaining amount that reaches the earth, some is reflected back. Snow, for example, can reflect about 80% of the energy received, while dark soil may absorb about 90%. The earth loses energy to the atmosphere by evaporation, transpiration, and wind currents. Schematically, the energy cycle is shown in Figure 2.8.

The quantity of pollutants in our atmosphere that have the potential to absorb or reflect radiation is of concern. Decreased solar radiation may reduce food production and may produce bad weather. In 1972, after nearly two decades of increases in world food production, poor weather reduced crops; for example, there was a 3 percent shortfall in the world grain harvest. From Figure 2.6, it is evident that the energy available for people decreased. In the early 1980's,

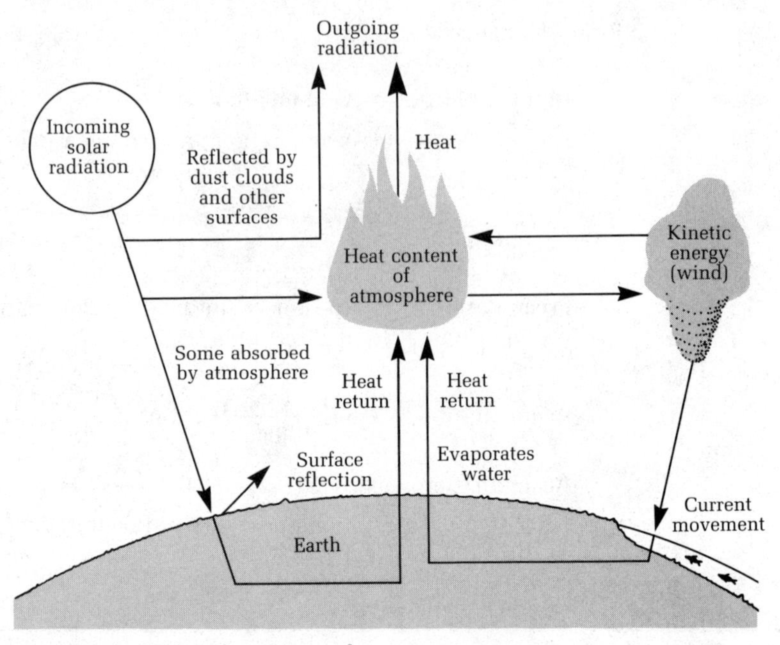

Figure 2.8 The solar energy cycle.

volcanic eruptions injected great quantities of dust and gases into the atmosphere, with yet uncertain effects.

2-4 Biogeochemical cycles

The elements listed in Table 2.1 are essential for life; thus they continuously cycle from reservoirs through food chains and chemical interactions and back to reservoirs. The cybernetic and cyclic movements of these elements are described by biogeochemical cycles. The carbon, oxygen, and nitrogen cycles have the atmosphere as the primary reservoir. The land is the primary reservoir for the phosphorus and sulfur cycles.

The carbon and oxygen cycles are closely related because carbon dioxide (CO_2) is an integral part of the respiration and photosynthesis processes. Photosynthesis, the process whereby sunlight is stored by green plants as chemically bound energy, can be expressed in equation form as

$$6CO_2 + 6H_2O \xrightarrow{\text{sunlight}} C_6H_{12}O_6 + 6O_2 \tag{2.2}$$

Carbon, hydrogen, and oxygen are converted from CO_2 and water

(H_2O) by green plants into complex organic molecules. These molecules are transferred through the food chain and provide energy and material for all forms of life. Respiration, the cellular process of burning food for energy, can be expressed in equation form as

$$C_6H_{12}O_6 + 6O_2 \longrightarrow 6CO_2 + 6H_2O + \text{Heat} \tag{2.3}$$

Another definition of respiration is inhalation of oxygen and exhalation of carbon dioxide. At each step in the food chain, both energy and material are returned to the environment. Waste material is eventually broken down by decomposers into carbon dioxide, water and other simple molecules. A small fraction of the carbon, hydrogen, and oxygen is incorporated over time into fossil fuels. Because this process takes many centuries to occur, fossil fuels are called nonrenewable.

Figure 2.9 shows the close relationship of the carbon and oxygen cycles. In the atmosphere carbon is found as carbon dioxide. This gas is formed from fuel combustion and respiration and is used by green plants in photosynthesis. Carbonates are formed in water when carbon dioxide dissolves. Some of these carbonates eventually become limestone or coral reefs. Oxygen is produced from water by green plants during photosynthesis and released. Oxygen is utilized

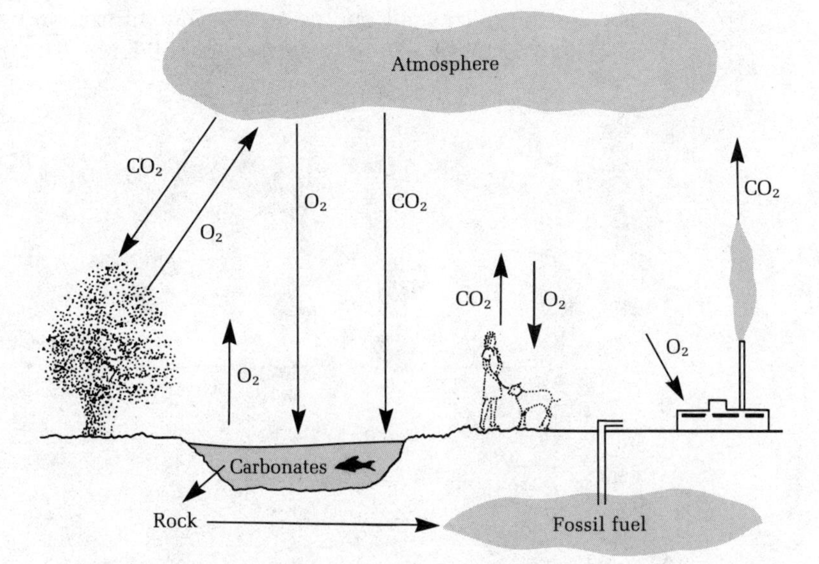

Figure 2.9 Simple carbon and oxygen cycles, showing the processes of photosynthesis and respiration with recycling.

during respiration and in the combustion of organic materials or oxidation of other materials to form oxides. It is a free gas in the atmosphere and a part of water.

Has the increased burning of fossil fuels caused an increase in CO_2 that may increase global temperatures? This question and others are related to the carbon and oxygen cycles. Further studies are necessary to help define important levels in the various transformations of both carbon and oxygen.

Nitrogen is contained in most important body molecules, such as protein, vitamins, enzymes, and hormones. It makes up approximately 78% of the atmosphere. The distribution and supply of nitrogen compounds is necessary to maintain human health. However, gaseous nitrogen (N_2) is not usable to most plants and animals. Only certain kinds of bacteria and some algae can "fix" nitrogen gas into nitrogen compounds which can eventually be incorporated into proteins. Small amounts of nitrogen also are fixed by electrical discharge during storms. Some plants (such as alfalfa and clover) stimulate nitrogen-fixing bacteria. The bacteria gather around the roots of the plants and are fed by the root secretions while they fix nitrogen. If a soil is low in nitrogen, these plants can be used to renew nitrogen supplies.

Other bacteria can convert ammonia to nitrates, which are used by most flowering plants. This process is called nitrification. Still another group of bacteria and some fungi convert nitrates into free nitrogen by denitrification. Thus the nitrogen cycle involves the removal of nitrogen gas from the atmosphere, utilization by plants and animals, and the return to the atmosphere (Figure 2.10).

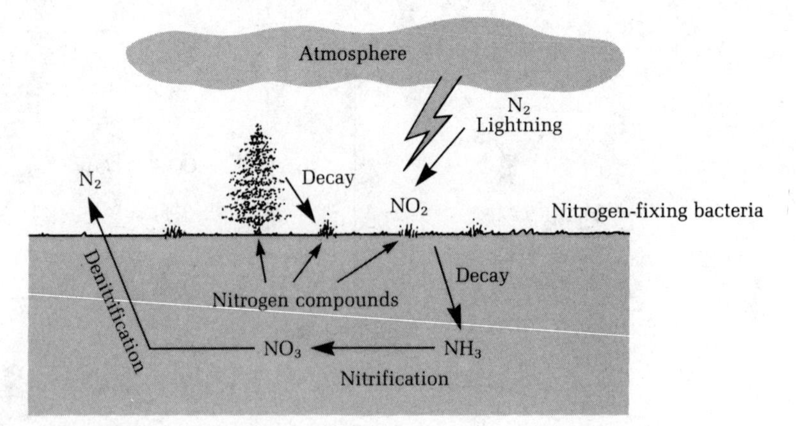

Figure 2.10 A simplified nitrogen cycle.

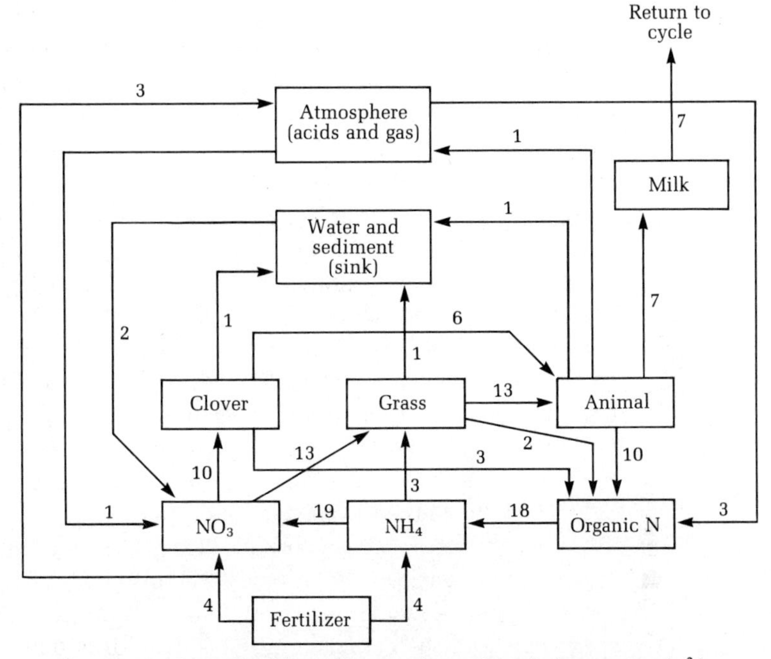

Figure 2.11 Generalized flow diagram of transfers of nitrogen, in $g/m^2/yr$, in cultivated temperate grassland grazed by diary cattle. (Derived from Deevy, 1970.)

A generalized flow diagram for nitrogen in a cultivated dairy pasture is shown in Figure 2.11. When cattle graze on plants, nitrogen is transferred to them. Some nitrogen is lost when it is tied up in the soil or leached to water bodies.

Example Problem 2.3

A farmer wishes to increase the nitrogen in the grass used to feed dairy cattle. The farmer plants clover, which will produce on the average over a year about 6 kg of nitrogen (as N or as NO_3^-) per square meter of clover. In addition, the farmer plans to flood the clover fields with runoff water containing various forms of nitrogen. If the farmer plants 1000 square meters with clover, what is the nitrogen uptake? The annual uptake of nitrogen gas (N_2) can be as high as 6000 kg ($6\ kg/m^2 \times 1000\ m^2$).

It should be noted in the previous example that the farmer is supplying runoff water, which most likely contains other elements. The farmer hopes to provide additional nutrients when runoff water is used. However, one should check the water for toxic materials that would kill the nitrogen-fixing organisms.

Some algae use nitrogen gas and are part of the carbon, oxygen, nitrogen, and phosphorus cycles. Any of these elements can control algal growth. Trace nutrients are also important and can be the limiting nutrient. Eutrophication is the nutrient enrichment of lakes. Associated with this enrichment is the excessive growth of plants and animals and depletion of dissolved oxygen. Most plans to control nutrient enrichment of water bodies address the limiting nutrient concept. The concept of the law of the minimum—that growth is limited by the scarcest essential nutrient—can also be expanded to include not only nutrients but also physical factors such as moisture, salinity, light intensity, and so forth.

Phosphorus is an important element of nucleic acids (DNA, RNA) and is a major part of bones. The major reservoir for phosphorus is phosphate rock or sediment. Plants absorb phosphorus through their roots and then animals ingest it. Sea birds that feed on marine life excrete large quantities of phosphorus. Certain large deposits of this excreta, or guano, are mined for phosphorus. Some of the phosphorus in water bodies is lost to deep sediments; however, recent evidence (Yousef and Harper, 1981) indicates some recycling of the phosphorus to the water column in lakes. Because of erosion and increased use of natural deposits for fertilizers, phosphorus may become an element in short supply, or at least the cost may increase relatively fast with time.

Civilization has affected these basic biogeochemical cycles. A relatively simple example is the burning of fossil fuels, which adds nitrogen oxides, sulfur oxides, carbon monoxide, and carbon dioxide to the atmosphere while using oxygen. The oxides are pollutants; oxygen utilization is small compared to the reservoir.

An increase in animal and human waste that reaches a water body raises the carbon, nitrogen, and phosphorus levels, which in turn increases productivity and oxygen demand and can result in a polluted condition due to excessive plant and animal growth. This kind of pollution is currently found in local environments and has regional effects where the degree of pollution is widespread. Erosion control for the conservation of soil, and urban runoff management for the reduction of toxic metals, nutrients, and organic matter must be done where pollution, eutrophic, or toxic potential exists.

Figure 2.12 Aquatic plants (hyacinths) used to remove select nutrients and other chemicals as a final stage in the treatment of domestic wastewater. The nutrients are recycled as animal feed.

Careful consideration must be given to an accounting of all the elements in relation to their cycles. If human use can be managed to allow the recycling of elements without loss of materials or excessive economic hardship, then the engineer will reconcile some of the current economic and environmental problems. An example of a possible solution for recycling wasted nutrients is the use of aquatic plants in wastewater treatment (Fig. 2.12).

2-5 Problems

2.1 Define the following:

Ecosystem
Influent and Effluent
Aerobic and Anaerobic
Eutrophication
Biosphere
Hydrologic cycle
Law of the minimum

2.2 A major input of mercury to a lake is an industry located on the shore. The industry recovers 80% of the mercury in the effluent. The remaining amount settles to the bottom of the lake. The industry recovers 10^4 kg of

Hg per year. If other sources of mercury into the lake are 10^2 kg of Hg per year. (1) what percentage of the total does the industry contribute? (2) Explain how the mercury can become available for human ingestion.

2.3 What is the rainfall excess over a 2 week period if the rainfall was 10 cm, infiltration was 3 cm, evapotranspiration was 2 cm, and initial abstraction was 0.5 cm?

2.4 Given a lake with a volume of 60×10^8 liters of water at the start of a year, what is the evaporation per year if rainfall on the lake was 6×10^8 L/yr, runoff was 80×10^8 L/yr, final yearly volume was 110×10^8 L, and net surface and groundwater discharge was 10×10^8 L/yr?

2.5 Explain in your own words, using examples not found in this text, the difference between a food chain and a food web.

2.6 Using Figure 2.5, what is the food the human should eat to minimize energy loss to the atmosphere?

2.7 Using examples not found in this text, explain how a human can be at once a primary, secondary, and tertiary consumer.

2.8 A farmer interested in raising fish on a commercial basis in a small pond is considering two species of fish:

carp, a herbivore, and largemouth bass, a carnivore.

Based on your knowledge of energy flow in a food web (chain), which species would you recommend to maximize fish biomass assuming no limits on food supply? Explain in a qualitative manner.

2.9 Define intensification and supply two examples. Why are humans vulnerable to the effects of intensification?

2.10 Discuss the effect of human activities on the nitrogen cycle.

2.11 In Figure 2.11, is there a build-up of nitrogen in soil and water environments? Explain your answer using a materials balance. Next, check the materials balance about each element (ex. grass, atmosphere) to prove there is a balance.

References

Anderson, J. M. 1981. *Ecology for Environmental Sciences: Biosphere, Ecosystems and Man.* New York: Halsted.

Deevy, E. S. 1970. Mineral cycles. *Scientific American, September,* 1970.

Miller, G. T. 1982. *Living in the Environment,* 3rd ed. Belmont, CA: Wadsworth.

Odum, E. P. 1971. *Fundamentals of Ecology.* Philadelphia: Saunders.

Yousef, Y. A. and Harper, H. H. 1981. *Inactivation of the Lake Sediment Release of Phosphorus.* EIES Project 11-1609-034, University of Central Florida, Orlando.

3

Materials and Energy Balances

3-1 Background

Materials and energy balances are key tools for engineers. The first step in understanding a process is to compile an overall materials and energy balance. This step alone may locate imbalances that may have existed unsuspected for some time; take the example of the engineer assigned to improve operations at a refinery's wastewater treatment plant. Inflow volume had increased steadily over the last month and outfall quality had deteriorated. A materials balance analysis led the engineer to suspect that the excess flow was coming from one part of the refinery. Upon investigation, he found a 2-inch fire hose emptying into the plant's industrial sewer system. It had been turned on a month before to dilute an acid spill and had never been turned off.

It might be asked why we should bother with materials and energy balances when we have sophisticated, automatic measuring instruments. Besides helping the engineer to understand in depth a process unit to which he or she has been assigned responsibility, materials and energy balances serve other important purposes. First, we may not be able to measure every flowing stream in a process unit because of cost or design. Second, we need an analytical method to check the accuracy (or at least the consistency) of the flow meters. Third, we use materials and energy balances to model a proposed change in the way an existing unit is operated. Fourth and most importantly, materials and energy balances are essential in design whether it be assessing the impact on the local environment of a proposed new plant, calculating the profit or loss expected from a proposed new chemical process, or actually sizing equipment to be purchased. It is impossible, of course, to measure parameters that do not yet exist.

3-2 Flowing streams of material

Industrial plants have at least two things in common: energy usage, and the flow of raw materials into and finished products out of the plant. A fundamental principle of engineering is that of conservation of mass: matter can neither be created nor destroyed. (While not strictly true for nuclear reactions, the principle is exact for ordinary physical and chemical processes.) Put another way, this principle tells us that if one or more streams of material are flowing into a region of space (that is, a process unit with definite boundaries), then material must be either flowing out of that region at the same rate or accumulating in the region.

How do we measure the rate of flow of streams of material? There are many different instruments to measure flow rate, based on several different properties of flowing materials. The most common property is that the flow of a fluid through an orifice results in a pressure drop across the orifice. From the Bernoulli law for a nearly incompressible fluid, we can easily show that the volumetric flow rate of the fluid is proportional to the square root of the pressure drop; that is,

$$F = k\sqrt{\Delta p} \tag{3.1}$$

where

F = volumetric flow rate
Δp = pressure drop across orifice
k = an empirical constant (may depend on orifice size, Reynolds number, and fluid conditions such as pressure and temperature)

This is the principle on which many fluid meters are based.

Regardless of whether we measure the volumetric flow rate of a gas, calculate the linear velocity of a liquid, or simply count the number of bags of a dry chemical received per hour, we must convert our material flow rates to proper units to make a materials balance. The only units that are *always* correct are units such as mass/time or moles/time. In certain special cases, it is permissible (and more convenient) to make a volumetric flow balance, but the assumption implicit in this approach must be understood: all streams must have identical, constant densities.

To aid in converting from one set of flow units to another, we often make use of the continuity equation

$$\begin{aligned} \dot{m} &= \rho u A \\ &= \rho F \end{aligned} \tag{3.2}$$

where

$\dot{m}$ = mass flow rate

ρ = fluid density

u = fluid linear velocity

A = area normal to the flow

The continuity equation is also useful for calculating linear or volumetric flow velocities at different points along a given conduit, as the mass flow rate does not change under steady conditions.

Example Problem 3.1

Water is flowing in a standard 4-inch pipe at 3.00 m/s. The pipe splits into two 2-inch pipes with an equal flow in each. Calculate the linear velocity and mass flow rate in one of the 2-inch pipes. The density of water is 1000 kg/m^3 and the inside diameters are as follows: 4 inch = 10.23 cm, 2-inch = 5.250 cm.

Solution First, calculate the mass flow rate in the 4-inch pipe:

$$\dot{m}_4 = 1000\,\frac{\text{kg}}{\text{m}^3}\,3.00\,\frac{\text{m}}{\text{s}} \times \frac{\pi}{4}\,(10.23)^2\ \text{cm}^2 \times \frac{1\ \text{m}^2}{(100)^2\ \text{cm}^2}$$

$$= 24.66\ \text{kg/s}$$

Exactly half of this flow goes into each 2-inch pipe. The volumetric flow is

$$F_2 = 12.33\,\frac{\text{kg}}{\text{s}} \times \frac{1}{1000\ \text{kg/m}^3}$$

$$= 0.01233\ \text{m}^3/\text{s}$$

and the linear velocity is

$$u_2 = 0.01233\ \text{m}^3/\text{s}\,\frac{1}{\pi(5.25)^2/4(100)^2}\ \text{m}^2$$

$$= 5.70\ \text{m/s}$$

Note the increase in velocity of the fluid even though the mass flow rate is constant.

3-3 The materials balance equation

A materials balance is simply an accounting of all materials into and out of an identifiable process area (see Figure 3.1).

Consider first an overall mass balance. We want to account for the total mass flow rates regardless of chemical type. Since mass can neither be created nor destroyed, we can say with certainty that *whatever flows into the process area must either flow out of it or accumulate within it.* In other words, the rate of accumulation of mass within the designated process area is equal to the inflow rate minus the outflow rate. Mathematically, we write

$$\frac{dM}{dt} = \dot{m}_{\text{in}} - \dot{m}_{\text{out}} \tag{3.3}$$

where

M = total mass within the boundaries of our system

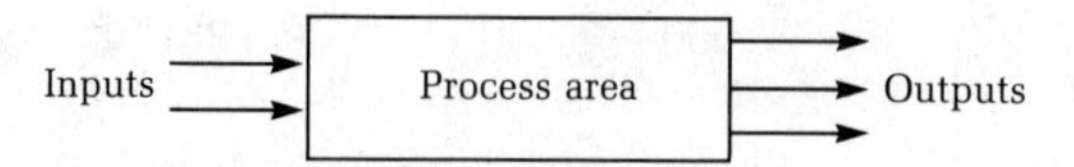

Figure 3.1 A materials balance diagram.

Usually several different chemical components are flowing through a process. The process may use a chemical reaction to change a raw material into a more valuable product, or it may be designed to use chemical or biological reactions to change harmful waste products into less harmful ones. An independent materials balance equation can be written for each component, or for all components but one, if an overall balance is also made. In the general case, a materials balance equation has the following form for component i:

$$\begin{pmatrix}\text{Accumulation}\\ \text{Rate}\end{pmatrix}_i = \begin{pmatrix}\text{Input}\\ \text{Rate}\end{pmatrix}_i - \begin{pmatrix}\text{Output}\\ \text{Rate}\end{pmatrix}_i + \begin{pmatrix}\text{Generation}\\ \text{Rate}\end{pmatrix}_i \tag{3.4}$$

Usually, we are interested in steady-state operations, which means zero accumulation. And if there are no chemical reactions generating (or destroying) component i, the balance simply becomes

$$(\text{Input Rate})_i = (\text{Output Rate})_i \tag{3.5}$$

3-4 Choosing the basis and the boundaries

A key step in making a material balance is picturing the region in space that is to be analyzed. Next the boundaries across which mass is flowing must be defined. Process engineers do this by drawing simplified flow diagrams. Blocks represent reactors, tanks, separation columns, and so forth and lines with arrows indicate material flow. For example, Figure 3.2 could be used to represent a process in which a mixed petroleum stream is catalytically desulfurized using hydrogen and then separated into products.

In Figure 3.2, the dotted line represents the boundaries for materials balance. In this case, it represents a whole processing unit rather than a reactor or a separations tower. Commonly we may draw up several material balances on a single unit and its individual equipment to better understand the process.

Since the material is flowing we may choose as a basis either a unit of time or a unit of material. The most general approach is to use the equations as written and solve in terms of flow rates. However, it may be more convenient to pick a specified time interval or amount of material as the basis and solve in terms of mass of material only.

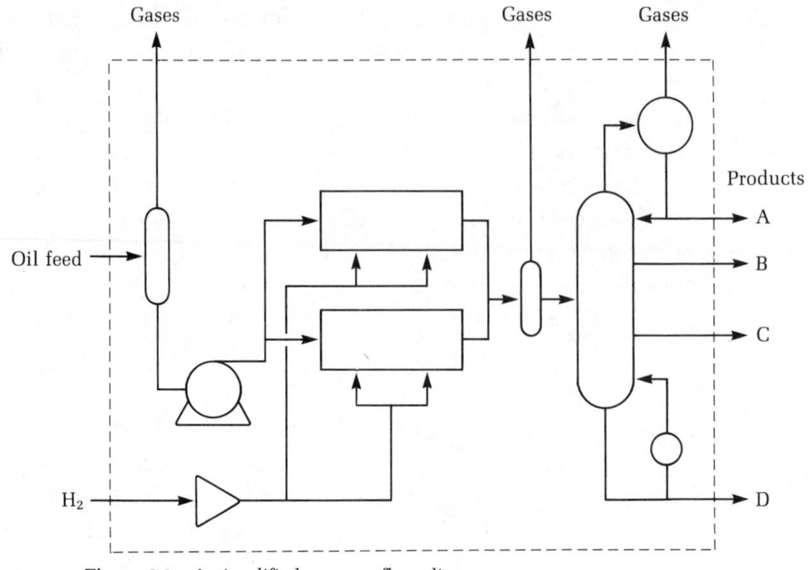

Figure 3.2 A simplified process flow diagram.

After drawing a diagram (with boundaries) and choosing a basis, the next step is to write down the general nonsteady-state equation represented by equation (3.4). Next label the known streams and write down given data. Once everything known is identified on the diagram, we are ready to solve the problem by simplifying the general equation as appropriate, substituting for terms in the equation, and solving. The following simple example illustrates this approach.

Example Problem 3.2

In some places, rainwater is collected for drinking water in barrels placed on the ground. If a certain rainfall is 4.0 cm/hr, how much water (in kg) will be caught in a barrel with a 50-cm diameter at the top in 2 hours?

Step 1. Draw the diagram. A diagram with boundaries is presented below.

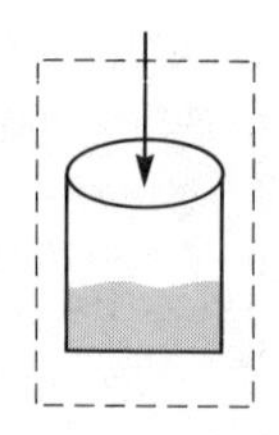

Step 2. Choose a basis: The basis we pick is 2 hours. (Note: A rainfall of 4.0 cm/hr means a flux of 4.0 cm/hr through an extended area, or flow rate of 4.0 cm^3/cm^2hr).

Step 3. Write the general equation:

$$\text{Accumulation Rate} = \text{Input Rate} - \text{Output Rate} + \text{Generation Rate}$$

Step 4. Simplify the general equation.

In this problem, there is no output stream and no generation of water by reaction. We can write:

$$\text{Accumulation Rate} = \text{Input Rate}$$

Step 5. Substitute known terms and solve the simplified equation.

$$\text{Accumulation Rate} = 4.0\ \text{cm/hr} \times \pi(50\ \text{cm})^2/4 \times 1.0\ \text{g/cm}^3 \times 1\ \text{kg}/1000\ \text{g}$$

$$\text{Accumulation Rate} = 7.85\ \text{kg/hr}$$

Or 15.7 kg accumulated during the basis period.

3-5 Analysis of steady-state flow processes

Steady state is an important condition for process engineers. By definition, when a process is operating at steady state, nothing changes with time. This means all flow rates, temperatures, pressures, liquid levels, and so forth are constant. While a true steady state is rarely achieved for extended periods of time, it is often approximated to a reasonable degree. Designs are usually based on steady-state conditions, and achieving a steady state is a goal of most process operators. Analyzing steady-state processes is simpler because we can neglect the accumulation rate term (at steady state there can be no accumulation, positive or negative). The accumulation rate term usually leads to differential equations; without it we are often left with algebraic equations.

Systems without chemical reaction

Let us demonstrate the materials-balance approach to a steady-state flow process in which there is no chemical reaction occurring.

Example Problem 3.3

A baghouse is being used to remove dust from an air exhaust stream flowing at 100.0 m^3/min. The dirty air contains 15.0 g/m^3 of particles, while the cleaned air from the baghouse contains 0.020 g/m^3. The industry's operating permit allows the exhaust stream to contain as much as 0.90 g/m^3. For various operating reasons, the industry wishes to bypass some of the dirty air around the baghouse and blend it back into the cleaned air so that the total exhaust stream meets the permissible limit. Assume no air leakage and negligible change in pressure or temperature of the air throughout the process. Calculate the flow rate of air through the baghouse and the mass of dust collected per day (in kg).

Step 1. Draw the diagram. The labeled diagram is presented in Figure 3.3. Using an overall material balance for dust on the whole process, we can calculate the removal rate of dust. However, to solve for the air-flow rates requires other material balances. Possible sets of boundaries are shown in Figure 3.4. Which boundaries should we choose to start solving this problem? Proper choice of a region is important and can make the problem easier to solve.

Step 2. Choose a basis. Since a flow rate of air is required we will work with rates in the units given. To calculate the total mass of dust collected, we choose a basis of 1 day.

Step 3.

$$\text{Accumulation Rate} = \text{Input Rate} - \text{Output Rate} + \text{Generation Rate} \tag{3.3.1}$$

Steps 4 and 5 (Dust Balance). For dust: from the overall balance,

$$0 = \text{Input Rate} - \text{Output Rate} \tag{3.3.2}$$

Note: There are two outputs streams with dust.

$$\begin{array}{c}\text{Output rate}\\ \text{from baghouse}\\ \text{(Z)}\end{array} = 100.0\,\frac{\text{m}^3}{\text{min}} \times 15.0\,\frac{\text{g}}{\text{m}^3} - 100.0\ \text{m}^3 \times 0.90\,\frac{\text{g}}{\text{m}^3}$$

$$= 1410\ \text{g/min} \tag{3.3.3}$$

Thus, daily output from the baghouse is 2030 kg.

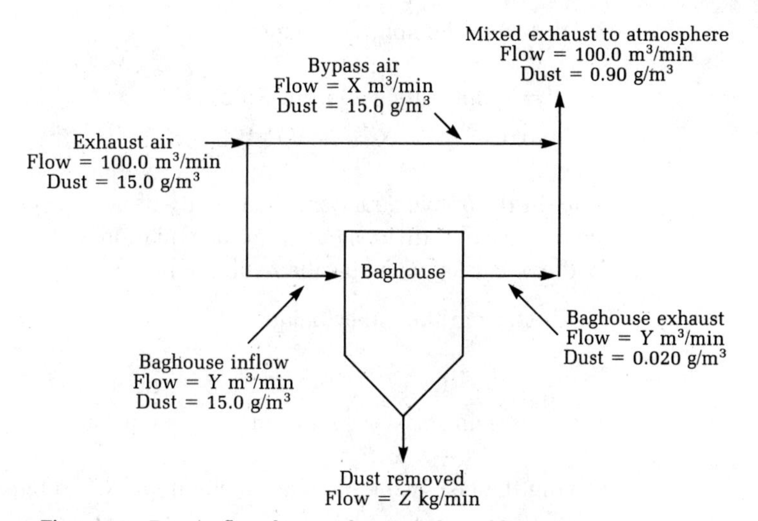

Figure 3.3 Process flow diagram for example problem 3.3.

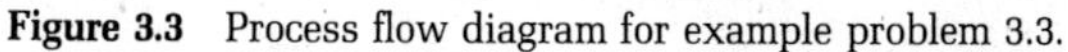

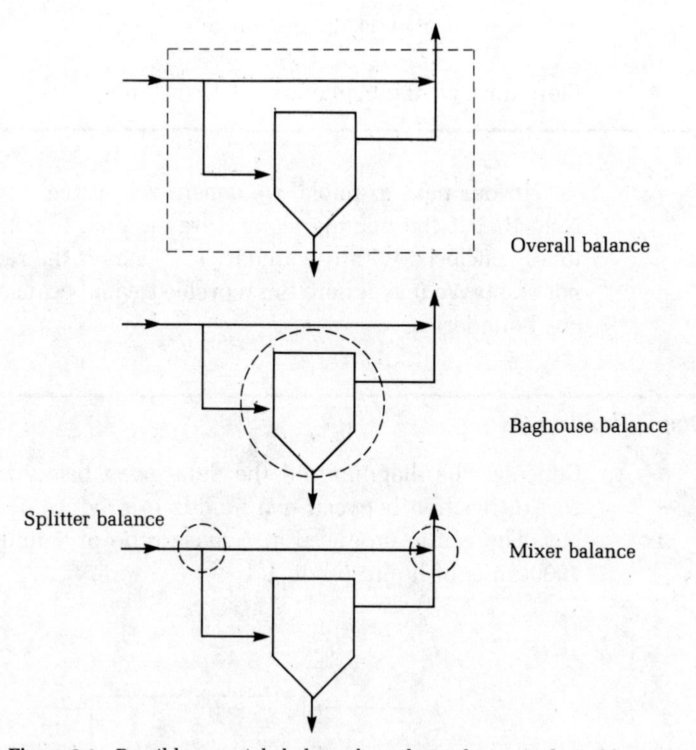

Figure 3.4 Possible materials balance boundaries for example problem 3.3.

For air: from the splitter balance,

$$0 = \text{Input Rate} - \text{Output Rate} \tag{3.3.4}$$

$$100.0\ \text{m}^3\rho_{\text{air}} = X\rho_{\text{air}} + Y\rho_{\text{air}} \tag{3.3.5}$$

Note: In the above equation, the density of air is constant throughout the process and divides out from the equation. For a constant density fluid, we can make a so-called volume balance.

For dust: from the mixer balance,

$$100.0\frac{\text{m}^3}{\text{min}}\frac{0.9\ \text{g}}{\text{m}^3} = X\frac{\text{m}^3}{\text{min}}\frac{15.0\ \text{g}}{\text{m}^3} + Y\frac{\text{m}^3}{\text{min}}\frac{0.02\ \text{g}}{\text{m}^3} \tag{3.3.6}$$

Solving the last two equations to eliminate X, the bypass flow:

$$90.0\ \text{g/min} = (100 - Y)15.0 + 0.02\ Y \tag{3.3.7}$$

$$14.98Y = 1410 \tag{3.3.8}$$

$$Y = 94.126\ \text{m}^3/\text{min} \tag{3.3.9}$$

Flow through the baghouse is 94.1 m^3/min.

In our next example, we consider a system with a recycle loop. Note that if the boundaries are drawn such that the recycle loop is totally enclosed within boundaries, it is as if the recycle stream does not exist. We may ignore the recycle stream because it does not cross our boundaries.

Example Problem 3.4

Consider the diagram and the data given below representing some sort of reaction between two liquids to produce a third liquid and a gas. The gas is produced at a mass ratio of 1 unit of gas for every 1000 units of liquid product.

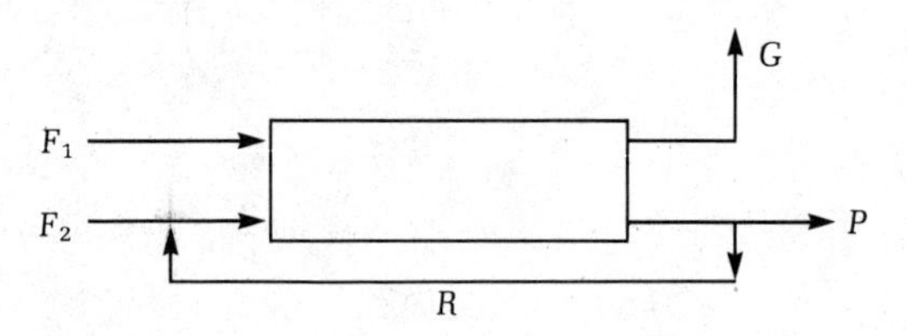

Stream	*Flowrate*	*Density*
Feed F_1	1000 L/min	1.5 kg/L
Feed F_2	6000 kg/min	1.2 kg/L
Product P	Not given	1.3 kg/L
Recycle R	0.5 times P	same as P
Gas G	Not given	1.0 kg/m³

Calculate P, G, and R in L/min. What is the total flow rate into the reactor in kg/min?

Solution First draw the boundaries around the whole system.

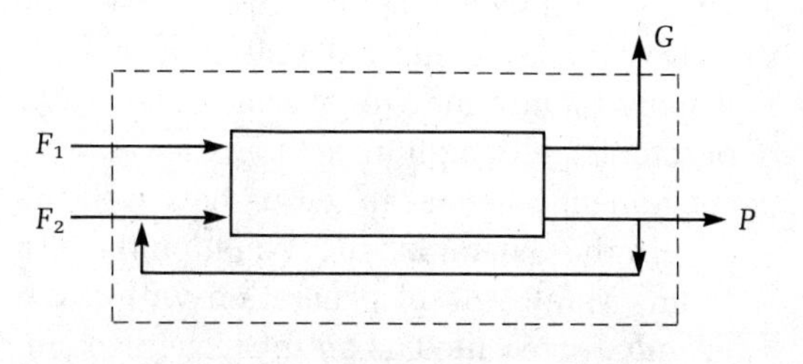

The general balance for total mass simplifies to Input Rate = Output Rate or $F_1 + F_2 = G + P$ (Note that R does not appear in this equation). Also, we know that $G = 0.001\ P$. Substituting, we get

$$1000\ \frac{\text{L}}{\text{min}} \times 1.5\ \text{kg/L} + 6000\ \text{kg/min} = 1.001\ P \tag{3.4.1}$$

$$P = 7492.5\ \text{kg/min} \tag{3.4.2}$$

Converting to volumetric flow rate, we get

$$P = 5763\ \text{L/min} \tag{3.4.3}$$

Recall that

$G = 0.001$ times P on a mass basis, therefore
$G = 7.49$ kg/min or, using the density of the gas,
$G = 7490$ L/min.

Finally,

$$R = 0.5P = 3746\ \text{kg/min} = 2882\ \text{L/min} \tag{3.4.4}$$

The fresh feed rate to the reactor is $F_1 + F_2$; the total feed rate is $F_1 + F_2 + R$, which equals 11,247 kg/min. Without assuming something about how the streams mix, we cannot say anything about the total volumetric feed rate. Even if there were no gas products, we could not have made a volume balance in this problem because the liquid densities are different.

Systems with chemical reaction

In most processes with a reaction step, we are not satisfied merely with a total mass balance. Indeed, we build a reactor specifically to convert at least one reactant into at least one product. Hence we usually require individual mole balances for each of the components of interest, although in some instances we need to know only enough component balances to tell us how well the reactor is working.

In the general balance equation the term "Generation Rate" refers to an overall rate of generation within the system boundaries. It is usually represented as an intrinsic reaction rate times the system volume. (If a component is being used up in the reaction, then its rate of generation is negative.) An intrinsic reaction rate is defined as the time rate of change in the number of moles N, due to reaction, of a given component per unit volume V. In the laboratory we can most easily obtain intrinsic reaction rates from constant-volume, batch, well-stirred reactors. For the reaction A → B, the rate of production r of B is

$$-r_A = r_B = \frac{d(N_B/V)}{dt} \tag{3.6}$$

which for a well-mixed constant-volume system is

$$r_B = \frac{dC_B}{dt} \tag{3.7}$$

For our simple A → B reaction, a simple kinetic model, widely used and usually adequate to predict r_A and r_B is

$$r_A = -kC_A{}^a \tag{3.8}$$

$$r_B = kC_A{}^a \tag{3.9}$$

where

k = a temperature-dependent rate constant

a = an empirically determined order or exponent

Note: The production or generation rate of A is negative because A is being used up.

The rate constant k is usually modeled with an Arrhenius equation:

$$k = k_o e^{-E/RT} \tag{3.10}$$

where:

k_o = frequency factor

E = activation energy

R = universal gas constant, in energy units

T = absolute temperature

Two common models of ideal chemical reactors are the completely mixed or continuous-stirred tank reactor (CSTR) and the plug flow reactor (PFR). We must approach the material balances of these two types differently.

The CSTR model (see Figure 3.5) is that of a tank in which the contents are rapidly and continuously mixed. There is no difference in concentration of any species anywhere in the tank. Reactants flow in and are immediately diluted to the final concentration in the tank. Because the outlet stream is continuously drawn from the tank and the contents of the tank have the same composition everywhere, the concentrations in the outlet stream are those in the tank. These are the concentrations at which all reactions occur. A steady-state material balance on a reacting component in a CSTR results in a simple algebraic expression, as shown in the next example problem.

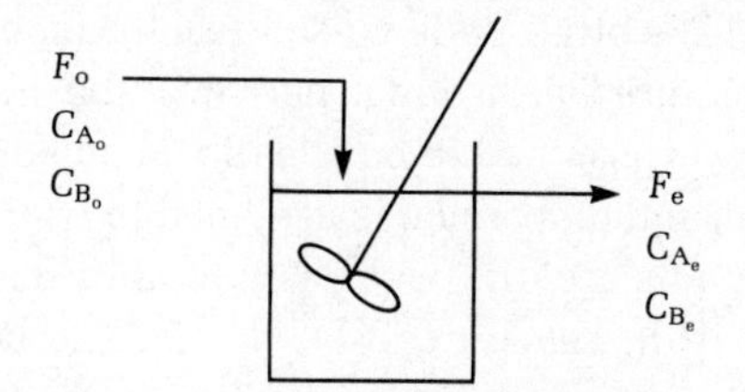

Figure 3.5 Diagram of a continuous-stirred tank reactor.

Example Problem 3.5

Calculate the CSTR volume required for 98% conversion of component A. The kinetics of conversion are

$$r_A = -kC_A \tag{3.5.1}$$

with $k = 0.10s^{-1}$.

The inflow rate is 75 L/s with an initial concentration of $C_{A_o} =$ 0.05 mol/L. There is no volume change on reaction.

Solution The general balance for component A is

$$\text{Accumulation Rate} = \text{Input Rate} - \text{Output Rate} + \text{Generation Rate}$$

At steady-state operation there is no accumulation, so

$$0 = F_o C_{A_o} - F_e C_{A_e} - kC_{A_e}V \tag{3.5.2}$$

Note that the concentration in the generation rate term is C_{A_e}, the exit concentration, because for a CSTR it equals the in-tank concentration. Solving for the volume of the reactor and noting that $F_e = F_o$, we get

$$V = \frac{F_o(C_{A_o} - C_{A_e})}{kC_{A_e}} = \frac{F}{k}\frac{(1 - C_{A_e}/C_{A_o})}{C_{A_e}/C_{A_o}} \tag{3.5.3}$$

Substituting, we find that

$$V = \frac{75 \text{ L/s}}{0.1 \text{ s}^{-1}}\frac{0.98}{0.02} = 36{,}750 \text{ L} \tag{3.5.4}$$

The other widely used reactor model, the plug flow reactor (PFR), is pictured as a long, narrow tube through which fluid is flowing. Flow is assumed to be one-dimensional, velocity is constant across the tube, and dispersion is assumed to be negligible. Temperature may not vary much with axial position, and the analysis is very much simplified if the reactor can be assumed to be isothermal. We shall use the material balance approach to develop the basic steady-state design equation for an isothermal plug flow reactor (Figure 3.6).

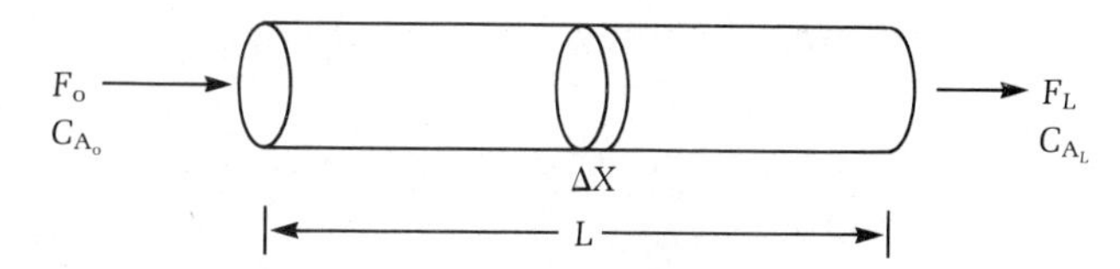

Figure 3.6 Schematic diagram of a plug flow reactor.

Consider the simple reaction A → B with kinetics as described by equations (3.7) to (3.10). Start by making a material balance on component A in a small volume increment of the reactor ($\Delta V = S\Delta x$, where S is the cross-sectional area) located at the arbitrary position x along the length of the reactor. The basic balance equation (subject to our assumptions) becomes

$$\text{Accumulation Rate} = F_x C_{A_x} - F_{x+\Delta x} C_{A_{x+\Delta x}} + r_A \Delta V \tag{3.11}$$

Assume that S remains constant throughout the reactor and that F, the volumetric flow rate, does not change with x. Further, assume that the reactor operates at steady state; therefore, accumulation is zero. Dividing through by $S\Delta x$ and noting that $F/S = u$, the linear velocity, we obtain

$$0 = \frac{u(C_{A_x} - C_{A_{x+\Delta x}})}{\Delta x} + r_A \tag{3.12}$$

Now let Δx approach zero and take the limit. Substituting equation (3.8) for r_A, equation (3.12) becomes

$$0 = -u\frac{dC_A}{dx} - kC_A{}^a \tag{3.13}$$

or

$$u\frac{dC_A}{dx} = -kC_A{}^a \tag{3.14}$$

which we can solve by integration. For a first-order reaction ($a = 1$), we write

$$\int_{C_{A_o}}^{C_{A_L}} \frac{dC_A}{C_A} = -\frac{k}{u}\int_0^L dx \tag{3.15}$$

which upon integration can be written as

$$\ln \frac{C_{A_L}}{C_{A_o}} = -k\frac{L}{u} \tag{3.16}$$

or

$$C_{A_L} = C_{A_o} e^{-k\tau} \tag{3.17}$$

where

τ = reactor residence time (L/u or V/F)

Example Problem 3.6

Rework example problem 3.5, except calculate the volume of a PFR required for the same 98% conversion.

From equation (3.16),

$$\ln\left(\frac{0.02C_{A_o}}{C_{A_o}}\right) = -0.1\ \text{s}^{-1}\tau \tag{3.6.1}$$

Solving for τ we get $\tau = 39.1$ s. But $\tau = V/F$, so

$$\begin{aligned} V &= F\tau \\ &= 75\ \text{L/s}\ (39.1\ \text{s}) \\ V &= 2932\ \text{L} \end{aligned} \tag{3.6.2}$$

Note that for a PFR, the volume required is much less than for a CSTR (especially for large conversion).

In the previous examples, we had a homogeneous reaction with very simple kinetics. In the next example problem we demonstrate that the material balance approach is also useful in more complicated situations.

Example Problem 3.7

The diagram below (Figure 3.7) represents a reactor-separator system through which is flowing water with a chemical contaminant (sub-

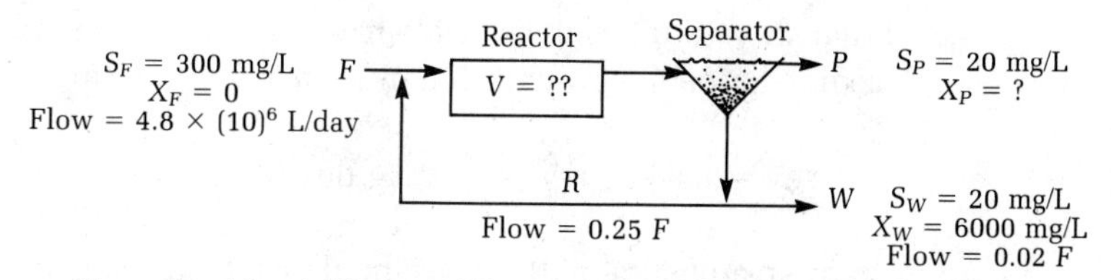

Figure 3.7 Material flows and characteristics of a reactor-separator system.

strate S). The substrate is biologically converted by bacteria to CO_2, H_2O, and more bacteria (or biomass, symbol X). The yield of biomass is 0.50 gram of biomass produced per gram of substrate converted.

Given the information on the diagram, and that the substrate intrinsic reaction rate is $-r_S = 0.03(S)^{1.5}$ mg/L-day, calculate (1) the conversion rate of substrate (kg/day), (2) the growth rate of biomass (kg/day), (3) the concentration of biomass in the effluent leaving the separator (mg/L), and (4) the volume of the reactor (liters). Assume the following:

1. Reactor is well mixed (assume CSTR behavior).
2. No reaction occurs in the separator or recycle line.
3. Densities of all streams are constant and are equal to that of pure water.
4. Steady-state conditions prevail.

Balances around the whole system for the three components of interest are

$$\text{Substrate:}\quad 0 = FS_F - PS_P - WS_W + r_S V \tag{3.7.1}$$

$$\text{Biomass:}\quad 0 = 0 - PX_P - WX_W + r_X V \tag{3.7.2}$$

$$\text{Total mass:}\quad 0 = F - P - W \tag{3.7.3}$$

From equation (3.7.3) and the fact that $W = 0.02F$, we determine that

$$P = 4.704 \times 10^6 \text{ L/day}$$
$$W = 9.6 \times 10^4 \text{ L/day}$$

Now, from equation (3.7.1), solve for the quantity $r_S V$, which is the conversion rate of substrate:

$$\begin{aligned} r_S V &= -FS_F + PS_P + WS_W \\ &= -1440 \text{ kg/day} + 94.1 \text{ kg/day} + 1.9 \text{ kg/day} \\ &= -1344 \text{ kg/day (Answer 1)} \end{aligned}$$

From the relation between biomass and substrate we can solve immediately for the growth rate of biomass:

$$r_X V = 0.50\,(-r_S V) = 672 \text{ kg/day (Answer 2)}$$

As at steady state nothing accumulates, biomass must be discharged at the same rate as it grows (exactly what equation 3.7.2 tells us). From equation (3.7.2), we find that

$$X_P = \frac{r_X V - W X_W}{P}$$

$$= 20.4 \text{ mg/L (Answer 3)}$$

Finally, combining $r_S V$ and the expression for r_S, we can solve for V:

$$r_S V = -1344 \text{ kg/day}$$

$$-r_S = 0.03(20)^{1.5} \text{ mg/L-day}$$

$$V = \frac{1344 \text{ kg/day}}{2.683 \text{ mg/L-day}} \times \frac{10^6 \text{ mg}}{1 \text{ kg}} = 5.01 \times 10^8 \text{ L (Answer 4)}$$

3-6 Transient processes

Up to now we have always dropped the accumulation term in our general material balance equation by assuming a steady-state condition. Although steady-state operations are desirable, and knowledge of the steady-state condition is useful for design understanding, the transient (non-steady-state) response of a process is also important. A study of the dynamic behavior of systems is a complete course in itself; however, we want to illustrate in a very simple fashion the use of the non-steady-state material balance equation with the following two examples.

Example Problem 3.8

A large oil tanker is off-loading 300,000 barrels of crude oil to a refinery. During this operation, both refinery personnel and the ship's crew must be careful not to overfill a shore tank nor to let the pumps run dry while pumping out a ship's compartment. The ship is discharging at a rate of 20,000 B/hr into a tank that is 100 feet in diameter, has a maximum filling height of 48 feet and starts with a

6-foot liquid level. How long do the refinery personnel have to prepare another tank to receive oil? One barrel is 42 gallons and 7.48 gallons equals 1.0 cubic foot.

Solution Accumulation Rate = Input Rate − Output Rate + Generation Rate

Accumulation Rate = Input Rate

$$\frac{dV}{dt} = \frac{20{,}000\ \text{B}}{\text{hr}} \times \frac{42\ \text{gal}}{\text{B}} \times \frac{1\ \text{ft}^3}{7.48\ \text{gal}}$$

$$= 1.123 \times 10^5\ \text{ft}^3/\text{hr}$$

Separating variables and integrating, we have

$$\Delta V = 112{,}300\ \text{ft}^3/\text{hr}\ (\Delta t)$$

But

$$\Delta V = \frac{\pi(100)^2}{4}\ \text{ft}^2\ (48 - 6)\ \text{ft}$$

$$= 330{,}000\ \text{ft}^3$$

So

$$\Delta t = \frac{330{,}000}{112{,}300} = 2.94\ \text{hours}$$

Example Problem 3.9

A company has been discharging its waste into a holding pit for a long time. For the last several years the pit has been full and overflowing into a local river. The waste concentration has been nearly constant at 10 mg/L of the pollutant of interest (as has the pit overflow stream). At this concentration there have been no adverse effects on the stream. Suddenly there is a process change and the company's waste stream goes up to 100 mg/L. Given that the waste stream flow rate is 100,000 L/day and the holding pit volume is 1 million liters, calculate the concentration of the pit's overflow stream after 10 days.

Solution Accumulation Rate = Input Rate − Output Rate + Generation Rate

For the pollutant in the pit, the balance is:

$$\frac{d(C_e V)}{dt} = FC_i - FC_e + 0 \tag{3.9.1}$$

where C_e is the concentration in the exit stream at any time. Thus

$$\frac{dC_e}{dt} + \frac{F}{V} C_e = \frac{FC_i}{V} \tag{3.9.2}$$

The hydraulic residence time of the pit τ is defined as V/F, so equation (3.9.2) becomes

$$\frac{dC_e}{dt} + \frac{1}{\tau} C_e = \frac{1}{\tau} C_i \tag{3.9.3}$$

Because both τ and C_i are constants with time, the solution to this linear, first-order differential equation is

$$C_e(t) = C_o + (C_i - C_o)(1 - e^{-t/\tau}) \tag{3.9.4}$$

where C_o is the concentration in the exit stream at the time when the inlet stream suddenly jumps to 100 mg/L. Note: This solution accords with our intuitive understanding of the situation. At $t = 0$, $C_e = C_o$ and at $t = \infty$, $C_e = C_i$. Substituting for our particular problem:

$$\begin{aligned} C_e &= 10 \text{ mg/L} + (100 - 10) \text{ mg/L } (1 - e^{-10/10}) \\ &= 10 + 90\ (0.632) \\ &= 67 \text{ mg/L after 10 days} \end{aligned}$$

3-7 Various forms of energy and power

Energy is often defined as work or the capacity to do useful work (however, we also classify useless heat as energy). Power, on the other hand, is the rate of doing work or the rate of expending energy. On a global scale, we have a continuous flow of energy into and through the environment. Although matter is always conserved, the earth is not a closed system with respect to energy. We depend on high-quality energy in the form of solar radiation flowing into the biosphere just as we must have the flow of low-quality thermal radiation away from the earth. Not only does the sun drive our hydrologic cycle, power our winds, and drive other physical processes, but through green plants it is the basis for energy flow up the food chain.

Energy is analogous to mass in that it cannot be created nor destroyed. Therefore, it can be balanced in the same way as mass. However, it must be recognized that energy can be transferred by radiation and by conduction through the walls of containers and pipelines as well as the usual means of bulk flow. We will make use of mass and energy balances throughout this course, just as process engineers make use of them throughout their careers.

To do work requires energy, but no process is 100% efficient in converting energy into useful work. There will always be waste heat produced in any natural or artificial energy conversion process. However, we can convert units among the many forms of energy and not incur losses, just as we can convert from pounds to grams. We can also mathematically convert from energy to power and back again by dividing or multiplying by a unit of time.

Energy has many forms: thermal, mechanical, kinetic, potential, electrical, and chemical, to name a few, and thousands of units of measure. In Table 3.1 we present a few of the units of measure of the different forms of energy along with some common conversion factors between units. Power may also take on many different units, although the most common deal with electrical power (kilowatts, megawatts) or with mechanical power (horsepower). In Table 3.2 we present several sets of units for power and some common conversion constants. The next example illustrates units conversion in energy.

Example Problem 3.10

A family of four uses about 15,000 kWh of electrical energy in a year's time. (1) How many gallons of gasoline is this equivalent to? (2) Where does this family use more energy, operating their house for a year or operating two cars for a year (say about 20,000 miles at 20 miles per gallon)? (3) If gasoline is $1.25/gal and electricity is 8.33 cents/kWh, where does the family spend more money?

Solution

1. 15,000kWh/yr × 1 gal gasoline/36.9kWh = 406.5gal/yr equivalent
2. 20,000 miles/yr × 1 gal/20 miles = 1000 gal/yr: the family expends more energy operating the cars
3. 15,000 kWh/yr × $.0833/kWh = $1250/yr vs. 1000 gal/yr × $1.25/gal = $1250/yr

---

As stated earlier, energy and power are related through time. The next example illustrates the relation between electrical energy and electrical power.

TABLE 3.1 Units of measure and conversion factors for various energy forms.

*Form**	*BTU*	*kWh*	*kJ*
Thermal:			
1 British thermal unit (BTU)	1.0	2.93×10^{-4}	1.055
1 kilocalorie (kcal)	3.968	1.163×10^{-3}	4.184
1 kilojoule (kJ)	0.948	2.778×10^{-4}	1.0
1 therm	100,000	29.3	105,500
Chemical Potential (typical higher heating values):			
1000 standard cubic feet (MSCF) of natural gas	1.03×10^{6}	302	1.08×10^{6}
1 barrel (B) of crude oil (42 gal)	6.0×10^{6}	1,760	6.33×10^{6}
1 gallon of gasoline	126,000	36.9	133,000
1 pound coal			
Pennsylvania anthracite	13,500	3.96	14,200
Illinois bituminous	11,500	3.37	12,100
North Dakota lignite	7,200	2.11	7,600
1 gallon of ethanol (anhydrous)	83,800	24.6	88,500
Electrical:			
1 kilowatt-hour (kWh)	3,412	1.0	3,600
1 watt-second	9.48×10^{-4}	2.778×10^{-7}	0.001
1 megawatt-day	8.19×10^{7}	24,000	8.64×10^{7}
Mechanical:			
1 horsepower-hour (HPhr)	2,545	0.746	2,685
1 foot-pound (ft-lb_f)	1.286×10^{-3}	3.766×10^{-7}	0.001356
1 newton-meter (N-m)	9.48×10^{-4}	2.778×10^{-7}	0.00100
Pressure-volume:			
1 liter-atmosphere (L-atm)	0.0961	2.82×10^{-5}	0.01014
1 cubic foot-atmosphere (ft^3-atm)	2.72×10^{-4}	7.97×10^{-8}	2.87×10^{-4}
1 cubic meter-pascal (m^3-Pa)	9.48×10^{-4}	2.778×10^{-7}	0.00100

*1 J = 1 N-m; 1 Pa = 1 N/m^2; 1 N = 1 kg-m/s^2.

Example Problem 3.11

(1) What is the average hourly electrical power delivery to the family in example problem 3.10? Actually, most of the electricity used is delivered to homes in relatively short periods of time—the peak hours are 7–9 A.M. and 4–7 P.M. on weekdays. Assume that peak power delivery to a home is 15 kW. (2) How much electrical energy is delivered if the 15-kW load is sustained for one hour? (3) For one day? (4) For one year?

Solution

1. 15,000 kWh/yr × 1 yr/365 days × 1 day/24 hrs = 1.71 kW
2. 15 kW × 1 hour = 15 kWh
3. 15 kW × 1 day × 24 hr/day = 360 kWh
4. 15 kW × 1 yr × 365 days/yr × 24 hr/day = 131,000 kWh

TABLE 3.2 Units of measure and conversion factors for various forms of power.

*Form**	*kW*	*HP*
1 watt (W)	0.001	0.00134
1 kilowatt (kW)	1.000	1.341
1 megawatt (MW)	1,000	1,341
1 foot-lb-force/s ($ft\text{-}lb_f/s$)	0.001356	0.001818
1 horsepower (HP)	0.746	1.00
1 kilojoule/hour (kJ/hr)	2.778×10^{-4}	3.725×10^{-4}
BTUhr	2.93×10^{-4}	3.93×10^{-4}
$ft\text{-}lb_f/min$	2.26×10^{-5}	3.03×10^{-5}
1 newton-meter/s	0.001	0.00134

*1 watt = 1 J/s; 1 J = 1 N-m; and 1 N = 1 $kg\text{-}m/s^2$ (also 1 lb_f = 4.45 N).

3-8 Flowing streams of energy

The first law of thermodynamics states that energy is conserved. Therefore, in the analysis of any process we must account for all energy inputs and outputs as well as energy accumulation. We can change the form of energy through an energy conversion process (for instance, when we burn natural gas we convert chemical energy into thermal energy), but basically, the energy that goes in must come out or accumulate. Thus, an energy balance is like a mass balance.

A flowing stream of material carries with it an associated energy flow. If we consider only thermal energy for the moment, a more useful property to engineers than energy is enthalpy. Enthalpy is a thermodynamic property of material which depends on temperature, pressure, and composition. Enthalpy has a precise mathematical thermodynamic definition, but here we will use the intuitive definition "heat content." Absolute values of enthalpy are not required; rather, we are interested in the change in enthalpy when the material passes from one set of conditions to another. In many situations, we can approximate the change in enthalpy of a fixed amount of material as

$$\Delta H = mC_p\Delta T \tag{3.18}$$

where

ΔH = enthalpy change

m = mass

C_p = specific heat at constant pressure

ΔT = final minus initial temperature

This equation can easily be extended to matter that is flowing from one place to another as well as matter that is stationary, as long as a constant flowing mass is considered. The assumptions inherent in this equation are that C_p is constant over the range of temperatures, that the effect of pressure is negligible (or pressure is constant), that any change in composition has negligible effect on enthalpy, and that there has been no change of phase during the process. Phase changes of a pure compound can be accounted for by the following equation:

$$\Delta H = m\lambda \tag{3.19}$$

where

λ = the heat of phase change (melting, boiling, condensing, freezing, or subliming)

Example Problem 3.12

A 30,000-L/day stream of waste material is to be anaerobically digested as part of its treatment process. The input stream must be heated from 15° to 40°C for the process to work properly. At what rate must we add heat to the stream? Assume that the stream has a density and specific heat similar to those of water.

Solution The heating process can be pictured as shown below. The general form of the energy balance is the same as that for a material balance:

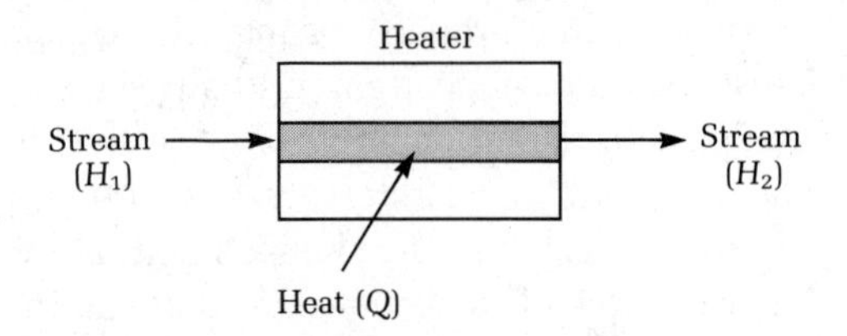

$$\text{Accumulation Rate} = \text{Input Rate} - \text{Output Rate} + \text{Generation Rate}$$

which for this problem simplifies to

$$0 = H_1 + Q - H_2 + 0 \tag{3.12.1}$$

or

$$\Delta H = Q \tag{3.12.2}$$

but

$$\Delta H = mC_p\Delta T$$

so

$$\begin{aligned}\Delta H &= 30{,}000 \text{ L/day} \times 1 \text{ kg/L} \times 4.18 \text{ kJ/kg °C} \times 25\text{°C} \\ &= 3.14 \times 10^6 \text{ kJ/day} \qquad (3.12.3)\end{aligned}$$

Thus,

$$Q = 3.14 \times 10^6 \text{ kJ/day}.$$

The way problem 3.12 is worded, there are no heat losses or heat transfer inefficiencies to consider. However, in a real situation, in order to get this much heat into the stream, we must use a real-world heat exchanger. In real-world processes there are always inefficiencies—the topic of the next section.

3-9 Real-world energy conversion processes

There are many different ways of stating the second law of thermodynamics. One form of stating it non-mathematically is to say that no process exists that is 100% efficient in using energy to do useful work, or in transferring energy from one useful form to another useful form. As an example, if we use electrical energy to lift an elevator full of people, not all of the electrical energy will be used to do the lifting. Some portion will be lost as heat in the motor wiring, as friction in the gears, and so on. Note that this is not a violation of the first law; we can still say that the total amount of energy is conserved. But the second law requires that some of the energy must be dissipated in a useless form (as low-level heat in this case). The following examples demonstrate this concept. Later in the book we will investigate the second law a bit more thoroughly.

Example Problem 3.13

Rework example problem 3.12, but now calculate the heat that must be fed into the heater in order to transfer the required amount of heat into the process stream. The heater has an efficiency of 85%.

Solution From the previous example problem, the heat required by the process stream was 3.14×10^6 kJ/day. This is the output of the heater. The input heat must be greater than this to allow for the less-than-perfect heat transfer efficiency. Therefore, the heat input is

$$\frac{3.14 \times 10^6 \text{ kJ/day}}{0.85} = 3.69 \times 10^6 \text{ kJ/day}$$

Example Problem 3.14

If we design a furnace-boiler to burn natural gas at the rate of 500 SCF/hr to generate steam, then transport the steam in an insulated pipe to the heater in example problem 3.13, is our design sufficient to supply the required heat (plus 50% capacity for future expansion)? Assume that the furnace-boiler is 87% efficient and the pipe transportation is 92% efficient.

Solution To calculate the overall efficiency of several steps in series, we simply multiply the efficiencies of each step.

The overall efficiency of the process is

$$(0.87)(0.92)(0.85) = 0.68$$

The heat available to the process is

$$0.50 \frac{\text{MSCF}}{\text{hr}} \times \frac{24 \text{ hr}}{\text{day}} \times 1.08 \times 10^6 \frac{\text{kJ}}{\text{MSCF}} \times 0.68$$
$$= 8.9 \times 10^6 \text{ kJ/day}$$

The process needs

$$1.50 \times 3.14 \times 10^6 \text{ kJ/day} = 4.71 \times 10^6 \text{ kJ/day}$$

Therefore, the design is adequate.

3-10 Combined energy and material balances

We do nothing different when we combine material and energy balance considerations; in fact, the two complement each other and assist the engineer in truly understanding a process. For example, combined energy and material balances are quite useful in gaining a preliminary understanding of the major environmental effects of large electricity-generating plants (we will deal with electricity generation in detail in Chapter 9). Since we are not introducing any new concepts, we will end this chapter with two examples demonstrating the use of combined energy and materials balances.

Example Problem 3.15

A 1000-MW coal-burning power plant is burning anthracite coal from Pennsylvania which has 6% ash and 2.5% sulfur. The plant has a thermal efficiency of 40%. Assume that the overall removal efficiencies for ash and SO_2 are 99.5% and 88% respectively. Calculate (1) the rate of heat emitted to the environment (kJ/s), (2) the rate of coal input to the furnace (kg/day), (3) the rate of ash emission to the atmosphere (kg/day), and (4) the rate of SO_2 emission to the atmosphere (kg/day).

Solution First, do an energy balance on the plant (recall that power is a rate of energy):

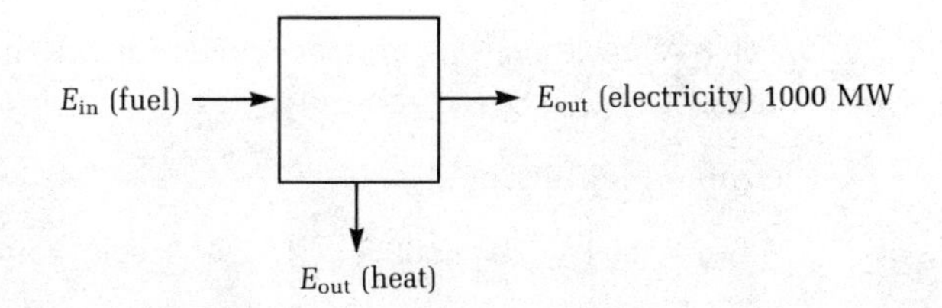

An efficiency of 40% means that only 40% of the input rate of energy is converted to useful electricity, so

$$E_{in} = \frac{1000 \text{ MW}}{0.40} = 2500 \text{ MW}$$

Consequently, the rate of heat emitted to the environment is

$$\text{Heat} = (1 - 0.40)\ 2500 \text{ MW} = 1500 \text{ MW}$$

or

$$\text{Heat} = 1500 \text{ MW} \times \frac{1000 \text{ kW}}{1 \text{ MW}} \times \frac{1 \text{ kJ/s}}{1 \text{ kW}}$$

$$= 1.5 \times 10^6 \text{ kJ/s (Answer 1)}$$

To calculate the coal input rate, we need the energy content. From Table 3.1, 1 pound of anthracite contains 14,200 kJ or 3.96 kWh.

$$\text{Coal Input} = 2500\ \text{MW} \times \frac{1000\ \text{kW}}{1\ \text{MW}} \times \frac{24\ \text{hr}}{1\ \text{day}} \times \frac{1\ \text{lb}}{3.96\ \text{kWh}} \times \frac{1\text{kg}}{2.2\text{lb}}$$

$$= 6.89 \times 10^6\ \text{kg/day (Answer 2)}$$

To get ash emissions, first calculate ash input:

$$\text{Ash Input} = 6.89 \times 10^6\ \text{kg/day} \times \frac{0.06\ \text{kg ash}}{1\ \text{kg coal}}$$

$$= 4.13 \times 10^5\ \text{kg/day}$$

Then multiply by 1.0 minus the removal efficiency:

$$\text{Ash emissions} = 4.13 \times 10^5\ \text{kg/day} \times (1.00 - 0.995)$$

$$= 2065\ \text{kg/day (Answer 3)}$$

For SO_2 emissions, we first assume that all the incoming sulfur is oxidized to SO_2: $S + O_2 \rightarrow SO_2$.

The rate of sulfur input is

$$S_{in} = 6.89 \times 10^6\ (0.025) = 1.72 \times 10^5\ \text{kg/day}$$

As the mass ratio of SO_2 to S is 2.0 to 1, and as the problem specifies 88% removal efficiency, SO_2 emissions are

$$SO_2 = 1.72 \times 10^5\ \text{kg/day} \times \frac{2\ \text{kg}\ SO_2}{1\ \text{kg S}} \times (1 - 0.88)$$

$$= 4.3 \times 10^4\ \text{kg/day (Answer 4)}$$

Example Problem 3.16

Figure 3.8 represents a simplified process flow diagram for making gasoline from coal.

Assume we use North Dakota lignite coal, with properties as listed in Table 3.1, and assume we are given the following data:

1. The process is 60% efficient in producing gasoline from coal on an energy-content basis.

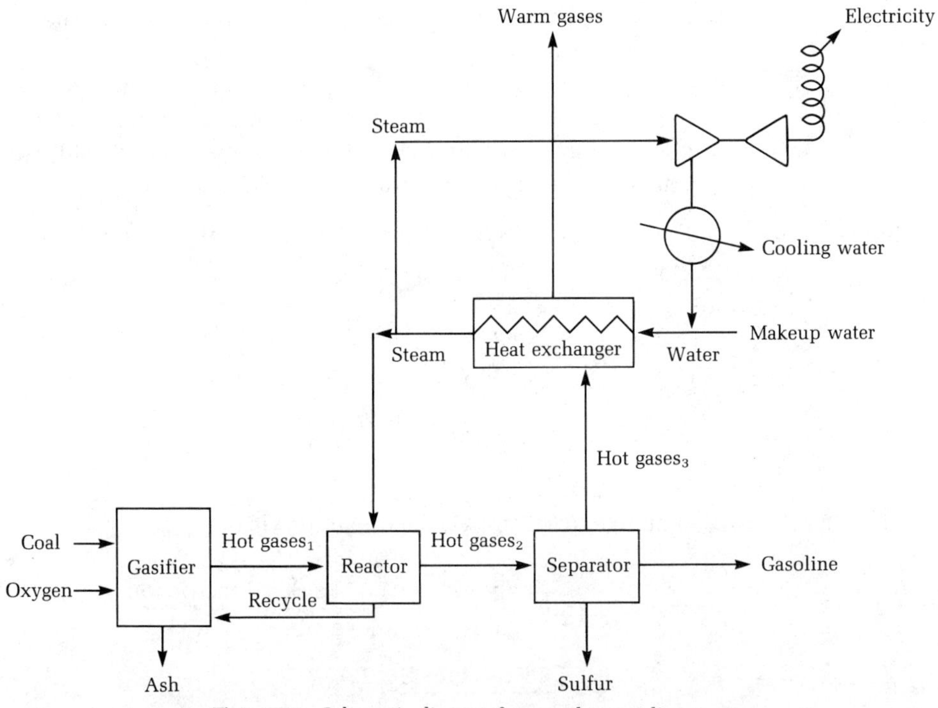

Figure 3.8 Schematic diagram for a coal-to-gasoline process.

2. The coal feed rate is 1.00×10^6 kg/day; coal contains 0.7% sulfur and 10.0% ash.
3. All the ash is removed in the gasifier.
4. All the sulfur is removed in the separator.
5. The steam feed rate into the reactor is in a ratio of 1 kg steam/10 kg coal; all of the steam that goes into the reactor is consumed.
6. Much of the energy available in the hot gases coming from the separator is recovered in the heat exchanger and is used to boil water to make steam.
7. Some of the steam produced in the heat exchanger is sent through a turbine-generator to make electricity.

Questions

1. Calculate the production rate of gasoline, in gal/day.
2. Calculate the production rate of sulfur, in metric tons per day (1 metric ton = 1000 kg).

3. Calculate the makeup rate of water, in gal/day (1 gal of water = 8.33 lb of water).
4. Assume that the only net energy input is due to coal, neglect the energy output due to ash and sulfur, and assume that all other energy losses are 20% of the net energy input due to coal. Calculate the production rate of electricity in kW.

Solutions

1. From the mass feed rate and energy contents of coal and gasoline,

$$1.00 \times 10^6 \frac{\text{kg coal}}{\text{day}} \times \frac{4.64 \text{ kWh}}{1 \text{ kg}} \times 0.60 \times \frac{1 \text{ gal gasoline}}{36.92 \text{ kWh}}$$

$$= 7.54 \times 10^4 \frac{\text{gal gasoline}}{\text{day}}$$

2. A simple material balance on sulfur yields

$$1.00 \times 10^6 \frac{\text{kg coal}}{\text{day}} \times \frac{0.007 \text{ kg S}}{1 \text{ kg coal}} \times \frac{1 \text{ metric ton}}{1000 \text{ kg}}$$

$$= \frac{7 \text{ metric tons S}}{\text{day}}$$

3. A material balance of the steam circuit indicates that makeup water equals steam into the reactor.

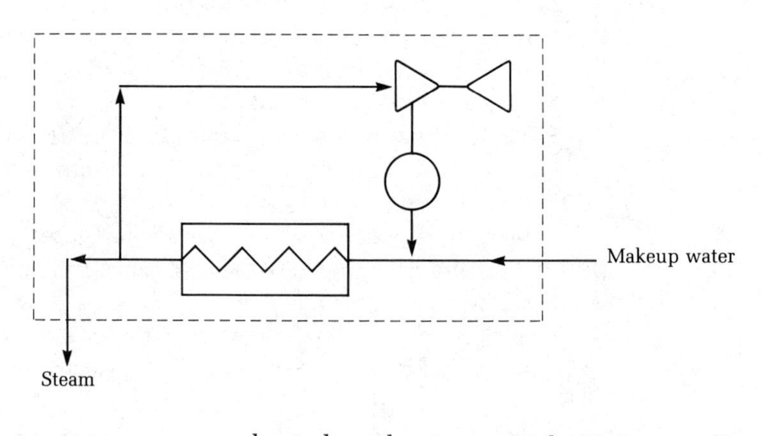

$$1.00 \times 10^6 \frac{\text{kg coal}}{\text{day}} \times \frac{1 \text{ kg steam}}{10 \text{ kg coal}} \times \frac{1 \text{ kg } H_2O}{1 \text{ kg steam}} \times \frac{2.2 \text{ lb}}{1 \text{ kg}}$$

$$\times \frac{1 \text{ gal}}{8.33 \text{ lb}} = 2.64 \times 10^4 \text{ gal } H_2O/\text{day}$$

4. We can draw an overall energy balance diagram for the process as shown below.

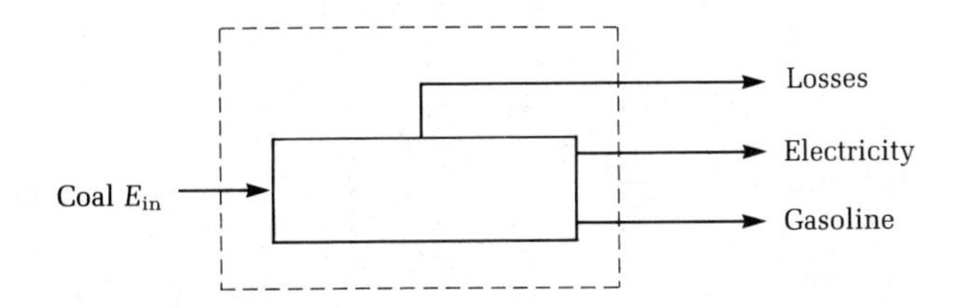

Notice that when we do this, all internal inefficiencies, recycle loops, and similar losses become included in the overall losses term. Therefore, we have

$$\text{Gasoline} = 0.60\ E_{in}$$

$$\text{Losses} = 0.20\ E_{in}$$

$$\text{Electricity} = E_{in}\ (1.0 - 0.60 - 0.20) = 0.20\ E_{in}$$

Thus, electricity production is

$$1.00 \times 10^6\ \frac{\text{kg coal}}{\text{day}} \times \frac{4.64\ \text{kWh}}{\text{kg}} \times \frac{1\ \text{day}}{24\ \text{hr}} \times 0.20$$

$$= 3.87 \times 10^4\ \text{kW}$$

In this chapter, you have been exposed to an extremely powerful and valuable engineering technique: making material and energy balances. You have progressed from simple situations (such as rainwater being caught in a barrel) to some rather complicated processes (such as electric generating plants). When you have learned the techniques presented in this chapter, not only are you ready to master the rest of this textbook, you also have made a lasting contribution to your success as an engineer.

3-11 Problems

3.1 Five million kilograms per day of coal is burned in an electric power plant. The coal has an ash (non-combustible portion) content of 12% by mass. Forty percent of the ash falls out the bottom of the furnace. The rest of the ash is carried out of the furnace with the hot gases into an electrostatic precipitator (ESP). The ESP is 99.5% efficient in removing the ash that comes into it. Draw a diagram representing this process and calculate the mass emissions rate of ash into the atmosphere from this plant.

3.2 Given the following diagrams and data, calculate the quantity of steam required in step 2.

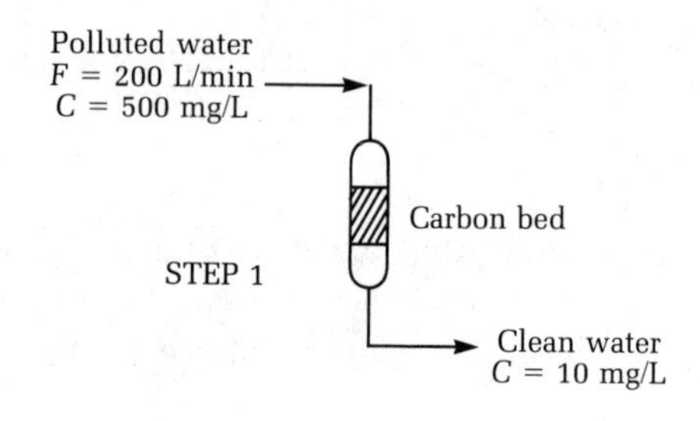

In step 1, we allow a polluted water stream to flow through a bed of activated carbon. Most of the pollutant molecules remain on the carbon and, thus, are removed from the water. This step lasts for 5 hours.

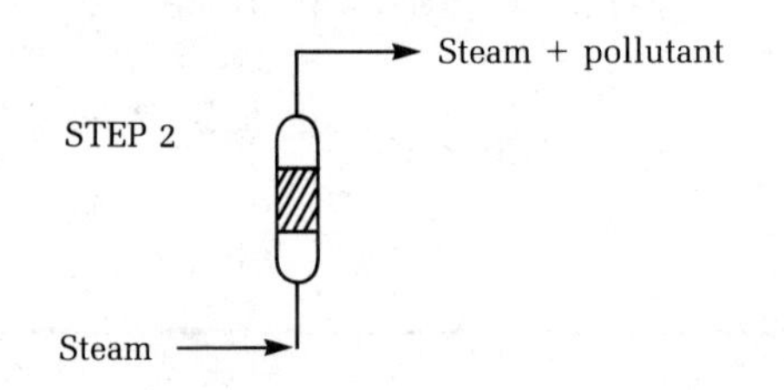

In step 2, after the bed gets loaded with pollutant, it is cleaned with steam. The steam removes all of the pollutant from the carbon. It has been found that a ratio of steam to pollutant of 7 kg steam/1 kg pollutant is necessary for complete cleaning. Calculate the quantity of steam required to do this job.

3.3 The following is a clarifier-thickener system to separate solids and liquids. Ignore the effect of solids content on density of the streams (that is, assume density of solids = density of water).

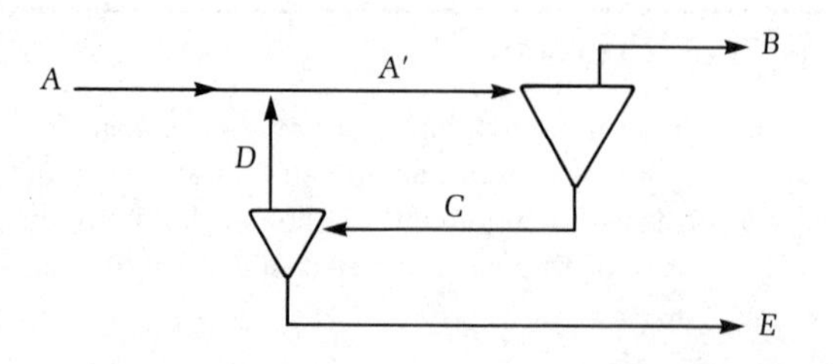

Complete the following material balance table.

Stream	Flow rate, L/s	Solids, mg/L
A	100	3000
B	95	15
C		6000
D	50	
E		
A′		

3.4 A 600-MW coal-burning power plant is burning Illinois bituminous coal with 8% ash content. The plant is 39% efficient. Thirty percent of the ash drops out in the furnace, and the electrostatic precipitator is 99.0% efficient.

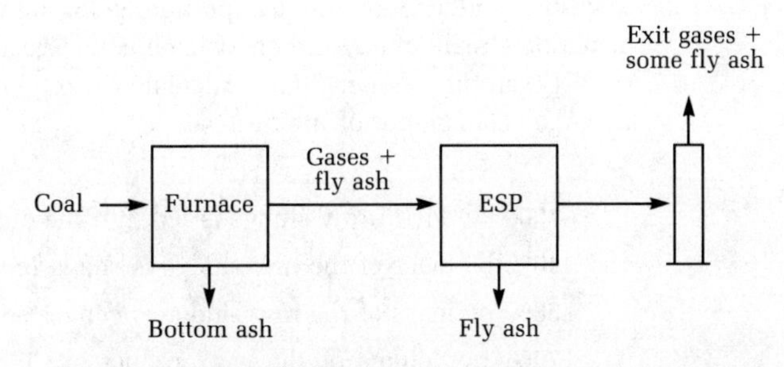

1. Draw an energy balance diagram for the plant and calculate the rate of heat emitted to the environment in joules per second.
2. Calculate the rate of coal input to the furnace, in kg/day.
3. Calculate the rate of ash emissions to the atmosphere, in kg/s.

3.5 Consider the following diagram:

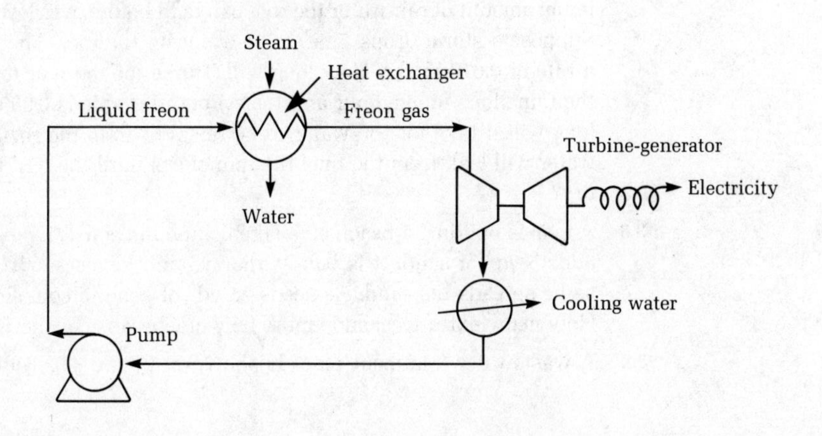

Given the following data, calculate the overall energy conversion efficiency of this process and the electricity production in kW.

Steam flows through the heat exchanger at the rate of 8000 lb/hr.

The enthalpy (heat content) of the steam is 1050 BTU/lb.

The exchanger is 75% efficient in transferring energy from the steam to the freon, changing the freon from liquid to vapor form.

The thermal energy of the freon vapor is converted to mechanical energy in the turbine with 55% efficiency.

The generator converts mechanical energy to electrical energy with 98% efficiency.

3.6 Draw a labeled diagram showing raw sludge being fed to an anaerobic digester which is producing three products: a gas, a supernatant, and digested sludge. Show the gas product going to a furnace along with a separate stream of air, and show combustion gases leaving the furnace.
Given the following data, calculate the total molar flow rate of CO_2 gas coming out of the furnace.

1000 kg/day of raw sludge is fed into the digester.

40% by mass of the raw sludge is converted to digested sludge.

58% by mass of the raw sludge is converted to supernatant.

70% by volume of the gas product is CH_4 (methane); the other 30% is CO_2.

The density of the gas product at 1 atm and 25°C is 1 kg/m^3.

Enough air is used to burn all of the methane to CO_2 and H_2O.

A value of *R*, the ideal gas constant, is 0.08205 L-atm/mole-K.

3.7 A rooftop holds rainwater puddles because of poor drainage. The maximum amount of rainwater the roof can hold before overflowing is 3.5 m^3. Suppose a storm drops 7 m^3/hr of water on the roof. The roof leaks at a rate of 0.016 m^3/hr. How long will it take for the roof to overflow? If the rain stops in one hour and the evaporation rate is 0.0004 m^3/hr how long will it take for the water to be removed from the roof? How much water will leak from the time the rain starts until the roof is completely dry?

3.8 A man is building a beach on a lake behind his house. Every weekend he adds 5 yd^3 of sand. It is windy, though, and he loses 2 ft^3/day of sand as he spreads the sand. He needs 75 yd^3 of sand to complete the beach. How many extra weekends must he work because of the wind?

3.9 A wastewater treatment plant is shown in Figure 3.9. Total suspended

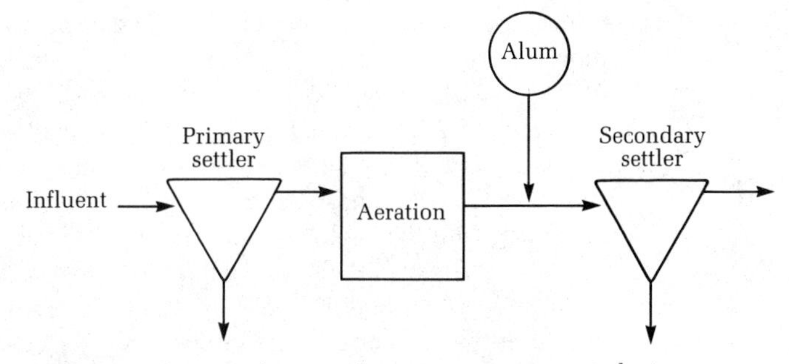

Figure 3.9 Schematic diagram of wastewater treatment plant.

solids (TSS) in the influent amount to 226 mg/L. Alum is added to help settling. Effluent requirements call for 90% reduction in TSS. If the primary settler removes 47% of initial TSS, find the amount of secondary removal required to achieve the 90% removal requirement. Assume all alum settles out.

3.10 The stack gas of a small plant contains 4.0 g particulates per m^3 of gas. The gas flow rate is 5 m^3/s. If an electrostatic precipitator removes 1000 kg of particulates per day, what is the emission rate of particulates? Give your answer in kg/day.

3.11 A 1000-megawatt plant operates at 40% efficiency. It consumes 10,000 tons of coal per day. The plant uses coal that is 4% sulfur and 10% ash. Virtually all the sulfur is converted to sulfur dioxide, and the plant is equipped with a 90% efficient SO_2 scrubber. Calculate the amount of sulfur dioxide emitted in gaseous form. Assume that one-third of the ash in the coal forms bottom ash, and the remainder forms fly ash. Pollution control devices capture 98% of the fly ash. Find the amount of bottom ash and fly ash collected in one day and the daily emissions of fly ash to the atmosphere.

3.12 A coal-burning plant uses 100 kg/hr of Pennsylvania anthracite coal. The plant utilizes an electrostatic precipitator to remove ash from the stack gases. The coal is 8% ash, and 40% of the ash falls out the bottom of the fire box. The efficiency of the precipitator is 96%. How much ash is released to the atmosphere?

3.13 A 100-MW power plant releases 18% of its energy as heat to the atmosphere and 45% as heat to the cooling towers. What is the efficiency of the power plant? What is the rate of heat input in BTU/hr, and how much heat is lost directly to the atmosphere in one day?

3.14 A CSTR reactor of 500 liters volume has a flow of 2000 liters per day. The reaction (which destroys the pollutant) is first-order with a rate constant of $k = 0.8$ day^{-1}. The influent concentration is 60 mg/L. Determine the effluent concentration of the reactant.

3.15 A company is in the business of rebuilding airplanes. Part of their operation involves sandblasting to remove old paint and rust. This process takes place in an enclosed building. Fine dust in the air is removed in a baghouse before exhausting the air outside. The baghouse cleans the air to 0.01 g/m^3, but regulations allow the exhaust to be 0.5 g/m^3. So, to save money, a portion of the dirty air from the operation is bypassed and mixed with the baghouse air to produce a final exhaust of 0.5 g/m^3. If 50,000 m^3/hr of air is used (with 1200 g/m^3 of sand) for the sandblasting operation and 99% of the sand falls to the floor, calculate the percent of the air that must go through the baghouse and the weight of the sand removed. Assume that all air (50,000 m^3/hr) exits through the final exhaust and contains the allowable limit of 0.5 g/m^3.

3.16 A city of 150,000 people discharges sewage (that has been treated to remove 90 percent of its suspended solids) into a small stream at point A (after Berthouex and Rudd, 1977). A critic has stated that what the city is doing is equivalent to discharging untreated sewage from 15,000 people; this comparison is shown in Figure 3.10. It is clear that the mass of suspended solids (SS) discharged is the same in each case. Compare the suspended solids concentration at point B. Is the criticism fair?

3.17 Calculate the minimum rate at which cooling water must be pumped through the condensers of a 800-MW nuclear power plant if the maximum cooling-water temperature rise is 15°F. The efficiency of the plant is 32 percent. Assume that no heat is lost to the atmosphere directly from the system. Give your answers in lb/hr and ft^3/s. Assuming that the plant is designed with five identical pipes in parallel to carry this cooling water, calculate the diameter of each pipe if the water velocity is 6.0 ft/s.

3.18 The wastewater from an industry has a salt, or total dissolved solids (TDS), concentration of 110,000 mg/L (after Berthouex and Rudd, 1977).

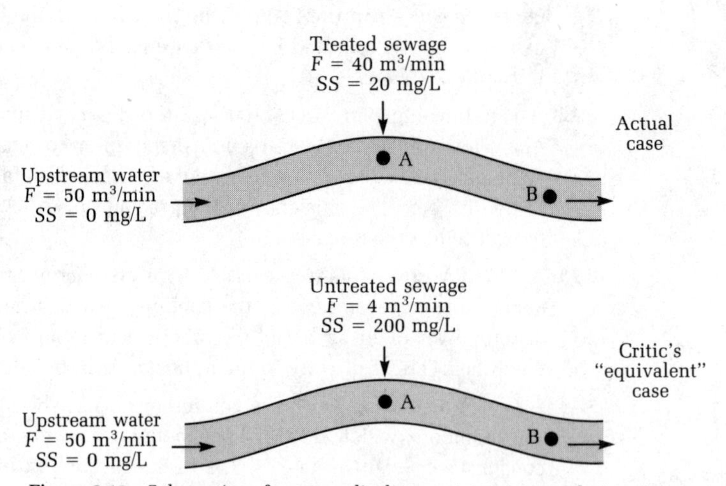

Figure 3.10 Schematics of sewage discharge; at top is actual situation, at bottom is critic's analogy.

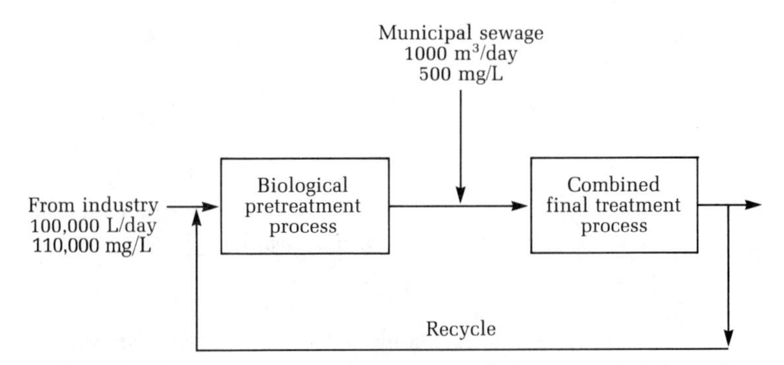

Figure 3.11 Wastewater treatment scheme including a recycle loop to dilute TDS in incoming stream.

This waste cannot be treated biologically in the pretreatment process to remove organic pollutants unless the TDS level is reduced to 20,000 mg/L by dilution because the high salt concentration interferes with the supply of oxygen to the treatment microorganisms. A treatment scheme that was proved in pilot plant tests is shown in Figure 3.11. How much recycle of the combined municipal-industry waste is required for the TDS concentration in the pretreatment process to be 20,000 mg/L?

3.19 You must prepare 2000 kg of a solution containing 14 weight percent ethyl alcohol in water. Two storage tanks are available, the first of which contains 5% ethanol in water and the second of which contains 25% ethanol in water. How much of each solution should you weigh out?

3.20 A gas containing equal parts (on a molar basis) of H_2, N_2, and H_2O is passed through a column of silica gel, which absorbs 96% of the water and none of the other gases (after Felder and Rousseau, 1978). The column was initially dry and had a mass of 2 kg. Following 5 hours of continuous operation, the column is reweighed and is found to have a mass of 2.18 kg. Calculate the molar flow rate (in mol/hr) of the feed gas and the mole fraction of water vapor in the product gas.

3.21 In order to meet a certain octane number specification, it is necessary to produce a gasoline containing 83 weight percent isooctane and 17 weight percent *n*-heptane. How many gallons of a high-octane gasoline containing 92 weight percent isooctane and 8 weight percent *n*-heptane must be blended with a straight-run gasoline containing 63 weight percent isooctane and 37 weight percent *n*-heptane to obtain 10,000 gal of the desired gasoline? The density of each of the liquids is 6.7 lb/gal.

3.22 The analysis of the waste gas from a burner fueled with a natural gas (essentially pure CH_4) is as follows (after Thompson and Ceckler, 1977):

N_2	75.24 mol percent
O_2	10.48 mol percent
CO_2	4.76 mol percent
H_2O	9.52 mol percent

What is the ratio of moles of air to moles of natural gas fed to the burner?

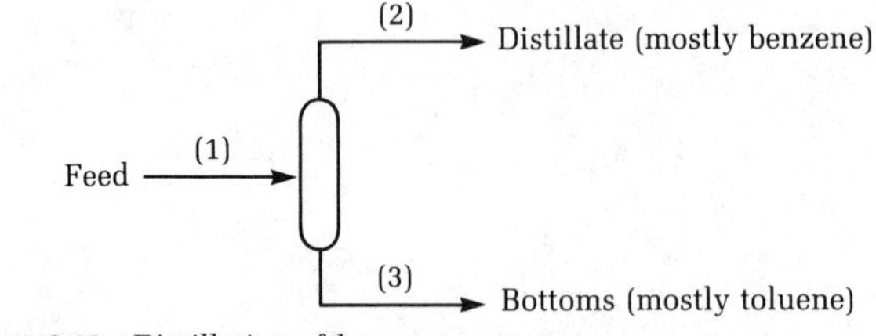

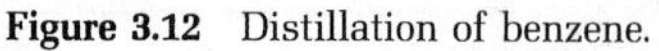

Figure 3.12 Distillation of benzene.

3.23 A mixture of benzene and toluene is separated by distillation, as shown in Figure 3-12 (after Myers and Seider, 1976). Toluene is slightly less volatile than benzene.

During continuous operation of the tower, the following flow rates and compositions are observed:

Stream number	*Flow rate (kg-mol/hr)*	*Mole fraction of benzene*
1	5.3	0.45
2	2.3	0.93

Calculate the flow rate and composition of the bottom stream. What is the flow rate of each stream in kilograms per hour?

3.24 A CSTR reactor is being fed a 100-L/min stream containing 0.10 mol/L of reactant A. The reaction proceeds according to $2A \rightarrow B$ with $-r_A = k[A]^3$. What is the volume of the reactor if k has a value of 3.0 $(L)^2/(mol)^2$ min and the outlet concentration of A is 0.05 mol/L? Calculate the production rate of B in mol/day.

3.25 A coal slurry is a mixture of crushed coal and water which can be pumped through a pipeline. A power plant needs 20 million kg/day of the coal (dry basis). A 50% coal-50% water slurry is pumped in the pipeline to the power plant. When it is received at the power plant, the coal is separated from the water (Figure 3.13). The separation process is inefficient: the separated water stream contains 2% (by weight) coal and the separated coal contains 20% (by weight) water. Fill in the following material balance table for the flow rates received at the power plant.

Mass flow rate (millions of kg/day)			
Stream	*Coal (dry)*	*Water*	*Total*
Pipeline slurry (stream 1)			
Separated water (stream 2)			
Separated coal (stream 3)	20		

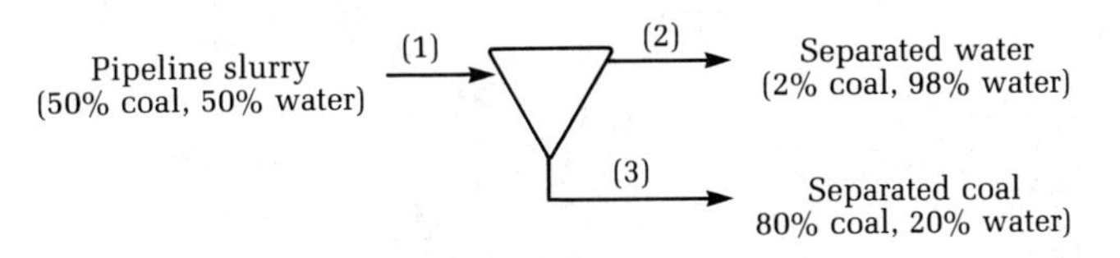

Figure 3.13 Separation of coal from slurry.

References

Berthouex, P. M. and Rudd, D. F. 1977. *Strategy of Pollution Control.* New York: Wiley.

Felder, R. M. and Rousseau, R. W. 1978. *Elementary Principles of Chemical Processes.* New York: Wiley.

Myers, A. L. and Seider, W. D. 1976. *Introduction to Chemical Engineering and Computer Calculations.* Englewood Cliffs, NJ: Prentice-Hall.

Thompson, E. V. and Ceckler, W. H. 1977. *Introduction to Chemical Engineering.* New York: McGraw-Hill.

4

Chemical Systems

4-1 Introduction

Of all the tools society has available for understanding, predicting, and minimizing its impact on the environment, chemistry is perhaps the best understood. That is not to imply that chemistry is a simple subject, but it is certainly better understood than the biological sciences. Moreover, some applications of some basic chemical principles allow us to gain valuable insight into the projected environmental impact of a given project. For example, a metal plating plant may sometimes discharge wastewater that contains a high concentration of chromium, a known heavy metal that is toxic to many microorganisms and has adverse effects on human health. The basic chemistry of chromium was developed before the concern over the environment surfaced in the late 1960's. It was developed because chromium is an excellent material for making a hard, attractive, and durable finish on metal. So, industry had already learned techniques for the removal of chromium when the need arose.

A large body of literature documents the adverse effects of chromium in the environment. This literature for the most part was not developed by and is not pertinent to the plating industry; however, given that a known heavy metal is in the waste stream, the means for removing that metal have been refined and are available to the engineer. Typically, reduction of Cr^{+6} to Cr^{+3} followed by lime precipitation of $Cr(OH)_3$ is used for chromium removal. Many of the chemical processes used by the environmental engineer to treat wastes utilizes concepts presented in introductory college chemistry courses. These processes, while challenging, are not overly complex and can provide any engineer with a technical appreciation of chemical pollution abatement.

Our purpose in this chapter is to provide an introduction to the chemical processes and systems that are the most common to pollution abatement: the gas laws, stoichiometry, acids and bases, oxidation reduction, precipitation reactions, and kinetics.

4-2 The gas laws

We will introduce the gas law by describing Boyle's Law and Charles' Law, which combine into the ideal gas law. Beyond these basic laws, gases are relatively straightforward because gaseous mixtures are considered homogeneous mixtures and contain equal numbers of moles if the volume, temperatures, and pressures are equal.

Boyle's Law states that the product of the pressure P and volume V of a combined gas is constant at a constant temperature. Mathematically stated, it is

$$P_1V_1 = P_2V_2 \tag{4.1}$$

where

P_1 = gaseous pressure of gas at V_1
V_1 = volume of gas at P_1
P_2 = gaseous pressure of gas at V_2
V_2 = volume of gas at P_2

Charles' Law deals with the effect of temperature on gas volume and states that the volume of a gas will expand or decrease by 1/273 for every degree Kelvin the absolute temperature of the gas is increased or decreased. The mathematical expression of Charles' Law is

$$V_T = V_O + \frac{T}{273} V_O \tag{4.2}$$

where

V_O = original gas volume
T = temperature, °C
V_T = final gas volume

This law predicts that a gas will have zero volume at $-273°C$. If the reference point is 0°K, the equation becomes

$$V_T = V_O \frac{T}{273} \tag{4.3}$$

This equation can be manipulated into the "perfect" or ideal gas law by multiplying each side by P_O, using Charles' Law to equate $P_O V_T$ and PV so that

$$PV = \frac{P_O V_O}{273} T \tag{4.4}$$

Since P_O and V_O are defined at standard conditions for 1 mole of gas, 1 atm pressure and 22.4 liters, equation (4.4) becomes

$$PV = RT \tag{4.5}$$

where R is the universal gas constant, and can easily be expanded to

$$PV = nRT \tag{4.6}$$

for n moles of gas. Common values of R are 0.08206 L-atm/mol-°K, 8.314 J/mol-degree and 1.987 cal/mol-°K. Many students find equating the ideal gas law as shown in equation (4.7) a convenient method of utilizing the gas equations for the same gas under changing pressure, volume, and temperature conditions:

$$\frac{P_1 V_1}{nRT_1} = \frac{P_2 V_2}{nRT_2} \tag{4.7}$$

The nR terms in the denominators are equal and will cancel.

Dalton's law

Ideal mixtures of different gases in a confined volume will exert individual pressures directly proportional to their molar fractions in that volume, and the total pressure will be the sum of those pressures:

$$P_T = \sum P_i \tag{4.8}$$

where P_i = partial pressure of any gas i in the volume

$$P_i = X_i P_T \tag{4.9}$$

where X_i = mole fraction i

$$X_i = \frac{n_i}{n} \tag{4.10}$$

where

n_i = moles of i
n = total moles

Henry's law

The amount of a gas that will dissolve in a solution is directly proportional to the partial pressure of that gas in contact with the solvent, typically water. This statement is known as Henry's Law and is the direct result of a chemical equilibrium between the gas and the liquid.

$$CO_2(\text{gas}) + H_2O \rightleftharpoons CO_2(\text{liq}) + H_2O \tag{4.11}$$

$$C = K_H P_i \tag{4.12}$$

where

P_i = partial pressure, in atm
K_H = equilibrium constant or Henry's Constant, in mol/L-atm
C = concentration of dissolved gas at equilibrium in mol/L

In different units the equation can be written as

$$C = \alpha P_g \tag{4.13}$$

where

P_g = partial pressure of gas, in atm
α = Henry's Constant, mg/L-atm
C = concentration of dissolved gas in mg/L

This is the normal form of Henry's Law. The two constants are different only in the units; both are derived from the equilibrium equation. Some values for K_H are given in Table 4.1.

When a gas is dissolved in solution, it can then react in solution. This sequence of reactions can be beneficially utilized to remove some environmentally harmful gases such as the oxides of sulfur. Sulfur removal can be accomplished by: (1) lime slurry scrubbing, (2) limestone slurry scrubbing, (3) magnesium oxide scrubbing, or (4) sodium base scrubbing, as shown below:

1. $Ca(OH)_2 + SO_2 \longrightarrow CaSO_3 + H_2O$
2. $CaCO_3 + SO_2 \longrightarrow CaSO_3 + CO_2(g)$
3. $Mg(OH)_2(s) + SO_2 \longrightarrow MgSO_3 + H_2O$
4. $Na_2SO_3 + H_2O + SO_2 \longrightarrow 2NaHSO_3$

Example Problem 4.1

A digester produces carbon dioxide and methane from the anaerobic decomposition of waste. Assume that 1000 kilograms of acetic acid (CH_3COOH) is decomposed to CO_2 and CH_4. Determine the volume of gas produced at 32°C and the required volume of water if 50% of each gas is captured in water that is open to the atmosphere at 25°C.

Reaction: $CH_3COOH \xrightarrow[\text{Digestion}]{\text{Anaerobic}} CO_2 + CH_4$

TABLE 4.1 Henry's constants for gases in H_2O at 25°C

Gas	*K_H(mol/L-atm)*
O_2	1.28×10^{-3}
CO_2	3.38×10^{-2}
H_2	7.90×10^{-4}
N_2	6.48×10^{-4}
CH_4	1.34×10^{-3}

Solution Volume of gas:

$$(1000 \text{ kg } CH_3COOH)\left(\frac{10^3 g}{kg}\right)\left(\frac{mol}{60g}\right)\left(\frac{mol\ CO_2}{mol}\right)$$

$$\times \left(\frac{22.4}{CO_2}\right)\left(\frac{273 + 32}{273}\right)$$

$$CO_2 \text{ gas} = 4.17 \times 10^5 \text{ L}$$

Since 1 mol CH_4 is produced with 1 mol CO_2 and Allogadro's number;

$$CH_4 \text{ gas} = 4.17 \times 10^5 \text{ L}$$

$$\text{Total gas volume} = 2 \times 4.17 \times 10^5 = 8.34 \times 10^5 \text{ L}$$

Volume of water:

$$K_{H(CH_4)} = 1.34 \times 10^{-3} \frac{mol}{L\text{-}atm};\ K_{H(CO_2)} = 3.38 \times 10^{-2} \frac{mol}{L\text{-}atm}$$

$$H_2O \text{ Volume} = \frac{\text{mol gas}}{K_H} = \frac{16{,}666.7}{1.34 \times 10^{-3}} + \frac{16{,}666.7}{3.38 \times 10^{-2}}$$

$$= 1.24 \times 10^7 + 0.05 \times 10^7 = 1.29 \times 10^7 \text{ L}$$

The moles of gas came from part of the CO_2 calculation and the stoichiometry (balance of constituents) of the reactions. Note that much more H_2O is required for CH_4 capture than for CO_2, so usually CH_4 is burned.

The automobile is a major source of nitrogen oxides, and to date no effective schemes are practiced on a national scale for removing NO_x from automobile exhaust. Schemes involving catalytic converters where the NO_2 is reduced to N_2 have been proposed, but they degrade the efficiency of automobiles and are not practiced nationally.

One major problem that has surfaced since 1970 is the decreasing pH of precipitation. This phenomenon has become known as acid rain and is thought to be caused by the absorption of sulfur and nitrogen oxides in rainfall. It occurs most predominantly in areas of heavily industralized population centers. In the last decade there has been a tenfold increase in the acidity of rainfall. The eastern United States is the section of the country that receives the most acidic rainfall.

Since pH is a logarithmic measure of the molar H^+ concentration, a tenfold increase represents a change of 1 pH unit. The average pH of rainfall in the northeastern United States is 4.25, contrasted to 7.0 in the western United States. This represents almost a thousand-fold increase in H^+ concentration in different areas of the country. The problems associated with acid rain are the eventual destruction of property susceptible to acid-base reaction, including the destruction of aquatic life in natural lakes. New York state has reported 83 lakes in which the pH has dropped to 4.5 or less; they are devoid of game fish. There is yet no plan to restore these lakes as the energy cost to the power plants producing the sulfur oxides would be increased dramatically if control measures were instigated. Canada has a worse problem with acid rain than the United States because the natural migration path of emissions from the United States is into Canada.

4-3 Stoichiometry

Introduction

The basis of stoichiometry is the atomic weights of elements (see Appendix C), which is a means by which the mass of one element can be compared to the mass of a second element. The basis of this comparison method is carbon, which is defined to have an atomic weight of exactly 12.

The mass of a molecule is simply the sum of the masses of atoms that bond together to form it. This simple constitutive relation is of importance to engineers because it allows a quantitative calculation for chemical reactions. Chemical reactions occur in moles, which are measured in gram-molecular weights. The moles involved in a chemical reaction can be determined by dividing the mass by the molecular weight in grams of each species.

$$\text{Moles X} = \frac{\text{mass X}}{\text{g-molecular weight X}} \tag{4.14}$$

The molecular weight of a substance is readily available from any periodic chart or table. The engineer must be sure to use the atomic mass (neutrons plus protons) of each element rather than the atomic number (protons only) when determining the molecular weight of a species.

Example Problem 4.2

How many moles of $Ca(HCO_3)_2$ are present in 100 grams of $Ca(HCO_3)_2$?

$$\text{molecular weight } Ca(HCO_3)_2 = 40 + (61)2$$
$$= 162$$

$$\text{gram-moles X} = \frac{100}{162} = 0.62 \text{ g-moles } Ca(HCO_3)_2$$

The student should note the term gram-moles. Because there are several different measurements of mass, it must be clearly understood what units are being used to measure mass. The metric system is the preferred system; however, chemical engineers in the past have used pound-moles to determine product and reactant mass. Results are satisfactory as long as the same mole units are carried throughout all calculations. The number of lb-moles of $Ca(HCO_3)_2$ in the previous example is 1.4×10^{-3}, which is exactly the same mass as 0.62 g-moles $Ca(HCO_3)_2$, only in different units. Usually in environmental applications, a wide range in masses of chemicals occurs, and the metric measurement of mass is the most convenient.

Another useful measure of mass is an equivalent, which is defined as that weight of any substance which contains 1 gram-atom of available hydrogen or its equivalent. Equivalence can be based on ion charge, proton transfer, or electron transfer depending on the reaction. It is possible for some substances to have different chemical reactions involving the same compound, thus different equivalent weights. The general equation for calculating the equivalent weight always requires dividing the molecular weight by a number n representative of the ion charge, proton transfer, or electron transfer as shown in equation (4.15).

$$\text{Eq. wt. X} = \frac{\text{Mol. wt. X}}{n} \tag{4.15}$$

Example Problem 4.3

Determine the number of equivalent weights of the specified substances in the following reactions:

1. 100 g $CaCl_2$.

$$CaCl_2 \longrightarrow Ca^{+2} + 2Cl^-$$

$$\text{eq. wt. } CaCl_2 = \frac{\text{mol. wt. } CaCl_2}{n} = \frac{40 + 71}{2} = 55.5 \text{ g}$$

$$\text{eq. } CaCl_2 = \frac{\text{mass } CaCl_2}{\text{eq. wt. } CaCl_2} = \frac{100}{55.5} = 1.80 \text{ eq.}$$

2. 10 g Fe.

$$Fe + 3H_2O = Fe(OH)_3 + 3H^+$$

Fe has been oxidized from 0 charge to +3, which involves a loss of 3 electrons.

$$\text{eq. wt. Fe} = \frac{\text{mol. wt}}{n} = \frac{55.85}{3} = 18.62 \text{ g}$$

$$\text{eq. Fe} = \frac{10}{18.62} = 0.54 \text{ eq.}$$

As previously noted, the number of equivalents is dependent on the particular chemical reaction involved. Equivalents have meaning in reality for the reactants of any reaction in that the number of equivalents of any reactant is equal to the number of equivalents of product produced.

Concentration

Most environmental engineering calculations require the manipulating of concentration, which is defined as mass solute per volume solution or in some cases mass solute per mass solution. Perhaps the molarity and the normality of a solution are the most familiar concentration terms for general applications. The molarity of a solution with regard to a specific substance is defined as the number of moles of a substance per liter of solution. The normality of a solution is the number of equivalents of a substance per liter of solution.

Example Problem 4.4

Determine the molarity and normality of 100 grams of $CaSO_4$ dissolved in enough water to make 1.00 liter of solution.

$$\text{Molarity} = \text{mol/L} = \frac{100 \text{ g/L}}{136 \text{ g/mol}} = 0.74 \text{ mol/L}$$

$$\text{Normality} = \text{eq/L} = \frac{100 \text{ g}}{136/2} \Big/ 1 \text{ L} = 1.47 \text{ eq/L}$$

The general relation between molarity and normality is

$$N = nM \tag{4.16}$$

where

N = normality
M = molarity
n = equivalent charge

Some people confuse the number of moles or equivalents with the concentration relationship of molarity or normality. The number of equivalents is equal to n times the number of moles, but an equivalent weight is determined by dividing the molecular structure by n. The above equation indicates that $N \geq M$ for all $n \geq 1$. This is by far the most common case. Students realize that the mass required to make a 1 M solution is greater than or equal to the mass required for a 1 N solution. Sometimes they interchange the mass with concentration and determine molarity to be greater than normality when the reverse is correct. The typical mistake is shown below from the previous sample:

$$\text{Normality} = \frac{\text{equivalents}}{\text{liter}} = \frac{(100/136)2}{1 \text{ liter}} = 1.47 \text{ eq (correct)}$$

$$\text{Molarity} = n \times N = 2 \times 147 = 2.94 \text{ M (wrong)}$$

Other concentration units of interest to the environmental engineer are milliequivalents per liter and millimoles per liter.

$$1 \text{ meq/L} = 1/1000 \text{ eq/L}$$
$$1 \text{ mmol/L} = 1/1000 \text{ mol/L}$$

This definition has also introduced milliequivalent and millimole as mass terms, which are more appropriate for studies of environmental reactions as they commonly deal with small quantities. Other units that measure lower concentrations and are helpful are mg/L and μg/L, which in aqueous solutions are taken to be the same as parts per million (ppm) or parts per billion (ppb), respectively.

Example Problem 4.5

Determine how many ppb 1 μg/L represents.

$$\text{mass/volume: } 1\ \mu\text{g/L}\left(\frac{10^{-6}\ \text{g}}{\text{g}}\right)\left(\frac{1\ \text{L}}{1000\ \text{g}}\right) = 10^{-9}\text{g/g}$$

or

$$1\ \mu\text{g/L} = 1 \text{ ppb by weight}$$

Note: This solution utilizes the density of water, which is correct for an aqueous solution. The same is not true for a gaseous mixture.

Since the ideal gas law predicts that 1 g-mole of gas will occupy 22.414 L at 0°C and 1 atm, the density of any gas can easily be found at STP: standard temperature and pressure. Furthermore, the ideal gas law provides the basis for development of a general equation that correlates mass per unit volume and volume per volume for any temperature and pressure. For most gaseous applications, the most common terms are $\mu g/m^3$ for M/V and ppm for V/V.

Example Problem 4.6

Develop a general equation for converting ppm to $\mu g/m^3$ for any gas at any temperature and pressure assuming the ideal gas law.

$$PV = nRT, \quad V = \frac{nRT}{P}$$

where

V = volume in liters

n = moles

R = 0.082054 L-atm/°K-mol

T = temperature in °K

P = pressure in atm

$$(\mathrm{X}\ \mu\mathrm{g/m^3})\left(\frac{10^{-3}\ \mathrm{m^3}}{1\ \mathrm{L}}\right)\left(\frac{RT}{P}\right) = (\mathrm{X}\ \mu\mathrm{g/m^3})\left(\frac{10^{-3}\ \mathrm{m^3}}{1\ \mathrm{L}}\right)\left(\frac{22.4\ \mathrm{L}}{\mathrm{mol}}\right) \times \left(\frac{\mathrm{mol}}{\text{g-mol. wt.}}\right)$$

For STP,

$$Y\ (ppm) = \mathrm{X}\ (10^{-3})\left(\frac{22.414}{\text{g-mol. wt.}}\right) \text{ where } \mathrm{X} = \mu\mathrm{g/m^3}$$

The above solution utilizes the volume of 1 mole of gas at STP and direct units conversion.

Concentrations may involve expressing chemical elements or compounds (species) in terms of other species but only on the basis of equivalents. This is commonly done in water-quality analyses, in which units are in terms of mg/L as $CaCO_3$ or mg/L of some quantitative analytical standard. Examples of this would be 40 mg/L Mg^{+2} as $CaCO_3$ or 10 mg/L BOD (biological oxygen demand).

To express the concentration of any chemical substance in terms of another, the equivalent weight of each of the two substances must be known before the conversions can be made. *Remember that one equivalent of reactant will produce one equivalent of product for a given chemical reaction.*

Example Problem 4.7

Express 10 mg/L of Mg^{+2} as $CaCO_3$.

Solution Here, the equivalent weights of Mg^{+2} and $CaCO_3$ are calculated by dividing the molecular weight of each by 2, the charge of the complete ionization.

$$\text{Eq. wt.} = \text{mol. wt.}/n$$
$$Mg^{+2}\ \text{Eq. wt.} = 24/2 = 12\ \text{g/eq;}$$
$$\text{Eq. wt. } CaCO_3 = 100/2 = 50\ \text{g/eq}$$

Since it is known that the equivalents of reactants are equal to the equivalents of products, a direct ratio can be taken that converts mg/L of any substance to mg/L of any other substance as long as the substances are properly related by a chemical reaction.

$$\text{mg/L } Mg^{+2} \text{ as } CaCO_3 = 10 \text{ mg/L } Mg^{+2}\left(\frac{\text{meq } Mg^{+2}}{12 \text{ mg } Mg^{+2}}\right) \times \left(\frac{50 \text{ mg } CaCO_3}{\text{meq } CaCO_3}\right)$$

Thus

$$10 \text{ mg/L of } Mg^{+2} \text{ as } CaCO_3 = 41.7 \text{ mg/L}$$

The equivalent weight of $CaCO_3$ is nearly always taken as 50 grams since the effective charge on calcium is usually 2 for all reactions involving $CaCO_3$ as a product. Likewise, the equivalent weight of any substance converted to calcium carbonate is determined by dividing the molecular weight by its stated charge. For example, the bicarbonate ion, HCO_3^-, has an equivalent weight of 61 grams. Therefore, it is typically not necessary to write the chemical reaction to determine mg/L as $CaCO_3$ unless special conditions are required. A general equation can be written for conversion to equivalent mg/L:

$$\text{mg/L X as Y} = \text{X mg/L}\left(\frac{\text{meq X}}{\text{mg X}}\right)\left(\frac{\text{mg Y}}{\text{meq Y}}\right) \tag{4.17}$$

For $CaCO_3$, typically

$$\text{mg/L X as } CaCO_3 = \text{X mg/L}\left(\frac{50}{\text{mg/meq X}}\right) \tag{4.18}$$

The general equation is applicable to any reaction and reading through the pages of *Standard Methods* provides many examples of such measurement. Ammonia (NH_3) and nitrates (NO_3^-) are reported as NH_3-N and NO_3-N. Chemical and biological oxygen demand (COD and BOD) are reported as equivalent O_2 mg/L. Because of the very nature of COD and BOD, it would be impossible to have any meaningful parameter unless equivalence was utilized. Differing elements such as carbon, nitrogen, and phosphorus, can be oxidized in this test, and totaling mg/L C, mg/L N, mg/L P, and so forth would have no meaning without using equivalence. The equivalent weight per liter of O_2 in any oxidation reaction is 8 g/L or 8000 mg/L. If the amount of O_2 reacted in known, the O_2 can be and is commonly reported as mg/L. The reduction of O_2 is

$$O_2 + 4H^+ + 4e^- \longrightarrow 2H_2O \tag{4.19}$$

$$\text{eq. wt. } O_2 = 32\text{ g}/4 = 8\text{ g/eq} = 8000\text{ mg/eq}$$

Example Problem 4.8

Determine the mg/L of O_2 required for a BOD test if 0.1 mmol of urea, $(NH_2)_2CO$, is completely oxidized in a 50-mL sample as shown below:

$$2(NH_2)_2CO + 8O_2 \longrightarrow 2CO_2 + 4HNO_3 + 2H_2O$$

$$\text{mg/L } O_2 = \left(\frac{0.1\text{ mmol }(NH_2)_2CO}{50\text{ mL}}\right)\left(\frac{8\text{ mmol }O_2}{2\text{ mmol }(NH_2)_2CO}\right)$$

$$\times\left(\frac{4\text{ meq }O_2}{\text{mmol }O_2}\right)\left(\frac{8\text{ mg }O_2}{\text{meq }O_2}\right)\left(\frac{1000\text{ mL}}{\text{L}}\right)$$

$$= 256\text{ mg/L}$$

In example problem 4.8, the use of the chemical reaction was expanded to allow the calculation of the O_2 required for complete oxidation by utilizing a stoichiometric conversion from urea to O_2 that was 8 mmol O_2/2 mmol urea. There was also a change in concentration units from mmol/mL to meq/L, which utilized the milliequivalent weight and a volume change from milliliters to liters. What is reported above is the stoichiometric O_2 requirement for the biological oxidation of C, H, and N in a given form in a given reaction, which provides a common basis for reporting the effect of potential pollution on the environment. BOD is seen then to be an indirect measurement of pollutants C, H, and N through measuring what is needed to satisfy the oxygen demand of the pollutants. This does not imply anything about toxicity or other measures of pollution.

Sometimes it is desirable to convert mg/L to lb/day for a given flow. The conversion factor is shown below for an aqueous solid:

$$\text{mg/L}\left(\frac{\text{g}}{10^6\ \text{mg}}\right)\left(\frac{\text{lb}}{454\ \text{g}}\right)\left(\frac{3.791}{\text{gal}}\right)\left(\frac{10^6\ \text{gal}}{\text{Mgal}}\right) = 8.345\,\frac{\text{lb/Mgal/day}}{\text{mg/L}}$$

This means that an engineer can determine the pounds of solids of interest in an aqueous stream for any given flow simply by multiplying the appropriate concentration in mg/L by 8.345 and the flow in millions of gallons per day (Mgal/day), or mgd.

Another useful tool is solids classification. A solids tree is shown in Figure 4.1. Total solids, TS, represents the sum of total dissolved solids and total suspended solids:

$$\text{TS} = \text{TSS} + \text{TDS} \tag{4.20}$$

Solids are usually measured in concentration units of mg/L. TSS is defined as solids that are removed by filtration. TDS is defined as solids that pass the filter. Obviously, the pore size of the filter is important to the determination of TSS and TDS. *Standard Methods* requires that a glass fiber or a No. 40 Whatman filter be used for solids classification. Actually, the TDS measured by this test not only includes the TDS, but also the colloidal solids. The approximate size range for TSS is over 1000 nm to 100 nm; colloidal solids range from 100 nm to 1 nm and TDS are all solids below 1 nm. (1 nm = 0.001 μm.)

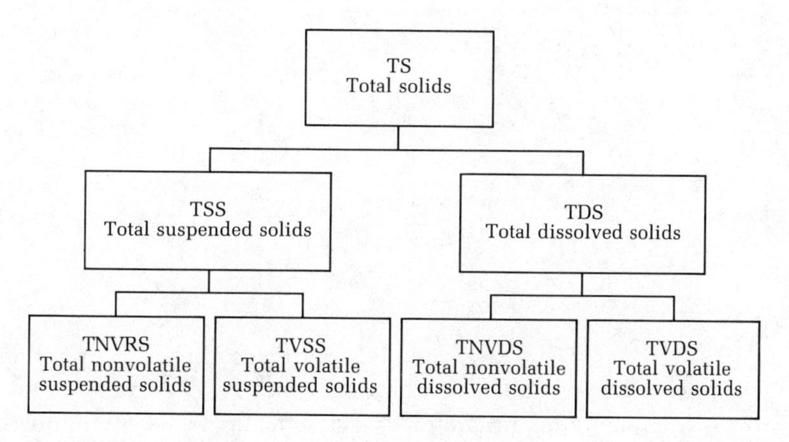

Figure 4.1 Solids tree for solids classification.

Many organics (natural color) exist in the colloidal state and cannot be removed by settling only. For that reason, TDS always includes true TDS and colloidal matter. Also shown in the solids tree is a volatile/nonvolatile solids classification.

$$\mathrm{TS} = \mathrm{TNVS} + \mathrm{TVS} \tag{4.21}$$

$$\mathrm{TSS} = \mathrm{NVTSS} + \mathrm{VTSS} \tag{4.22}$$

$$\mathrm{TDS} = \mathrm{NVTDS} + \mathrm{VTDS} \tag{4.23}$$

No chemical measurement is more useful to the process environmental engineer than solids classification. Sludge production from domestic or industrial wastewater and drinking water can be measured quantitatively by TSS concentrations and flow for a given time period by

$$T = QC\,(8.345) \tag{4.24}$$

where

T = lb solids/day
Q = flow, mgd
C = affected solids concentration, mg/L
8.345 = conversion factor for mg/L to lb/mgd

and

$$V = \frac{T}{(Sp)(\rho)} + \frac{T(1 - X)}{(X)(\rho)} \tag{4.25}$$

where

V = volume TSS/day (sludge)
T = lb solids/days
X = solids fraction
Sp = solid specific gravity
ρ = density of water

There are many different types of chemical reactions that are of interest to the environmental engineer. An example of a common precipitation reaction in water softening is shown in the reaction

$$\mathrm{Ca(HCO_3)_2 + Ca(OH)_2 \longrightarrow 2CaCO_3\downarrow + 2H_2O} \qquad (4.26)$$

In this reaction, calcium hardness is removed from solution by lime addition, which precipitates $CaCO_3$ (denoted by the downward arrow). Using this reaction as an example, mass calculations will demonstrate using stoichiometry.

Example Problem 4.9

Given a flow of 7.5 mgd, determine how much $Ca(OH)_2$ is required each day to reduce the Ca^{+2} concentration from 150 mg/L as $CaCO_3$ to 100 mg/L as $CaCO_3$ by the above equation. Also, determine the pounds of $CaCO_3$ and the sludge volume produced each day if the solids are settled to 1% by weight and the specific gravity of $CaCO_3$ is 2.8. Determine the annual cost of lime and sludge disposal if lime costs \$80/ton and sludge disposal costs \$1/yd^3.

Solution Since the calcium concentration is given on a meq basis as $CaCO_3$, the lime dose is calculated using milliequivalents. However, the student could convert to a molar basis using reaction (4.26) and solve if desired.

Reactions:

$$\mathrm{Ca(HCO_3)_2 + Ca(OH)_2 \longrightarrow 2CaCO_3 + 2H_2O}$$

$$150 - 100 = 50 \text{ mg/L } \mathrm{Ca^{+2}} \text{ as } \mathrm{CaCO_3} \text{ to remove}$$

Lime dose:

$$50 \text{ mg/L } \mathrm{Ca^{+2}}\left(\frac{\text{meq}}{50 \text{ mg } \mathrm{CaCO_3}}\right)\left(\frac{\text{meq } \mathrm{Ca(OH)_2}}{\text{meq } \mathrm{Ca(HCO_3)_2}}\right)$$

$$\times\left(\frac{37 \text{ mg } \mathrm{Ca(OH)_2}}{\text{meq}}\right) = 37 \text{ mg/L } \mathrm{Ca(OH)_2}$$

lb lime/day:

$$T = QC\,(8.345)$$

$$= 7.5 \text{ mgd}(37 \text{ mg/L})\left(\frac{8.345 \text{ lb/mgd}}{\text{mg/L}}\right) = 2316 \text{ lb } \mathrm{Ca(OH)_2}\text{/day}$$

Sludge concentration:

$$50 \text{ mg/L Ca}^{+2}\left(\frac{\text{meq}}{50 \text{ mg CaCO}_3}\right)\left(\frac{2 \text{ meq CaCO}_3}{\text{meq Ca(HCO}_3)_2}\right)\left(\frac{50 \text{ mg}}{\text{meq}}\right)$$

$$= 100 \text{ mg/L CaCO}_3$$

$$T = QC\,(8.345)$$

$$= (7.5)(100)(8.345) = 6259 \text{ lb CaCO}_3/\text{day}$$

Sludge volume:

$$V = \frac{T}{Sp\,\rho} + \frac{T(1 - X)}{X\rho}$$

$$= \frac{6259}{(2.8)(62.4)} + \frac{6259(1 - 0.01)}{(0.01)(62.4)}$$

$$= 36 + 9930 = 9966 \text{ ft}^3/\text{day}$$

Sludge cost:

$$\left(9966\,\frac{\text{ft}^3}{\text{day}}\right)\left(\frac{\text{yd}^3}{27 \text{ ft}^3}\right)\left(\frac{\$1}{\text{yd}^3}\right)\left(\frac{365 \text{ day}}{\text{year}}\right) = \$134{,}725/\text{year}$$

4-4 Chemical reaction rates

Now that the basic gas laws have been introduced in addition to the basic laws of stoichiometry, it is appropriate to discuss the fundamentals of chemical reaction with respect to time. If a reaction is time-dependent, such as biochemical oxidation or chemical disinfection, it is referred to as a kinetic reaction. If the reaction is instantaneous for engineering applications, it can be modeled by stoichiometry using the equilibrium constant. In this chapter, zero-, first-, and second-order rate constants will be evaluated and discussed.

For the basic chemical reaction given:

$$aA + bB = cC + dD \tag{4.27}$$

In equation (4.27), the law of mass action states, that a moles of A and b moles of B combined to form a product, in this case c moles of C and d moles of D. The rate of formation of C or D is

$$r = k(A)^a(B)^b \tag{4.28}$$

The reaction is of a order in A and b order in B with an overall order of $(a + b)$. The order of a reaction may be any number and in its simplest form is an integer. Let us develop simple rate equations for zero-, first-, and second-order kinetics.

A zero-order process is independent of any concentration. Mathematically stated, the change of A with respect to time is constant:

$$\frac{-dA}{dt} = k \tag{4.29}$$

Integration of (4.29) produces (4.30) where A_o is the initial concentration of A:

$$A_o - A = kt \tag{4.30}$$

Probably the most common reaction rate utilized in environmental engineering is a first-order reaction

$$A \longrightarrow P \tag{4.31}$$

For a first-order reaction,

$$\frac{dA}{dt} = -kA \tag{4.32}$$

Integrating,

$$\ln A/A_o = -kt \tag{4.33}$$

or

$$A = A_o e^{-kt} \tag{4.34}$$

This reaction can be easily solved for the production rate:

$$\frac{dP}{dt} = \frac{-dA}{dt} \tag{4.35}$$

The loss of one mole of A in this elementary reaction means an increase in one mole of P, so that utimately at the end of a complete reaction,

$$A_u = 0 \tag{4.36}$$

$$P_u = A_o \tag{4.37}$$

$$P_t = P_u - P_r \tag{4.38}$$

$$P_t = A_o - A_o e^{-kt} \tag{4.39}$$

$$P_t = A_o(1 - e^{-kt}) \tag{4.40}$$

$$P_t = P_u(1 - e^{-kt}) \tag{4.41}$$

where

P_t = product formed at time t
A_u, P_u = ultimate reactant or product
P_r = product remaining to be formed

The ultimate value of A is zero for a complete reaction. The ultimate value of P is A_o, the initial value of A. At any time t, the product formed, P_t, is shown in equation (4.38). Equations (4.39) through (4.41) develop the relation for the product formed at any time t. Equations (4.41) and (4.34) are useful forms for first-order reactions if reactant or product concentrations or mass at any time t are desired.

A second-order kinetic rate equation would be developed for

$$2A \longrightarrow \text{Products} \tag{4.42}$$

For second-order conditions, the product formed would be

$$\frac{-dA}{dt} = kA^2 \tag{4.43}$$

This can be rearranged and solved:

$$\frac{-dA}{A^2} = kdt \tag{4.44}$$

$$\frac{1}{A} - \frac{1}{A_o} = kt \tag{4.45}$$

The student should note that if the reactants were different, then the kinetic equations derived from the rate law would also be different. For example, the second-order equation below has the integrated solution shown beneath it, where X is the number of moles of A or B reacted. The student may derive this equation for a review of elemental kinetic applications.

$$A + B \longrightarrow \text{Products} \tag{4.46}$$

$$(A_o - B_o)^{-1} \ln \frac{B_o(A_o - X)}{A_o(B_o - X)} = k_2 t \tag{4.47}$$

A useful concept for any rate equation is the concept of half-lives or doubling time, which is defined as the amount of time required for 50% of the reactant present to react or for the amount of product present to double. Obviously, when the reactions are just beginning, product doubling time is meaningless because $P = 0$ at that instant. Conversely, when no reactant is present, reactant half-life is also meaningless. The most common example of half-life might be the loss of 50% of the initial material by a first-order reaction, or find t for $A_o = 0.5A_o$:

$$\ln \frac{A}{A_o} = -kt \qquad \text{let } A = 0.5A_o \qquad \ln \frac{0.5A_o}{A_o} = -kt$$

$$t_{1/2} = \frac{\ln 0.5}{-k} = \frac{0.693}{k} \tag{4.48}$$

The student can easily derive the corresponding equation for the doubling time for a first-order equation. Note that time is always positive.

$$t_2 = \frac{\ln 2}{k} = +\frac{0.693}{k} \tag{4.49}$$

The first-order decay constants are negative whereas the first-order growth (products) constants are positive, which combine with equations (4.48) or (4.49) to give positive times. See Table 4.2 for a summary of some reactions.

One technique common to applied science or engineering design is to plot all equations in final form from clarity. The plot of the rate

TABLE 4.2 Summary table for elemental kinetic equations.

Chemical equation	Order	Rate equation	Integrated form	$t_{1/2}$
$A \longrightarrow P$	0	$\frac{-dA}{dt} = k$	$A_o - A = kt$	$\frac{0.5A_o}{k}$
$A \longrightarrow P$	1	$\frac{-dA}{dt} = kA$	$\ln \frac{A}{A_o} = -kt$	$\frac{0.693}{k}$
$2A \longrightarrow P$	2	$\frac{-dA}{dt} = kA^2$	$\frac{1}{A} - \frac{1}{A_o} = kt$	$\frac{1}{A_o k}$

equations and the integrated rate equations as a function of time are shown in Figure 4.2. The half-life of a first-order equation is constant, whereas the half-life of the second-order equation increases by a proportional amount in the order of a geometric progression. Plotting the correct integrated kinetic equation versus time will give a straight line of slope $-k$, the rate constant. A common technique for approximating the order of a reaction is to record the decrease in reactant concentration with time and plot the appropriate form versus time. Direct observation or a statistical calculation can be made for the goodness of fit and the appropriate rate equation selected that provides the best estimate over the experimental range.

Example Problem 4.10

The following data were collected during the degradation of R to P ($R \rightarrow P$). Approximate the order and the rate constant of this reaction.

Time min)	0	5	10	30	60	120	240
R, mmol/L	10	2.5	1.1	0.3	0.2	0.1	0
P, mmol/L	0	7.5	8.9	9.7	9.8	9.9	10.0
1/R	0.1	0.4	0.9	3.3	5.0	10.0	—

A plot of the data in example problem 4.10, assuming various orders, can be constructed (see Figure 4.3) and a visual inspection made to determine the best approximation of order and for the value of k.

Two points should be noted from this example. First, although the second-order equation provides the best fit of the experimental data

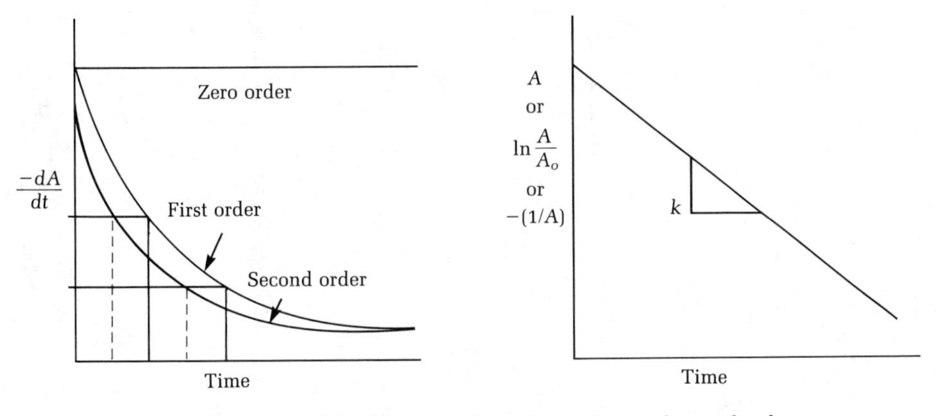

Figure 4.2 Plots of rate equations of zero, first, and second order.

from visual inspection, no equation provides a perfect fit. The reason may be that the laboratory procedures were not accurate to less than 0.1 mmol/L. Therefore, some data collected and fitted to any equation may be slightly displaced from that equation.

The second point concerns the units of k, the rate constant. The units for the zero-order equation rate constant are mmol/L-min, the first-order k units are min^{-1}, and the second-order k units are L/mmol-min. There will be different units for the rate constant as long as the order of the rate equation changes. Inspection of the rate equations will show that these rate constants must have different units in order for the equations to be dimensionally correct.

Temperature effects on reaction rate

All chemical reactions are affected by temperature. This effect is predicted by the Arrhenius equation

$$k = Ae^{-E_a/RT} \tag{4.50}$$

where

k = reaction rate constant
A = pre-exponential factor
E_a = activation energy
R = universal gas constant
T = absolute temperature

This equation can be used to determine E_a and A by plotting ln

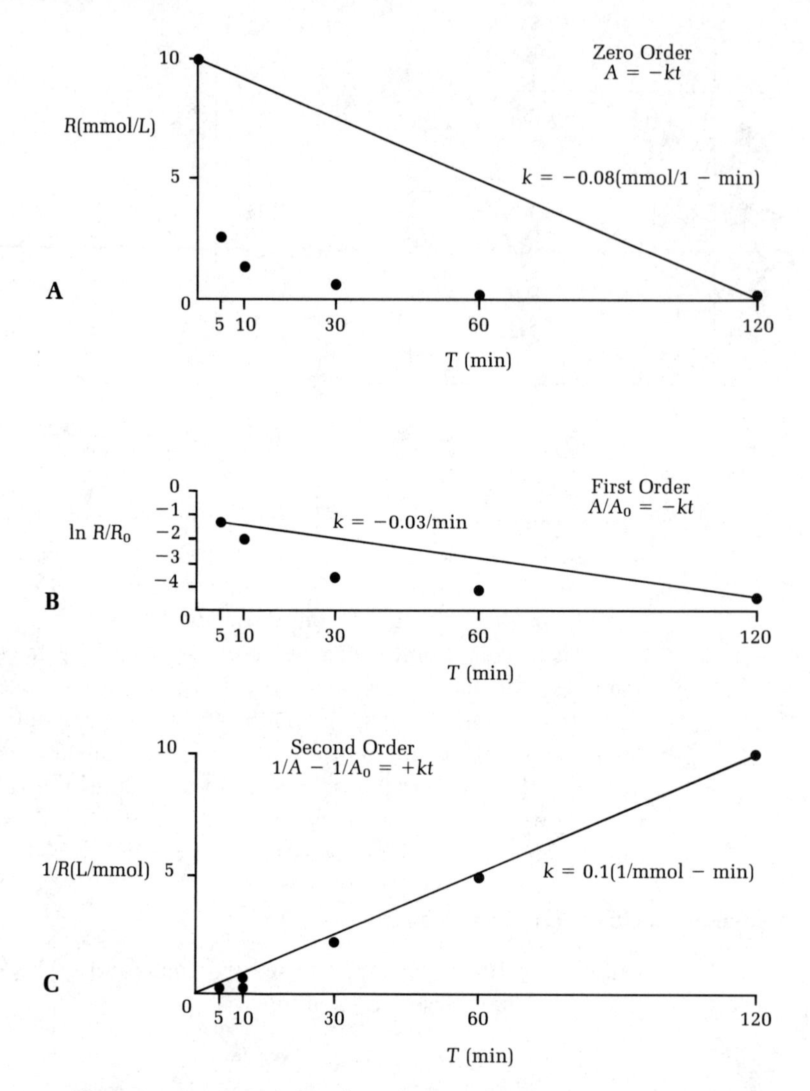

Figure 4.3 Data points plotted against equations of various order. A, zero-order; B, first-order; C, second-order.

k versus $1/T$. E_a becomes the slope and A is determined from the intercept. However, this procedure is unusual for most environmental engineering applications. Typically, relations such as equation (4.51) are given in the literature to determine k as a function of temperature:

$$k = k_{20}(1.024)^{T-20} \quad k_{20} = \text{rate constant at } 20^\circ\text{C} \tag{4.51}$$

where

k = first-order BOD rate constant for any temperature

T = temperature, °C

These relations have been developed from the Arrhenius equation by determining rate constants at different temperatures, plotting them as a function of temperature, fitting a semi-log function to them, and reporting the resulting equation in the literature.

Thermodynamics

In this section we will introduce free energy and chemical equilibrium. These concepts are useful to the environmental engineer in judging the possibility of a chemical reaction proceeding and, therefore, being utilized in some process. Thermodynamics deals with the behavior of materials in bulk. It does not speak to the rate of change but to the final position of a reacting system, or equilibrium.

Gibbs free energy, ΔG, can be developed from the laws of thermodynamics but will not be discussed here. It is sufficient here simply to state that Gibbs free energy has been measured for many compounds and elements (see Table 4.3), which is useful if the feasibility

TABLE 4.3 Gibbs free energy ($\Delta G°$) for selected species at 25°C.

Species	*State*	$\Delta G°$ *(kcal/mole)*
Ca^{+2}	Liquid	−132.18
$CaCO_3$	Solid	−269.78
$Ca(OH)_2$	Solid	−214.33
CO_2	Gas	−94.26
CO_2	Liquid	−92.31
CO_3^{-2}	Liquid	−126.22
HCO_3^-	Liquid	−140.31
H_2CO_3	Liquid	−149.00
H_2O	Gas	−54.64
H_2O	Liquid	−56.690
H^+	Liquid	0
OH^-	Liquid	−37.595
H_2S	Gas	−7.89
H_2S	Liquid	−6.54
HS^-	Liquid	3.01
S^{-2}	Liquid	20
NH_3	Gas	−3.98
NH_3	Liquid	−6.37
NH_4^-	Liquid	−19.00
NO_2^-	Liquid	−8.25
NO_3^-	Liquid	−26.41

of a given reaction involving these reactions and elements is of interest. A good example of how ΔG can be used is shown in Figure 4.4 for a reversible reaction of R going to P. At the bottom of the curve where $dG = 0$, the reaction is at equilibrium and no net change in R or P occurs. On the left or R side of $dG = 0$, dG is negative and the formation of P is favored. On the P side of $dG = 0$, the equilibrium point, dG is still less than zero, but now the formation of R, the reverse reaction, is favored.

Many reactions of significance in water and wastewater treatment occur at constant temperature and pressure, and the change in Gibbs free energy is given. Consequently, the statement can be made that a reaction will proceed *only if the* ΔG *for that reaction is negative.* Many first-time readers may miss the significance of ΔG and not realize that a positive ΔG indicates that a reaction will occur, but in the reverse direction as written. The ΔG referred to up to this point is any free energy at any temperature or pressure; however, in practice ΔG is the standard ΔG° at STP. For a given reaction, ΔG is usually determined from either of two equations for environmental engineering applications:

$$\Delta G^\circ_{\text{reaction}} = \Delta G^\circ_{\text{product}} - \Delta G^\circ_{\text{reactants}} \tag{4.52}$$

or, for the reaction

$$aA + bB \rightleftharpoons cC + dD$$

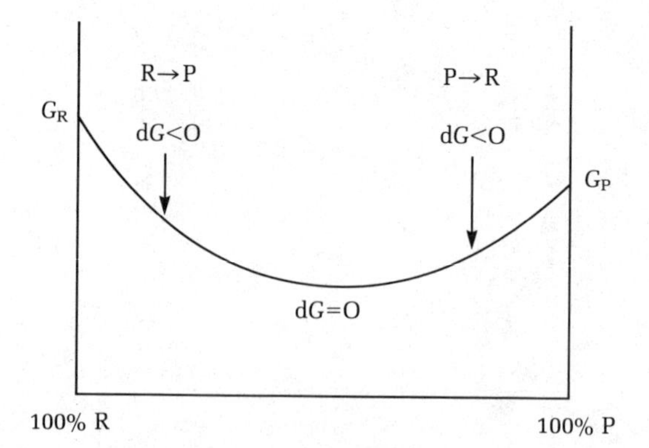

Figure 4.4 Gibbs free energy in the reversible reaction R ⇄ P.

ΔG can be determined by

$$\Delta G = \Delta G^{\circ} + RT \ln \frac{\text{Product}}{\text{Reactant}}$$

$$= \Delta G^{\circ} + RT \ln \frac{(C)^c(D)^d}{(A)^a(B)^b} \tag{4.53}$$

The natural-log term represents the activity, but will be approximated by the molar concentration of the reactants and products at the completion of the reaction. Equation (4.53) can be representative of equilibrium concentrations, $\Delta G = 0$, and solved for the equilibrium constant.

$$\Delta G = 0 = \Delta G^{\circ} + RT \ln \frac{\text{Product}}{\text{Reactant}}$$

$$\Delta G^{\circ} = -RT \ln \frac{\text{Product}}{\text{Reactant}}$$

$$\ln \frac{\text{Product}}{\text{Reactant}} = -\frac{\Delta G^{\circ}}{RT} = \ln K$$

$$K = e^{-\Delta G^{\circ}/RT} = 10^{-\Delta G^{\circ}/2.3RT} \tag{4.54}$$

or

$$K = \frac{\text{Product}}{\text{Reactant}} = \frac{(C)^c(D)^d}{(A)^a(B)^b} \tag{4.55}$$

By now we have discussed three approaches to engineering chemical reactions. First was the stoichiometric approach, in which a direct reaction was given and the molar ratios of the products and reactants were used to predict final product quantities. Complete reaction was assumed. Second, the transition from a non-steady state to a steady state was presented by kinetic analysis for elementary zero-, first-, and second-order reactions. Here, the rate of reaction and the amount of product could be predicted at any point in time. Third the equilibrium constant was developed for a reversible reaction. It is useful in predicting quantities of reactants and products remaining in solution after a reaction is complete. It is a necessary tool when trace elements are removed to low concentrations. The equilibrium constant applies to any reversible reaction. Some common symbols for the equilibrium constant are K_{sp} for solubility product, K_a for acid ionization, K_b for base hydrolysis, and E° for redox reactions. However, it should be noted that all the different K values come from the same thermodynamics, even though they are identified for special reactions.

Solubility products

One of the most useful forms of the equilibrium constant is the solubility constant K_{sp}, which represents the maximum molar concentration of any species resulting from a precipitation reaction which is at equilibrium. The reaction of Ca^{+2} with CO_3^{-2} is typical of solubility relations:

$$CaCO_3 \rightleftharpoons Ca^{+2} + CO_3^{-2}$$

$$K_{sp} = \frac{\text{Products}}{\text{Reactant}} = \frac{(Ca^{+2})(CO_3^{-2})}{(CaCO_3)} = \frac{(Ca^{+2})(CO_3^{-2})}{1}$$

$$= 10^{-8.3} \tag{4.56}$$

When utilizing or calculating the equilibrium constant, the activity, as approximated here by the molar concentration, of any solid or water in an *aqueous* solution can be taken as unity.

This means that the product of molar concentrations of (Ca^{+2}) and (CO_3^{-2}) will always be $10^{-8.3}$ when in equilibrium with solid $CaCO_3$. This obviously enables us to calculate Ca^{+2} or CO_3^{-2} if either is given and the system is at equilibrium. These calculations are widely used by engineers working with reactions that approach equilibrium rapidly such as coagulation, softening, or precipitation of some heavy metals.

A useful plot of equation (4.56) can be made if the log of the solubility product is plotted in a log-log relationship (Figure 4.5). As Eq. (4.56) is plotted on a log-log scale, it defines two domains, one where $CaCO_3$ solid exists and one where Ca^{+2} and CO_3^{-2} exist. The line then presents a phase change from solid to liquid in aqueous solution. The molar concentration of any value of Ca^{+2} can be read directly from Figure 4.5 for any value of CO_3^{-2}. The terms Ca^{+2} and CO_3^{-2} represent free calcium in the $+2$ state and carbonate in the -2 state, and should not be incorrectly equated to total calcium or total carbonate soluble species. Some common solubility products are given as pK_{sp} at 25°C in Table 4.4, and more complete information is given in Appendix C.

TABLE 4.4 Common solubility products.

Reaction	*pK^sp at 25°C*	*Application*
$CaCO_3 = Ca^{+2} + CO^{-2}$	8.3	Water softening
$CaCO_3 = Ca^{+2} + 2OH^-$	5.3	Water softening
$Mg(OH)_2 = Mg^+ + 2OH^-$	10.8	Water softening
$Al(OH)_3 = Al^{+3} + 3OH^-$	33.0	Coagulation
$Fe(OH)_3 = Fe^{+3} + 3OH^-$	37.2	Coagulation
$Cu(OH)_2 = Cu^{+2} + 2OH^-$	18.7	Metal removal
$Zn(OH)_2 = Zn^{+2} + 2OH^-$	16.5	Metal removal
$Ni(OH)_2 = Ni^{+2} + 2OH^-$	15.7	Metal removal
$Cr(OH)_3 = Cr^{+3} + 3OH^-$	30.2	Metal removal
$Mn(OH)_3 = Mn^{+3} + 3OH^-$	36.0	Metal removal
$CaF_2 = Ca^{+2} + 2F^-$	10.5	Fluoridation
$BaSO_4 = Ba^{+2} + SO_4^{-2}$	10.0	Analysis

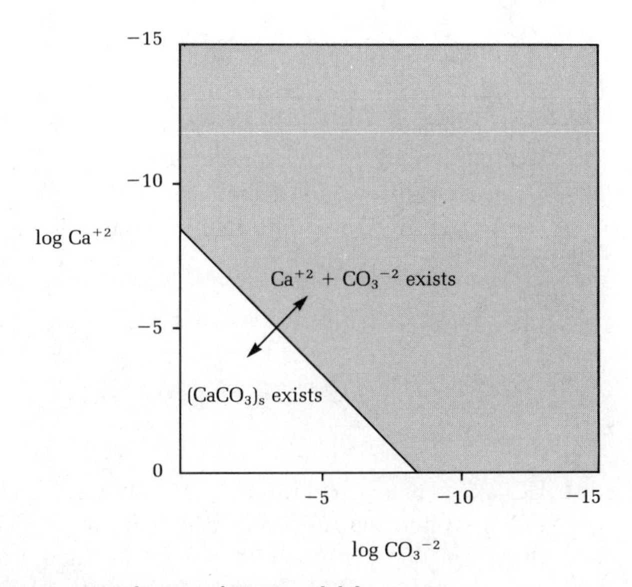

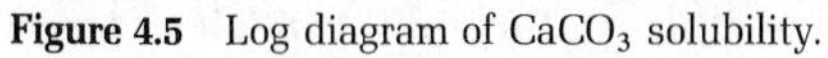

Figure 4.5 Log diagram of $CaCO_3$ solubility.

4-5 Acids and bases

In this chapter an acid is defined as any substance that can donate a proton, and a base is defined as any substance that can receive a proton. This discussion will further assume an aqueous medium where water is the controlling solvent. The first important reaction for water is the ion product.

$$H_2O = H^+ + OH^-$$

$$K_w = \frac{(H^+)(OH^-)}{H_2O} = (H^+)(OH^-) = 10^{-14} \tag{4.57}$$

This is another example of the equilibrium constant that states that the molar product of protons and hydroxide ions will always equal 10^{-14}. Most equilibrium constants are exponential and very small because of the direction in which the reaction is written, therefore, the negative log is a useful convenience:

$$pX = -\log x \tag{4.58}$$

This concept is used to define the term pH, which has come into common use to describe shampoos, chlorination agents for swimming pools, and other products. As defined here,

$$pH = -\log H^+ \tag{4.59}$$

Any acid has to donate an H^+ if it is to be an acid. A pH scale with some common substances is presented in Appendix C. Acids can be weak or strong. That is, some strong acids will donate protons almost always whereas weak acids can only donate protons under special conditions. By the equilibrium constant, HCl is seen to be a strong acid.

$$HCl = H^+ + Cl^-$$

$$K_a = \frac{(H^+)(Cl^-)}{HCl} = 10^{+3}, \; pK_a = -3$$

HCl exists in an "acid ratio" of 1000:1, which means, considering only HCl in water, that for every one molecule of HCl that is not ionized there are 1000 that are, that is, have donated a proton. Acetic acid is a weak acid: it will donate a proton but to a lesser extent.

$$CH_3COOH = CH_3COO^- + H^+$$

$$K_a = \frac{(CH_3COO^-)(H^+)}{CH_3COOH} = 10^{-4.7}, \; pK_a = 4.7$$

As the equilibrium constant is much smaller than K_a for HCl, acetic acid is much less likely to donate a proton. The "acid ratio" here would be 1:50,000. That is, considering only acetic acid in aqueous solution, for every CH_3COOH molecule that ionized there are 50,000 that did not. The stronger acid will always have a larger equilibrium constant K_a or a lower pK than a weak acid.

pH is a major indicator in water quality or water treatment. Many state and federal regulations are based on pH of the finished or final effluent. Because of the significance of pH, several different types of acid and base reactions to determine pH will be discussed. These will include strong and weak acids and bases. pH is significant in neutralization or precipitation reactions involving functional groups affected by proton transfer.

Strong acids

Perhaps neutralization reactions are most common involving strong acids or bases. Neutralization can be defined by the stoichiometry of the acid-base reaction. Equation (4.60) shows the reaction of hydrochloric acid with lime:

$$2HCl + Ca(OH)_2 = 2H_2O + CaCl_2 \tag{4.60}$$

This is a typical acid-base reaction in which an acid, HCl, is neutralized by $Ca(OH)_2$ to form water plus a salt, $CaCl_2$. The neutralization point will be reached when all of the moles or equivalents of acid have been reacted with base according to equation (4.60). The amount of base required by this reaction can be determined by

$$V_A N_A = V_B N_B \tag{4.61}$$

where

V_A = volume of acid, liters
N_A = equivalent of acid
V_B = volume of base, liters
N_B = equivalent of base

The neutral pH is defined as the pH that will result when all of the acid or base has been neutralized. In an aqueous system, this will usually be pH 7 because the protons are controlled by the ionization of H_2O at the neutralization point, so

$$H^+ = OH^-$$
$$K_w = (H^+)(OH^-) = (H^+)^2 = (10^{-14})$$
$$pH = 7$$

Until that point is reached, the pH is controlled by the ionization of the strong acid remaining in solution as shown by

$$pH = pC \tag{4.62}$$

where

C = equivalents of strong acid remaining in solution

Weak acids

The neutralization of weak acids is also a common treatment problem. Again, the pH of neutralization can be determined by adding an equivalent amount of base to the weak acid solution and determining the resulting pH.

Equation (4.61) can be used at any time to determine the mass of equivalents to add for neutralization. The weak acid CH_3COOH, which is commonly written as HAc for brevity, can be used to illustrate the three areas of neutralization for a weak acid. The first is a weak acid alone in water.

$$HAc \rightleftarrows H^+ + Ac^-$$

$$K_a = \frac{(H^+)(Ac^-)}{HAc} = 10^{-4.7}$$

The number of protons equals the number of acetate ions donated to solution, and both species come from the original HAc:

$$H^+ = Ac^-$$

$$\frac{(H^+)^2}{(C - H^+)} = 10^{-4.7}$$

where C is the initial amount of acid. If C is given, then

$$(H^+)^2 - 10^{-4.7}(C - H^+) = 0$$

The only unknown in this equation is H^+, which can be solved by quadratic equation. However, usually the first-order H^+ term is insignificant and the equation can be modified to

$$(H^+)^2 = (10^{-4.7})C$$

which can be expressed in the general form

$$pH = \frac{pC + pK}{2} \tag{4.63}$$

where

pC = the negative log of molar concentration, HAc

The second point of neutralization is the midpoint, where half of the HAc has been neutralized. Using NaOH, the reaction is

$$HAc + NaOH = H_2O + Na^+ + Ac^- \tag{4.64}$$

Since 50% of the HAc has been neutralized, the HAc remaining equals the Ac^- at the halfway point. If this relationship is substituted directly into K for HAc, the following equation is derived:

$$K_a = \frac{(H^+)(Ac^-)}{HAc} = \frac{(H^+)(Ac^-)}{(Ac^-)} = H^+$$

$$pK = pH \tag{4.65}$$

The third point is that of complete neutralization, when the original amount of acid has been neutralized by the OH^- added.

$$HAc + OH^- = H_2O + Ac^- \tag{4.66}$$

The Ac^- produced at the neutralization point equals the original HAc concentration denoted as pc in 4.67. The pH at neutralization is given by

$$K_a = \frac{(H^+)(Ac^-)}{(HAc)}$$

$$H^+ = \frac{K_a HAc}{(Ac^-)} = \frac{K_a(OH^-)}{(Ac^-)} = \frac{K_a K_w}{(Ac^-)(H^+)}$$

$$(H^+)^2 = \frac{K_a K_w}{Ac^-}$$

$$pH = \tfrac{1}{2}(pK_a + pK_c - pc) \tag{4.67}$$

Another useful equilibrium constant is K_b, which represents the equilibrium constant for the reaction of a salt and H_2O to produce an acid and a base.

$$Ac^- + H_2O \rightleftharpoons HAc + OH^-$$

$$K_b = \frac{(OH^-(HAc)}{Ac^-}$$

$$K_b = \frac{(OH^-)(Acid)}{(Salt)} \tag{4.68}$$

This equation can be used in conjunction with K_w to relate K_a by equation (4.63):

$$K_b = \frac{K_w(HAc)}{(H^+)(Ac^-)} = \frac{K_w}{K_a}$$

$$K_aK_b = K_w \tag{4.69}$$

This equation can be used to determine pH when a weak base is neutralized with a strong acid. Three general equations were developed representing the initial state, midpoint, and conclusion of the neutralization of a weak acid with a strong base and can be developed for the neutralization of a weak base and a strong acid. These equations are summarized in Table 4.5. Equilibrium constants for some acids and bases are given in Appendix C.

The only area of pH control not discussed is the regions between the equations. The pH values there can be determined by the log form of the equilibrium constant for a weak acid as in

$$pH = pK + \log\frac{(Salt)}{(Acid)} \tag{4.70}$$

where the salt and acid are the molar concentrations remaining after addition of base. When the addition of base causes the formed salt to equal the acid remaining, then equation (4.70) becomes equation (4.65).

The only other acids to be discussed here will be the diprotic acids. The pH of diprotic weak acid solutions can be determined in the same manner as monoprotic acid solutions. That is, the first ionization constant of a diprotic acid is used until the first proton has been completely donated from the diprotic acid, then the second

TABLE 4.5 Summary equations for determination of pH during weak acid or base neutralization with strong base or acid.

Stage	*Weak base/Strong acid*	*Weak acid/Strong base*
Initial	$pH = pK_w - 1/2\ pK_b - 1/2\ pC$	$pH = 1/(pK_a + pC)$
Midpoint	$pH = pK_w - pK_b$	$pH = pK_a$
Conclusion	$pH = (pK_w - pK_b + pC)$	$pH = 1/2(pK_a + pK_w - pC)$

ionization constant is used until neutralization is complete. The only difference arrives at the end of the neutralization of the first acid in which the pH equals one-half of the sum of the both pK values. For example, the weak acid H_2CO_3 has two ionization constants:

$$H_2CO_3 \rightleftharpoons H^+ + HCO_3^- \quad pK = 6.3 \tag{4.71}$$

$$HCO_3^- \rightleftharpoons H^+ + CO_3^{-2} \quad pK = 10.3 \tag{4.72}$$

The first ionization constant, 6.3, would be used in the identical manner as in Table 4.5 for pH calculations, if a strong base were added to solution. The pH at the point where all of the H_2CO_3 has been converted to HCO_3^- results in the moles of H_2CO_3 remaining equaling the moles of CO_3^{-2} formed. Therefore

$$K_1 = \frac{(H^+)(HCO_3^-)}{(H_2CO_3)} \qquad K_2 = \frac{(H^+)(CO_3^{-2})}{(HCO_3^-)} \tag{4.73}$$

$$HCO_3^{-2} = \frac{K_1(H_2CO_3)}{H^+} = \frac{(CO_3^{-2})(H^+)}{K_2} \tag{4.74}$$

$$K_1K_2 = \frac{(CO_3^-)}{(H_2CO_3)}(H)^{+2} = (H^+)^2 \tag{4.75}$$

$$H^+ = (K_1K_2)^{\frac{1}{2}} \tag{4.76}$$

Therefore

$$pH = \frac{pK_1 + pK_2}{2} \tag{4.77}$$

Example Problem 4.11

50 mL of a 0.1 N diprotic acid, H_2A, $pK_1 = 4.0$, $pK_2 = 8.0$, is titrated with 0.05 N NaOH. Determine the pH after 0, 25, 50, 75, 100, and 150 mL of base has been added.

Solution The first step in working this problem is to determine the effect of each addition of base on the regions within the neutralization reactions.

Volume 0.05 N NaOH required for neutralization:

$$V_A N_A = V_B N_B \text{ (law of equivalence)}$$
$$(50 \text{ mL})(0.1) = (0.05) V_B$$
$$V_B = 100 \text{ mL of } 0.05 \text{ N NaOH}$$

is required for complete neutralization. Note the switch from pK_1 to pK_2 when all H_2A is neutralized.

mL base added	*Region*
0	Beginning
25	Prior to first neutralization
50	Neutralization first proton donor, H_2A
75	Prior to second neutralization
100	Complete neutralization, all A is A^{-2}
150	After complete neutralization

0 mL base

$$pH = \tfrac{1}{2}(pK_1 + pC)$$
$$pH = \tfrac{1}{2}(4.0 + 2) = 3$$

25 mL base

$$pH = pK_1 + \log \frac{(HA^-)}{(H_2A)}$$

$$pH = \frac{(0.05)(25)}{50 + 25 \text{ mL}} \Bigg/ \frac{(0.1)(50) - 0.05(25)}{50 + 25 \text{ mL}} = 3.52$$

The calculations for (H_2A) and (HA^-) may at first seem complex, but they are not. Since the neutralization of H_2A is

$$H_2A + OH^- = HA^- + H_2O$$

an equal number of equivalents of HA^- will be produced as OH^- added until all of the H_2A is neutralized. So, where the HA^- is increased by 1.25 meq in 75 ml, the H_2A is reduced by 1.25 meq in

75 mL. Note that the 75 mL is cancelled in the fraction. This will always be true for weak acid or base pH calculations in similar instances.

50 mL base

$$pH = \tfrac{1}{2}(pK_1 + pK_2)$$
$$= \tfrac{1}{2}(4.0 + 8.0) = 6.0$$

75 mL base

$$pH = pK_1 + \log \frac{(A^{-2})}{(HA^{-})}$$

$$pH = 8.0 + \log \frac{0.05(25)}{75 + 50} \Bigg/ \frac{(0.1)(50) - (0.05)25}{75 + 50} = 7.52$$

Now the reaction is:

$$HA^{-} + OH^{-} = A^{-2} + H_2O$$

The A^{-2} produced at 75 mL base addition is (75 − 50)(0.05)/125 mL, since the first 50 mL of base was required to neutralize H_2O. Correspondingly, only the 25 mL of base past the 50-mL H_2A neutralization requirement affected the HA^{-} in solution.

100 mL base

$$pH = \tfrac{1}{2}(pK_2 + pK_w - pC)$$
$$= \tfrac{1}{2}(8.0 + 14.0 - 1.5) = 10.25$$

150 mL base

$$pH = pK_w = pOH$$
$$= 14 - \log \frac{(50)(0.05)}{200 \text{ mL}} = 12.1$$

This pH is determined by K_w since there is no more acid to react with. The OH^{-} from the excess NaOH simply is diluted to a volume of 200 mL.

Oxidation-reduction

Redox reactions are among the most significant to the environmental engineer. They are important to nutrient uptake in waste treatment or natural ecosystems, disinfection in potable water treatment, and corrosion in all areas. One area for professional employment in almost all engineering disciplines is corrosion engineering. It has been estimated that losses due to corrosion are $8 billion annually in the United States, $700 million specifically occurring in potable water distribution systems. Corrosion, the oxidative loss of electrons by a metal, renders metal weak, unusable, or undesirable for its purpose. Almost everyone is familar with red stains on the sides of homes near faucets or stains on freshly laundered clothing. These stains are usually the product of a corrosive or poorly treated water in the distribution system.

Oxidation results if a material loses electrons. If iron goes from Fe^0, a zero charge, to Fe^{+2}, a plus-two charge, then Fe^0 was oxidized to Fe^{+2} because Fe^0 lost two electrons. If Cu^{+2} gained two electrons, then it was reduced from a plus-two state to a zero state. One simple way to remember the electron transfer relationship for a redox reaction is "Loses Electrons, Oxidized; Gains Electrons, Reduced, or "*LEO* the Lion goes *GER!*"

All redox reactions are thermodynamically based; that is, the reactions proceed toward the state of greatest thermodynamic stability. Moreover, whether or not a proposed redox equation will occur can be determined by calculating ε, the cell potential. If ε is positive, the reaction will proceed as written from left to right. If ε is negative, then the reaction will proceed not as written but in the reverse direction from right to left. If ε is zero, then the reaction will be at equilibrium and no net increase in products or reactants will occur. The equation for ε, known as the Nernst equation, for the general reaction is

$$aA + bB \rightleftharpoons cC + dD$$

$$\varepsilon = \varepsilon^{\circ} - \frac{RT}{nF} \frac{(C)^c(D)^d}{(A)^a(B)^b} \tag{4.78}$$

where

ε° = standard half cell, volts

F = Faraday's constant, 23.062 cal/V eq.

R = gas constant, 1.987 cal/°K-mol

T = temperature, °K

n = electrons transferred, eq/mol

However, the cell potential can be related to Gibbs free energy by

$$\Delta G = -nF\varepsilon \tag{4.79}$$

Consequently, there is a relationship among Gibbs free energy, the cell potential, and the equilibrium constant. If $\varepsilon = 0$ equation (4.78) reduces to

$$\varepsilon^{\circ} = \frac{RT}{nF} \ln \frac{\text{Product}}{\text{Reactant}} = \frac{RT}{nF} \ln K$$

$$K = e^{nF\varepsilon^{\circ}/RT} = 10^{nF\varepsilon^{\circ}/2.3RT} \tag{4.80}$$

Equation (4.80) is comparable to equation (4.54) involving Gibbs free energy and the equilibrium constant and can be rederived by substituting equation (4.79) into (4.80):

$$K = 10^{nF\varepsilon^{\circ}/2.3RT} = 10^{-\Delta G/2.3RT} \tag{4.81}$$

The significance of these relationships to the environmental engineer is that any potential chemical reaction can be evaluated if the free energies, cell potential, or equilibrium constants are known. Consequently, means of treatment or pollution abatement can be evaluated.

The use of equation (4.78) requires that half cells be evaluated at STP and then coupled together in an electrochemical chemical cell. Half cell potentials are given in Table 4.6. These potentials are measures against a standard hydrogen half cell which is assigned an ε° of 0 V. This would be written $|H^+|\frac{1}{2}H_2 \cdot Pt|$ in a standard format. The Pt refers to a platinum electrode.

$$H^+ + e \longrightarrow \tfrac{1}{2}H_2\,(Pt) \quad \varepsilon^{\circ} = 0\ V \tag{4.82}$$

To determine whether any two elements will enter into a redox reaction, the first step is to write a balanced chemical reaction. The second step is to calculate ε from equation (4.78) and determine from its sign whether the reaction will proceed. Some very basic facts should be kept in mind concerning redox or corrosion. This is a basic

TABLE 4.6 Standard electrode potentials at 25°C.

Reaction	ε^o, *volts*
$H^+ + e^- = 1/2\ H_2(g)$	0
$Fe^{+3} + e^- = Fe^{+2}$	+0.77
$Fe^{+2} + 2e^- = Fe^0$	−0.44
$Cr_2O_7^{-2} + 14H^+ + 6e^- = 2Cr^{+3} + 7H_2O$	+1.33
$Cu^{+2} + 2e^- = Cu(s)$	+0.34
$Cu^{2+} + e^- = Cu^+$	+0.16
$Zn^{2+} + 2e^- = Zn(s)$	−0.76
$Cd^{2+} + 2e^- = Cd(s)$	−0.40
$Sn^{2+} + 2e^- = Sn(s)$	−0.14
$Pb^{2+} + 2e^- = Pb(s)$	−0.13
$NO_3^- + 2H^+ + 2e^- = NO_2^- + H_2O$	+0.84
$NO_3^- + 10H^+ + 8e^- = NH_4^- + 3H_2O$	+0.88
$N_2(g) + 8H^+ + 6e^- = NH_4^+ + 2H_2O$	+0.28
$O_3(g) + 2H^+ + 2e^- = O_2(g) + H_2O$	+2.07
$O_2(ag) + 4H^+ + 4e^- = 2H_2O$	+1.27
$Mg^{+2} + 2e^- = Mg(s)$	−2.37
$Al^{+3} + 3e^- = Al(s)$	−1.68

chemical reaction that requires electron transfer via some circuit. Therefore corrosion can be stopped if the circuit is disrupted in any way. If any substance is to be oxidized, another substance must be reduced. A very common form of corrosion is galvanic corrosion, which is brought about by (1) a coupling of two dissimilar metals, (2) a connection between the two metals that will allow electron flow, and (3) a difference in the overall potential of the galvanic cell. Properly changing any of these conditions could stop the corrosive reaction.

A simple illustration of corrosion of an iron pipe carrying water is shown in Figure 4.6. In this example, Fe in the zero state is oxidized to Fe^{+2}. The two electrons are used by dissolved oxygen and H^+ to complete the cell. The further oxidation of Fe^{+2} to Fe^{+3} can be

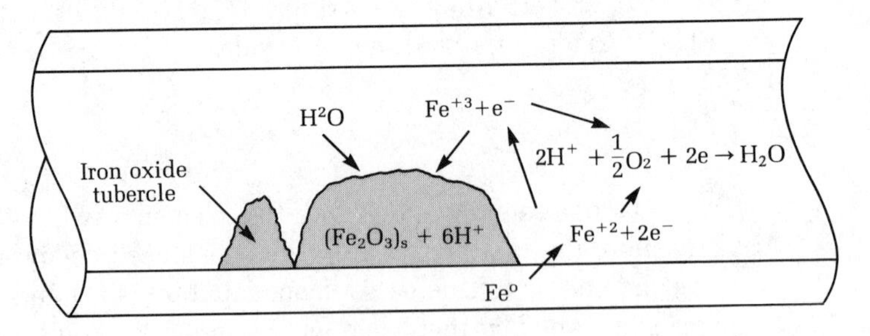

Figure 4.6 Common pathways for iron corrosion in a water distribution system.

easily facilitated by the presence of dissolved oxygen. Once Fe^{+3} is produced, it will react quickly with water to form Fe_2O_3 solid, producing protons, $6H^+$, which in turn completes the required oxygen half cell for the initial oxidation of iron. The actual thermodynamics are favorable for these reactions and the red stain, Fe_2O_3, is common to many households. Two good ways of reducing this reaction would be to deoxygenate the water or seal the inside of the pipe.

Example Problem 4.12

An engineer wishes to determine the suitability of using magnesium ingots to protect an oil derrick made of aluminum in the sea at 25°C. To write the reaction, the assumption is made that aluminum will be reduced from Al^{+3} to Al^0 and that magnesium will be oxidized from Mg^0 to Mg^{+2}.

$$Mg^0 = Mg^{+2} + 2e^- \qquad \varepsilon^o = +2.37 \text{ V}$$

$$Al^{+3} + 3e^- = Al^0 \qquad \varepsilon^o = -1.68 \text{ V}$$

A common denominator of 6 is required to balance the overall reaction.

$$\begin{array}{ll} 3\,Mg^0 = 3Mg^{+2} + 6e^- & \varepsilon^o = -2.37 \text{ V} \\ 2\,Al^{+3} + 6e^- = 2Al^0 & \varepsilon^o = +1.68 \text{ V} \\ \hline 3\,Mg^0 + 2Al^{+3} = 3Mg^{+2} + 2Al^0 & \varepsilon^o = +0.69 \text{ V} \end{array}$$

An ε^o of +0.69 V has been calculated, so for a standard cell condition, the derrick would be protected as long as it was connected to a magnesium ingot. However, this derrick is in the ocean where dissolved Mg is 0.055 mol/L and the Al^{+3} can be approximated to be a maximum of 10^{-6} mol/L.

$$\varepsilon = \varepsilon^o + \frac{RT}{2.3\,nF} \log \frac{\text{Product}}{\text{Reactant}}$$

$$= 0.69 \text{ V} + \frac{(1.987 \text{ cal/°K-mol})(25 + 273)}{(2.3)(6 \text{ eq/mol})(23.062 \text{ cal/V-eq})} \log \frac{(Mg^{+2})^3(Al)^2}{(Al^{+3})^2(Mg)^3}$$

$$= 0.69 \text{ V} + (0.002) \log \frac{(0.055)^3(1)^2}{(10^{-6})^2(1)^3}$$

$$= 0.69 \text{ V} + (0.0164) = +0.070 \text{ V}$$

So, the oil derrick is well protected. This technique of protection is known as cathodic protection because it uses a sacrificial anode, the magnesium ingot, to protect the aluminum derrick, which is turned into the cathode by the connection to magnesium. Oxidation or corrosion always occurs at the anode, so any cathode is automatically protected.

4-6 Problems

4.1 A sulfuric acid manufacturing plant produces 100 tons of H_2SO_4 per day which has a total volume of 21,175 gallons. What is the molarity and normality of the H_2SO_4 produced? How many milliequivalents of acid are produced?

4.2 A mining operation left mine waste which contained solid pyrite, FeS_2, in northeastern Oklahoma. One conservative estimate placed the total mine waste at 4×10^{10} yd^3. If the mine waste was 0.10% solid FeS_2 by weight in water and the given reaction occurred, how many pounds of protons would be produced if all of the FeS_2 leached from the waste? Assume specific gravity of FeS_2 is 5.

$$4FeS_2 + 15O_2 + 14H_2O \longrightarrow 16H^+ + 8SO_4^{-2} + 4Fe(OH)_3$$

How many pounds of NaOH and $Ca(OH)_2$ would be required to neutralize the waste?

4.3 How much $CaCO_3$ in kilograms would be required to neutralize the SO_2 produced from one ton of coal that contained 2% by weight S if all of the sulfur is converted to SO_2? Assume

$$CaCO_3 + SO_2 \longrightarrow CaSO_3 + CO_2(g)$$

4.4 A city has an area of 100 square miles and receives 50 inches per year of rainfall at pH 7.0. Determine how much coal with 2% S by weight must be burned to reduce the pH of the rain to 4.0 if all of the sulfur reacts as follows:

$$S + O_2 = SO_2$$
$$SO_2 + \tfrac{1}{2}O_2 + H_2O = H_2SO_4$$

4.5 If a combustion engine is running in a 20 ft × 20 ft × 10 ft room and the given reaction is occurring, how much C_9H_{28} would have to combust to reach a CO concentration of 10,000 ppm? What would the CO concentration be in mg/L?

$$C_9H_{28} + 9.5\,O_2 = 9CO + 10H_2O$$

4.6 A waste sludge is determined to be 1% solids of $Al(OH)_3$, which has a specific gravity of 2.5. How much of the sludge is solids and how much is water if 10 mgd of incoming water is treated by 40 mg/L $Al_2(SO_4)_3 \cdot 14H_2O$ and the $Al(OH)_3$ is produced by

$$Al^{+3} + H_2O = Al(OH)_3 + 3H^+$$

4.7 Assume that 1,000 mol/L of a substance A can undergo zero-, first-, or second-order degradation with reaction rate constants of -0.1 mole/L-min, -0.10 min^{-1}, and 0.1 L/mol-min, respectively. Plot the degradation of A as a function of time and report the half-life of each reaction.

4.8 A first-order reaction is 30% complete after 10 minutes. How long before it is 90% complete, and what is the rate constant?

4.9 A gas decomposed and the following data were recorded. Determine the order, half-life, rate constant, and average existence time of the gas.

Gas, mol/L	2.33	1.91	1.36	1.11	0.72	0.55
Time, s	0	319	867	1198	1877	2315

4.10 Determine from the Gibbs free energy data at 25°C whether the following reaction will occur if the residual Ca^{+2} concentration and HCO_3^- concentration are each 100 mg/L as $CaCO_3$.

$$Ca(OH)_2 + Ca^{+2} + 2HCO_3^- = 2CaCO_3 + 2H_2O$$

4.11 Determine the equilibrium constant for $CaCO_3$ at 25°C, 10°C, and 40°C. What effect does temperature have on the $CaCO_3$ solubility?

$$K_{sp} = 10^{-8.3} \text{ at } 25°C$$

$$Ca^{+2} + CO_3^{-2} = CaCO_3$$

4.12 Determine pK_1 and pK_2 for H_2S from Gibbs free energy data.

$$H_2S = H^+ + HS^-$$

$$HS^- = H^+ + S^{-2}$$

4.13 A stream flowing at 10^5 L/s containing 10^{-3} mol/L of Ca^{+2} and 10^{-4} mol/L of Ba^{+2} is mixed with a second stream flowing at 10^4 L/s containing 10^{-4} mol/L CO_3^{-2} and 10^{-5} mol/L SO_4^{-2}. How many kilograms will precipitate over a 24-hr period?

4.14 A copper pipe is placed in the soil and joined to an aluminum fitting. The water in the soil contains 10^{-6} mol/L Al^{+3} and 10^{-4} mol/L Cu^{+2}. Will there be any corrosion? If so, which element will corrode? How could this be stopped? Identify the cathode and anode.

4.15 A 200-acre lake with an average depth of 15 ft contains how many kilograms of oxygen and nitrogen if the water is in contact with the atmosphere which contains 79% N_2 and 21% O_2?

4.16 10 mgd of a waste containing an acidic liquor is discharged to a river. The waste has a pH of 4.0 and is produced from only the acid HX and water. How many moles of HX in the form of HX was opposed to X^- does the waste contain per day if pK = 3.0 for HX? How much $Ca(OH)_2$ would be required to neutralize this waste to pH 6.0 before discharge?

4.17 If the waste in problem 4.16 is produced from only water and a diprotic acid, $pK_1 = 2.0$ and pK_2 8.0, how much $Ca(OH)_2$ per day would be required to raise the pH to 7.0 before discharge?

References

American Public Health Association. 1980. *Standard Methods for the Examination of Water and Wastewater,* 15th ed. Washington, DC.

Fogler, H. S. 1974. *The Elements of Chemical Kinetics and Reactor Calculations.* Englewood Cliffs, NJ: Prentice-Hall.

Himmelblau, D. M. 1974. *Basic Principles and Calculations in Chemical Engineering,* 3rd ed. Englewood Cliffs, NJ: Prentice-Hall.

Perkins, H. C. 1974. *Air Pollution.* New York: McGraw-Hill.

Prince, N. C. and Dweka, R. 1979. *Principles and Problems in Physical Chemistry for Biochemists,* 2nd ed. Briston, U.K.: Oxford Science.

Sawyer, C. N. and McCarty, P. L. 1978. *Chemistry for Environmental Engineers,* 3rd ed. New York: McGraw-Hill.

Snoeyink, V. L. and Jenkins, D. 1980. *Water Chemistry.* New York: Wiley.

Wark, K. and Warner, C. F. 1976. *Air Pollution.* New York: Dun-Donnelly.

Wilkinson, F. 1980. *Chemical Kinetics and Reaction Mechanisms.* New York: Van Nostrand-Reinhold.

Williams, V. R. and Williams, H. B. 1967. *Basic Physical Chemistry for the Life Sciences.* San Francisco: W. H. Freeman.

5

Biological Processes

5-1 Introduction

In the previous chapter, we saw how the knowledge of chemical principles can be effective in the development of processes used to improve the quality of our environment. Similarly, biochemical reactions, which occur in association with living organisms, play an essential role in a variety of environmental control processes. Specific biochemical reactions are part of various biological processes to treat wastewater and to maintain a well-balanced ecosystem. Microorganisms are dependent on the nutrient pool, and at the same time they are constantly recycling nutrients to replenish the pool in their ecosystems (Berthouex and Rudd, 1977).

Selected microorganisms may be effective in decomposing certain pollutants and removing undesirable nutrients and toxic elements. Also, well-managed algal or plant systems may be used to remove nutrients, heavy metals, or radioactive matter from aqueous environments. This chapter covers the basic knowledge of the biological processes related to degradation of organic matter, photosynthesis, respiration, and other important processes. Material and energy balance concepts are applied to biological processes and an attempt is made to familiarize students with the role of microorganisms both as pollutants and as pollution-control agents. Also, the usefulness of material balance and energy balance in quantifying biological processes is emphasized.

5-2 Classification of organisms

Microorganisms can be divided into three groups on the basis of their nutritional requirements for carbon and energy sources. These groups are (1) heterotrophs, (2) photoautotrophs, and (3) chemoautotrophs.

Heterotrophic microorganisms utilize organic carbon compounds as building blocks and energy source. Most organisms, including all animals, are heterotrophic and can use a wide diversity of organic compounds for their growth. Different organic compounds are used by different microorganisms. Photoautotrophs utilize carbon dioxide or bicarbonates as their sole carbon source and sunlight as their energy source. Examples are algae and photosynthetic bacteria. Chemoautotrophic or chemosynthetic organisms utilize CO_2 as their source of carbon and their energy comes from the oxidation of reduced inorganic compounds. Examples are sulfur and nitrifying bacteria. The nitrifying bacteria oxidize ammonium to nitrate:

$$NH_4^+ \xrightarrow{\text{Nitrosomonas}} NO_2^- \xrightarrow{\text{Nitrobacter}} NO_3^-$$

The photoautotrophs together with the chemoautotrophs are the primary producers of organic matter in the biosphere.

Microorganisms are also classified on the basis of their oxygen utilization. Organisms that use molecular oxygen or air are known as aerobes and they survive in aerobic conditions. Organisms that do not use molecular oxygen and die in the presence of air are known as anaerobes. Facultative aerobes or anaerobes can survive in the presence or absence of molecular oxygen. Detailed discussion on classification of microorganisms can be found in biological textbooks.

5-3 Nutrients

Nutrients such as the nitrogen and phosphorus associated with urban and agricultural runoff, as well as effluent from most sewage treatment plants, stimulate the growth of algae and water weeds. These nutrients accelerate the process of eutrophication in lakes and streams (see Chapter 2). The nutrient pool must contain these elements essential for growth, since the least abundant nutrient needed for growth will control the rate as stated by the law of the minimum.

The basic nutrients are carbon, nitrogen, and phosphorus. The carbon source can be inorganic (from carbon dioxide or bicarbonates), which is utilized by autotrophic organisms such as plants and algae. Organic carbon is utilized by heterotrophic organisms such as animals and heterotrophic bacteria during respiration and decomposition of organic wastes. Nitrogen in the form of ammonia and nitrates is available for plant and animal growth; however, some organisms such as blue-green algae are able to fix nitrogen gas from the atmosphere. On the average, about 15% of the mass of protein is nitrogen. Phosphorus reaches the aquatic system through fertilization, surface runoff, effluent disposal, and internal cycling. These nutrients are recycled through the ecosystem and tend to sustain the productivity of the system. Availability of excess nutrients will enhance the growth of algae and plants, which are partially consumed by animals and partially decomposed upon death. The decay organisms, bacteria and fungi, recycle the nutrients back into the ecosystem. The decomposed matter, as well as respiration by plant and animal life, tends to depress the dissolved oxygen content in the aquatic system and may produce anaerobic conditions, particularly at night. Many fish kills have been reported in streams and lakes due to lack of dissolved oxygen.

Example Problem 5.1

If the bacterial cell can be simulated by $C_5H_7O_2N$, what are the relative nutrient requirements?

Solution

$$\text{Mol. wt. of } C_5H_7O_2N = 5(12) + 7(1) + 2(16) + 1(14)$$

$$= 60 + 7 + 32 + 14 = 113$$

$$C = \tfrac{60}{113}(100) = 53\%$$

$$H = \tfrac{7}{113}(100) = 6.3\%$$

$$O = \tfrac{32}{113}(100) = 28.3\%$$

$$N = \tfrac{14}{113}(100) = 12.4\%$$

$$100.0\%$$

Therefore, the nutrient requirements are 53% for C and 12.4% for N.

Example Problem 5.2

If algal protoplasm can be simulated by $C_{106}H_{263}O_{110}N_{16}P$, what are the relative nutrient requirements?

Solution

$$\begin{aligned}
C &= 106 \times 12 = 1272 \text{ g/g-mol} \\
H &= 263 \times 1 = 263 \text{ g/g-mol} \\
O &= 110 \times 16 = 1760 \text{ g/g-mol} \\
N &= 16 \times 14 = 224 \text{ g/g-mol} \\
P &= 1 \times 31 = 31 \text{ g/g-mol} \\
& \qquad\qquad\quad 3550
\end{aligned}$$

Therefore, the nutrient requirements are 35.8% for C, 6.3% for N, and 0.9% for P.

5-4 Bacterial growth

Bacteria, like other living organisms, must obtain from their environment all of the nutrient materials necessary for their growth and reproduction. The nutrients must be sufficient for synthesis of cytoplasmic material and serve as an energy source for cell growth and biosynthetic reactions. The principal elements of living matter are C, H, N, O, S, and P.

Most of the dry weight of microorganisms consists of C, H, N, and O, which are released as gaseous compounds upon combustion. The remaining ash elements constitute 2–14% of the dry weight of the cell. On the average, carbon accounts for about 50% of the dry weight, oxygen for about 30%, nitrogen for 7–14%, and hydrogen for approximately 7%. Phosphorus accounts for about 50% of the total weight of the ash (Benefield and Randall, 1980). A widely used molecular formula to describe the composition of dry bacterial cells as given by Lawrence and McCarty (1970) is $C_{60}H_{87}O_{23}N_{12}P$. This structure has a formula weight of 1374. Carbon, nitrogen, and phosphorus fractions compose 52.4%, 12.2%, and 2.3%, respectively.

When a small number of viable bacterial cells are placed in a closed vessel containing an ample supply of food in a suitable environment, conditions may be established in which unrestricted growth takes place. However, the growth of organisms does not go on to infinity and the rate of growth may follow a growth pattern like that shown in Figure 5.1. The curve shown in this figure may be divided into six well-defined phases for many species of bacteria: the lag

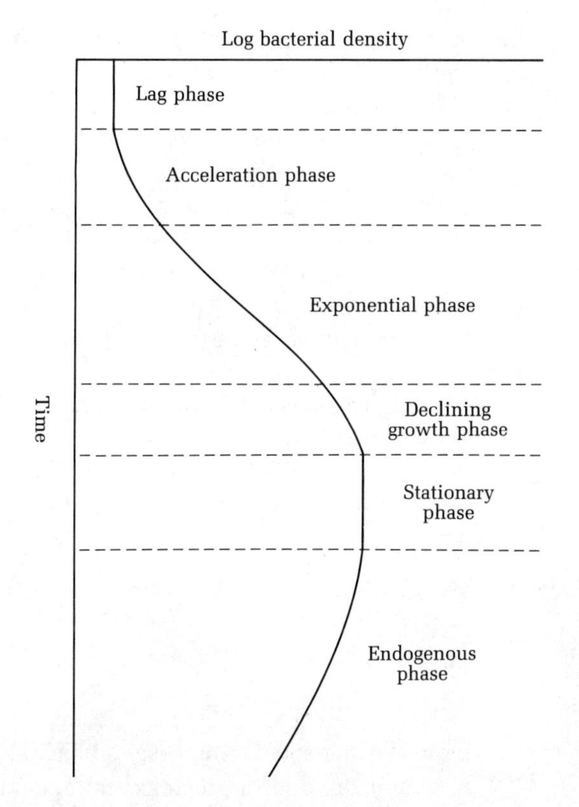

Figure 5.1 Characteristic growth curve of microorganisms.

phase, acceleration phase, exponential phase, declining growth phase, stationary phase, and endogenous phase. It must be emphasized that this growth cycle is a result of bacterial interaction with the environment in a closed system, and it is possible to maintain exponential growth for long periods of time. Suppose you start with one organism that doubles itself every 30 minutes; the number of bacterial cells after 24 hours will amount to 2^{48}, or 2.8×10^{14} cells if growth follows the exponential phase. This growth pattern is unrealistic in natural systems, since the food supply and environmental conditions are limiting factors. Growth is generally influenced by temperature, pH, oxygen requirements, food supply, limiting nutrient, amount of toxic matter present, and predator-prey relationships.

Growth is measured by cell counts or turbidity measurements. Cell counts, usually made by plating out (distributing) sample dilutions on nutrient media such as agar in a petri dish, reflect the viable cells which produce colonies. However, cell counts and turbidity measurements do not provide satisfactory information for wastewater treatment processes, which deal with quantities of material. Therefore, engineering measurements are made on the basis of mass concentration (milligrams per liter or parts per million). These concentrations are measured by weighing material retained on a specified filter paper dried at 105°C and reported as suspended solids or mixed liquor suspended solids (MLSS). The material may include organic constituents which volatilize at 550°C and are known as volatile suspended solids (VSS or MLVSS) and fixed fractions (FSS or MLFSS), which remain as ash or nonvolatile.

Measurement of bacterial growth on a mass increase basis is useful for establishing the material balance and rate determinations of biomass. Also, stoichiometric information should assist in developing relationships between the rates of growth of organisms and utilization of the limiting nutrient. The ratio of the growth rate to limiting nutrient is known as cell yield Y, and can be expressed as:

$$Y = \frac{\Delta x/\Delta t}{\Delta s/\Delta t} \tag{5.1}$$

where

Δx = the incremental increase in biomass which results from the utilization of an incremental amount of substrate Δs during an incremental time Δt

Also,

$$Y = \frac{\dfrac{dx/dt}{x}}{\dfrac{ds/dt}{x}} = \frac{\mu}{q} \tag{5.2}$$

where

μ = specific growth rate, which represents the rate of growth per unit amount of biomass

q = specific substrate utilization rate, which represents the rate of nutrient utilization per unit amount of biomass

s = substrate concentration

x = bacterial cell concentration

In biological processes used for wastewater treatment, the bacterial cells in the system are not in the exponential growth phase but generally in the declining growth phase. Also, the rate of growth is reduced due to energy consumption for cell maintenance and other factors such as death and predation. Usually, these factors are accounted for by allowing for endogenous decay or the decrease in cell mass per unit organisms present. The endogenous decay coefficient k_d of the biomass lost to endogenous respiration per unit of biomass per unit of time can be determined from laboratory experiments. The net growth rate would be equal to

$$\mu - k_d = qY - k_d \tag{5.3}$$

Monod (1949) described both the exponential and declining growth-rate region in Figure 5.1 by the expression

$$\mu = \frac{dx/dt}{x} = \frac{\mu_{max}(S)}{K_s + S} \tag{5.4}$$

where

μ_{max} = maximum value for μ at saturation concentration of growth-limiting substrate

S = growth-limiting substrate concentration

K_s = saturation constant, equal to the substrate concentration when μ is equal to $\mu_{max}/2$

Equation (5.4) is illustrated in Figure 5.2. When the value of S is much

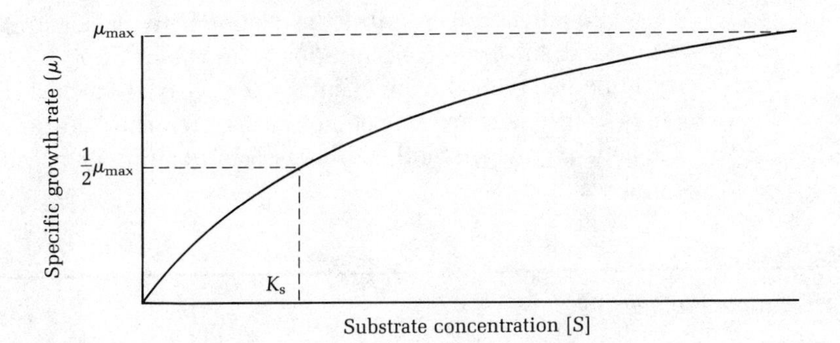

Figure 5.2 Growth rate versus limiting nutrient concentration.

greater than the value of K_s, equation (5.4) is reduced to

$$\mu = \mu_{max} \tag{5.5}$$

Equation (5.5) follows zero-order kinetics since the rate of reaction is constant and is independent of the substrate concentration. When the substrate concentration is much smaller than K_s, equation (5.4) is reduced to

$$\mu = \frac{\mu_{max}}{K_s}(S) \tag{5.6}$$

Since both μ_{max} and K_s in equation (5.6) are constants, the rate of reaction is proportional to the substrate concentration or limiting nutrient, and follows first-order kinetics. At intermediate ranges, mixed-order reaction rates may occur, requiring fractional exponents to describe the order of reaction.

Substituting from equation (5.2) for q and using μ_{max}/Y for q_m, equation (5.4) can be rewritten as

$$q = \frac{\mu}{Y} = \frac{\mu_{max}}{Y} S \Big/ K_s + S$$

$$= \frac{q_m S}{K_s + S} \tag{5.7}$$

where

q_m = the maximum specific substrate utilization rate

Equation (5.7) describes the relation between substrate utilization rate and substrate concentration. The values of Y, k_d, K_s, and μ_{max} can be determined experimentally for selected biochemical reactions and controlled environmental parameters. They are known as biokinetic parameters and can be used in quantifying specific biochemical reactions.

Example Problem 5.3

Examine the stoichiometry of the reaction of casein with oxygen, including the synthesis of cell material as follows:

$$C_8H_{12}O_3N_2 + 3O_2 \xrightarrow{\text{Bacteria}} C_5H_7O_2N + NH_3 + 3CO_2 + H_2O$$

(casein) (bacterial cells)

Mol. wts.:

184 + 96 = 113 + 17 + 132 + 18

Therefore:

1. Oxygen required per gram casein oxidized = $\frac{96}{184}$ = 0.522 g.
2. Oxygen required per gram of organic carbon = $\frac{96}{96}$ = 1.0 g.
3. Bacterial cells produced per gram of casein = $\frac{113}{184}$ = 0.61 g, which is the stoichiometric cell yield Y.
4. Bacterial cell produced per gram of oxygen consumed = $\frac{113}{96}$ = 1.177 g.

The stoichiometric amount of oxygen consumed by bacteria to oxidize one gram of casein is 0.522 g; that is, biochemical oxygen demand is 522 mg oxygen for 1000 mg casein. It must be realized that the cell yield and the biochemical oxygen demand are dependent on other environmental parameters such as pH, temperature, availability of other micronutrients, and presence of toxic elements.

Example Problem 5.4

A bacterial growth experiment conducted at 20°C in a completely mixed reactor produced the following results:

μ(day^{-1})	16	12	10	8	7
S(mg/L)	30	15	10	7	6

Determine the maximum specific growth rate μ_{max} and the saturation constant K_s.

Solution Equation (5.4) can be rewritten in the form

$$\frac{1}{\mu} = \frac{K_s}{\mu_{max}}\left(\frac{1}{S}\right) + \frac{1}{\mu_{max}}$$

This equation represents a straight-line relation between $1/\mu$ and $1/S$, with slope = K_s/μ_{max} and intercept = $1/\mu_{max}$:

$\frac{1}{\mu}$ (day)	0.063	0.083	0.100	0.125	0.143
$\frac{1}{S}$ (L/mg)	0.033	0.067	0.100	0.143	0.167

$$\text{Intercept} = 1/\mu_{max}$$
$$= 0.04$$
$$\mu_{max} = 25 \text{ day}^{-1}$$
$$\text{Slope} = K_s/\mu_{max}$$
$$= \frac{0.055}{0.10} = 0.55$$
$$K_s = (0.55)(25) = 14 \text{ mg/L}$$

These relations are plotted below on a graph.

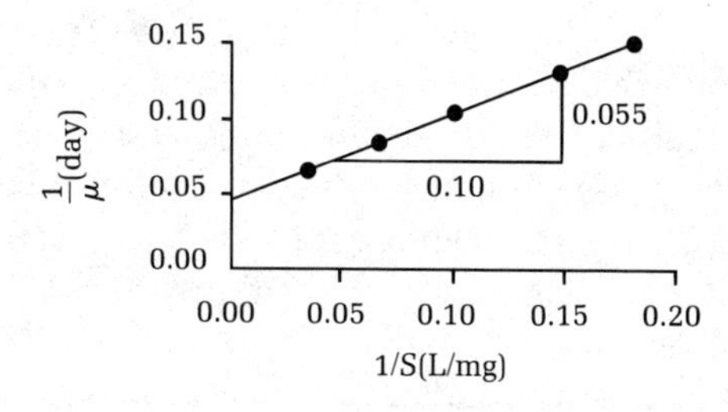

5-5 Decomposition of wastes

Heterotrophic organisms, especially bacteria, use organic wastes as food and in the process break down the complex organics into simple organic and inorganic materials. Decomposition can occur in the presence of molecular oxygen (aerobic decomposition) or in its absence (anaerobic decomposition). For example, simple oxidation of the carbohydrate glucose is described by

$$C_6H_{12}O_6 + O_2 \xrightarrow[\text{bacteria}]{\text{aerobic}} CO_2 + H_2O + \text{more cells}$$

$$C_6H_{12}O_6 \xrightarrow[\text{bacteria}]{\text{anaerobic}} CO_2 + CH_4 + \text{more cells}$$

The reaction products from aerobic decomposition of wastes are not offensive; however, dissolved oxygen is removed from the water. Typical anaerobic decomposition products include ammonia, methane, hydrogen sulfide, carbon dioxide, and water. The amount of organic content in water and wastewater systems is measured by its biochemical oxygen demand (BOD), chemical oxygen demand (COD), or total organic carbon (TOC).

BOD

Biochemical oxygen demand (BOD) is the amount of oxygen required by bacteria to stabilize decomposable organic matter under aerobic conditions. It is used, among other things, (1) to measure the strength of a waste in designing wastewater treatment facilities, (2) to measure the amount of organic pollution in a stream, and (3) to assess industries for discharging waste to municipal systems.

The complex reactions involved can be summarized as Organic material + O_2 + Microorganisms results in More microorganisms + CO_2 + H_2O + Residual organic material + NH_4^+.

Nitrogen-containing organic compounds can be degraded in the BOD test, and the ammonia released, together with ammonia already present, can be oxidized to NO_2 and NO_3 with significant oxygen consumption.

$$2NH_3 + 3O_2 + \xrightarrow[\textit{Nitrosomonas}]{\text{nitrite-forming autotrophs}} 2NO_2^- + 2H^+ + 2H_2O$$

$$2NO_2^- + O_2 + \xrightarrow[\text{Nitrobacter}]{\text{nitrate-forming autotrophs}} 2NO_3^-$$

A plot of the time versus oxygen used is shown in Figure 5.3.

BOD values are measured using dilution water prepared according to the American Public Health Association's *Standard Methods for Examination of Water and Wastes.* Biodegradable carbonaceous material may completely break down in 20 days, but for practical reasons

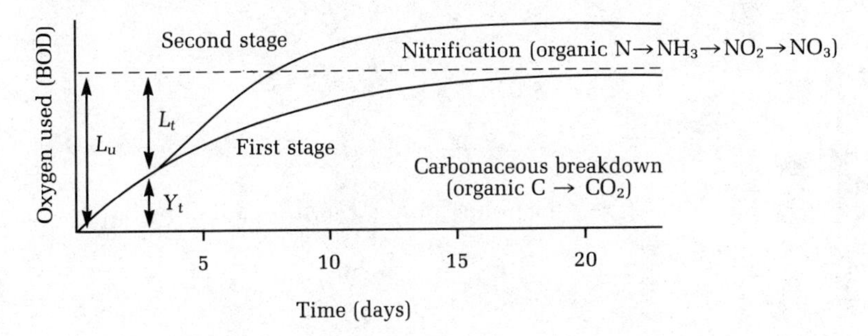

Figure 5.3 Decomposition of organic matter by bacteria.

a standard time of 5 days at 20°C has been selected for the test. BOD at 5 days is a certain percentage of BOD at 20 days or L_u (mg/L). Also, 20°C (68°F) was selected because it is a median temperature of natural streams and discharged wastes.

The amount of organic material present in water is proportional to the amount of oxygen required to oxidize this organic material; that is, the BOD remaining in water is proportional to the amount of organic matter present.

The rate of change of biochemical oxygen demand (BOD) remaining can be simulated by a pseudo first-order reaction:

$$-\frac{dL}{dt} = K_d L_t \tag{5.8}$$

where

L_t = BOD remaining (mg/L)

K_d = First-order reaction constant (time^{-1})

By integration of equation (5.8),

$$-\int_{L_u}^{L_t} \frac{dL}{dt} = \int_0^t K_d\, dt$$

$$-[\ln L]_{L_u}^{L_t} = K_d[t]_0^t$$

$$-(\ln L_u - \ln L_t) = -K_d t$$

$$\ln \frac{L_t}{L_u} = -K_d t$$

$$\frac{L_t}{L_u} = e - K_d t$$

$$L_t = L_u e - K_d t \tag{5.9}$$

where

L_u = BOD remaining at zero time or BOD ultimate

Also

$$L_t = L_u 10 - K_d' t \tag{5.10}$$

where

$$K'_d = 0.434K_d$$

BOD satisfied at time t is designated by Y_t as shown in Figure 5.3.

$$\begin{aligned} Y_t &= L_u - L_t \\ &= L_u - L_u 10^{-K'_d t} = L_u - L_u e - K_d t \\ &= L_u(1 - 10^{-K'_d t}) = L_u(1 - e - K_d t \end{aligned} \tag{5.11}$$

Y_5 represents the 5-day BOD.

Generally, decomposition of organic wastes by bacteria is more rapid in warm weather than in cold weather. The temperature effect on the carbonaceous rate constant can be represented by

$$K_{d_T} = K_{d_{20}}\ \theta^{(T-20)} \tag{5.12}$$

where

$\theta = 1.047$

$T =$ temperature

Other factors affecting the BOD test include seed organisms, toxicity, and pH. Acclimated bacteria in settled sewage effluent are used as seed organisms for the BOD determination.

Example Problem 5.5

The 5-day BOD for some waste has been found to be 150 mg/L with a reaction rate constant $K_d = 0.30\ \text{day}^{-1}$ at 30°C. Determine the ultimate BOD and the rate constant at 20°C.

Solution

(1) $$L_u = \frac{BOD_5}{1 - e^{-5K_d}} = \frac{150}{1 - e^{-5 \times 0.30}}$$

$$= \frac{150}{1 - 0.223} = 193\ \text{mg/L}$$

(2) $$K_{d_{20}} = \frac{0.30}{1.047^{(30-20)}} = 0.19\ \text{day}^{-1}$$

Measurement of BOD_5

The standard test for biochemical oxygen demand is the 5-day BOD at 20°C. The BOD of each wastewater sample can be measured using dilution water prepared according to standard methods with settled sewage seed of 2 ml/L. The dilution is made with aerated distilled water that contains buffers and nutrients. The necessary dilutions for each sample can be calculated from prior estimates of the BOD_5 for the solutions to be tested. Suppose the estimated BOD_5 for the wastewater is 200 mg/L at the incubation temperature of 20°C. The saturation level of DO (dissolved oxygen) in the dilution water is 9.2 mg/L. Clearly, we have to dilute the sample so that the measured BOD_5 of the mixture is less than 9.2 mg/L, preferably in the range of 5 to 7 mg/L, to ensure that some oxygen will remain after incubation. Determine the initial DO for each dilution water in triplicate. Assume that this is the initial DO of the mixtures of sample and dilution water, because the samples do not have any initial dissolved oxygen demand (IDOD).

Set up proper dilutions and incubate at 20°C for 5 days. The BOD_5 for the mixture is the difference between the initial and final DO concentrations. The BOD_5 for the wastewater can be calculated by multiplying the dissolved oxygen consumed times the dilution factor. Usually, 300-mL glass bottles are used in BOD tests. If 5 mL of wastewater is used and the bottle is filled with dilution water, the dilution factor will become $300/5 = 60$ for this experiment. If the initial DO is 9.0 mg/L and the final DO is 5 mg/L, or 4.0 mg/L oxygen is consumed by the mixture, then BOD_5 at 20°C for the wastewater sample is 4×60 or 240 mg/L.

Other parameters for measurement of organic content

The organic matter content in water and wastewater can also be quantified by the amount of oxygen required to chemically oxidize this matter—the chemical oxygen demand (COD)—or by its total organic carbon (TOC) content. If the organic matter is easily biodegradable by bacteria, the COD and the ultimate BOD may show similar values, otherwise there will be large differences between ultimate BOD and COD for non-biodegradable or poorly biodegradable organic matter. Generally, mathematical relations can be developed between BOD, COD, and TOC for different wastewaters.

Example Problem 5.6

Calculate the COD and TOC and ultimate BOD for glucose, knowing that glucose is easily degradable.

Solution $C_6H_{12}O_6 + 6O_2 \longrightarrow 6CO_2 + 6H_2O$

Mol. wt. of $C_6H_{12}O_6 = 6(12) + 12(1) + 6(16) = 180$

Oxygen required: $6O_2 = 6(2 \times 16) = 192$

Organic carbon: $6 \times 12 = 72$

or

$$\text{COD} = \text{ultimate BOD} = \frac{192}{180} = 1.07 \frac{\text{g } O_2}{\text{g glucose}}$$

$$\text{TOC} = \frac{72}{180} = 0.4 \frac{\text{g carbon}}{\text{g glucose}}$$

5-6 Photosynthesis and respiration

The oxygenation of a stream or a lake may be greatly influenced by the photosynthetic oxygen production of plants and algae. Chlorophyll in these plants enables them to synthesize cell material by photosynthesis using energy from the sun and CO_2 and bicarbonates from the stream, generating oxygen as a by-product:

$$6CO_2 + 6H_2O \xrightarrow{\text{sunlight}} C_6H_{12}O_6 + 6O_2$$

The carbohydrates such as glucose can be converted into proteins or cellulose and complex compounds. The number of photosynthetic organisms in water varies, and the oxygen production rate is affected by the availability and intensity of sunlight, temperature, and the availability of a carbon source and other nutrients.

Respiration is the reverse process of photosynthesis. All aerobic organisms derive their energy from the oxidation of organic matter, the reverse of the equation above:

$$C_6H_{12}O_6 + 6O_2 \longrightarrow 6CO_2 + 6H_2O + \text{Energy}$$

Respiration continuously removes oxygen from water, while photosynthesis adds oxygen during daylight hours. The effects of gross photosynthesis minus respiration result in the net photosynthesis for a specified period of time. This value can be determined by the clear-and-black-bottle technique.

In the clear-and-black-bottle technique, water samples are collected at various depths. Part of the samples are placed in transparent bottles and part in opaque bottles and resuspended in the streams at the depths from which they were obtained. The samples are left in position for a day or several days. The change in final dissolved oxygen in the clear bottles, divided by the number of days they are

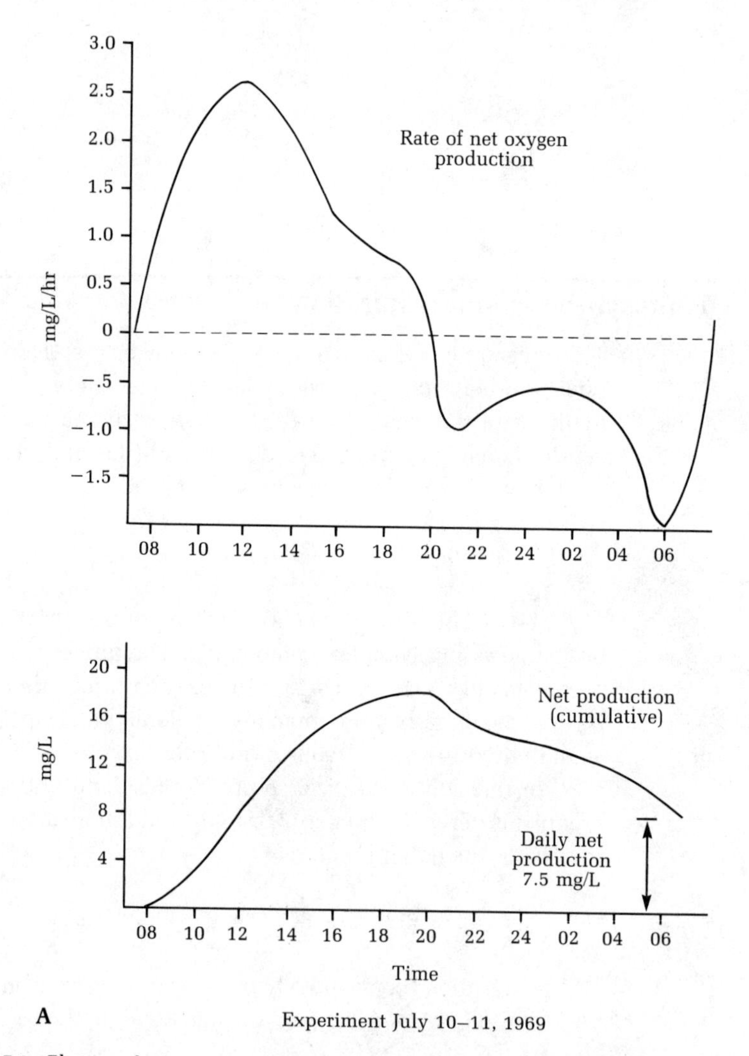

Figure 5.4 Photosynthetic oxygen production in a stream model (after Padden, 1970). A, July 10–11, 1969; B (*on right page*), November 13–14, 1969.

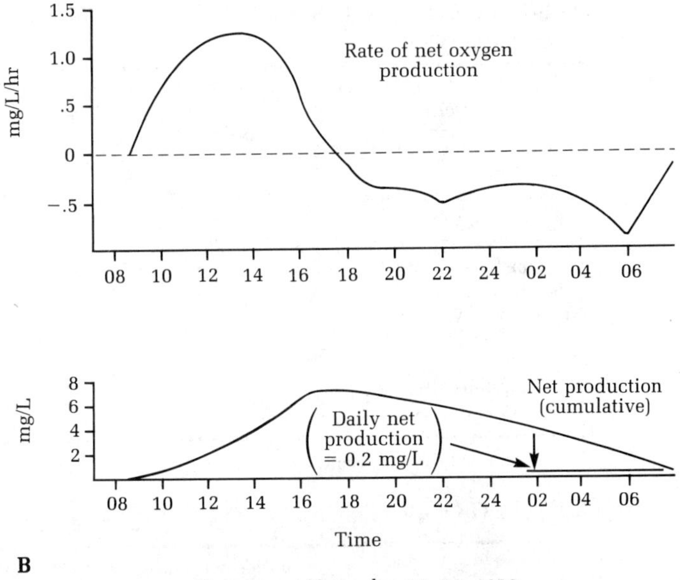

B

Experiment November 13–14, 1969

in the water, serves as a measure of the net production of oxygen per day by the phytoplankton. Whether or not this is a good method is debatable. The glass bottle may filter out some of the effective energy, and the behavior of photosynthetic organisms in a stagnant environment may be considerably different than in the turbulent environment of a stream. The reduction of dissolved oxygen in the black bottles, divided by the number of days they are in the water, serves as a measure of respiration. The gross photosynthesis is equal to the net photosynthesis plus respiration. The diurnal pattern of net photosynthetic oxygen production is shown from results obtained by Padden (1970) on a research flume as represented by Figure 5.4.

A sine-curve function is used to estimate the diurnal variability if the oxygen balance is to be mathematically modeled. Typical average values of net photosynthetic and respiration activity are listed in Table 5.1. This table shows the average values in g oxygen/m^2-day; these rates divided by the stream water depth in meters result in values with units of mg/L-day.

5-7 Oxygen balance in aquatic ecosystems

The dissolved oxygen concentration is an important criterion used to evaluate the quality of surface waters. The effect of oxidizable matter on streams, the evaluation of water condition for fish and other orga-

TABLE 5.1 Average values of net photosynthetic dissolved oxygen production

Warm type	*Average net production* (g/m^2–*day*)	*Average respiration* (g/m^2–*day*)
Truckee River, California, bottom—attached algae	9	11.4
Tidal creek diatom bloom (62–109 × 10^6 diatoms/L)	6	—
Delaware estuary, summer	3.7	—
Duwamish River estuary, Seattle	0.5–2.0	—
Neuse River system, North Carolina	0.3–2.4	—
River Ivel	3.2–17.6	6.7–15.4
North Carolina streams	9.8	21.5
Laboratory streams	3.4–4.0	2.4–2.9

Compiled by Hadi Elmi.

TABLE 5.2 Solubility of oxygen in water (mg/L).

	Chloride concentration in water (mg/L)					
Temp. (°)	0	5000	10,000	15,000	20,000	*Difference per 100 mg* chloride
0	14.6	13.8	13.0	12.1	11.3	0.017
1	14.2	13.4	12.6	11.8	11.0	0.016
2	13.8	13.1	12.3	11.5	10.8	0.015
3	13.5	12.7	12.0	11.2	10.5	0.015
4	13.1	12.4	11.7	11.0	10.3	0.014
5	12.8	12.1	11.4	10.7	10.0	0.014
6	12.5	11.8	11.1	10.5	9.8	0.014
7	12.2	11.5	10.9	10.2	9.6	0.013
8	11.9	11.2	10.6	10.0	9.4	0.013
9	11.6	11.0	10.4	9.8	9.2	0.012
10	11.3	10.7	10.1	9.6	9.0	0.012
11	11.1	10.5	9.9	9.4	8.8	0.011
12	10.8	10.3	9.7	9.2	8.6	0.011
13	10.6	10.1	9.5	9.0	8.5	0.011
14	10.4	9.9	9.3	8.8	8.3	0.010
15	10.2	9.7	9.1	8.6	8.1	0.010
16	10.0	9.5	9.0	8.5	8.0	0.010
17	9.7	9.3	8.8	8.3	7.8	0.010
18	9.5	9.1	8.6	8.2	7.7	0.009
19	9.4	8.9	8.5	8.0	7.6	0.009
20	9.2	8.7	8.3	7.9	7.4	0.009
21	9.0	8.6	8.1	7.7	7.3	0.009
22	8.8	8.4	8.0	7.6	7.1	0.008
23	8.7	8.3	7.9	7.4	7.0	0.008
24	8.5	8.1	7.7	7.3	6.9	0.008
25	8.4	8.0	7.6	7.2	6.7	0.008
26	8.2	7.8	7.4	7.0	6.6	0.008
27	8.1	7.7	7.3	6.9	6.5	0.008
28	7.9	7.5	7.1	6.8	6.4	0.008
29	7.8	7.4	7.0	6.6	6.3	0.008
30	7.6	7.3	6.9	6.5	6.1	0.008

nisms, and the progress of self-purification can all be estimated by means of DO. For example, McKee and Wolf (1963) recommended that the DO required for a well-rounded warm-water fish population should remain above 5.0 ppm (mg/L) for at least 16 hours of the day and during the other 8 hours should not drop below 3.0 ppm. Generally, game fish such as trout require more oxygen than coarser species such as carp. Raw surface water criteria and drinking water standards for public water supplies recommend permissible levels for DO concentration equal to or greater than 4 ppm, with a desirable level of concentration near saturation. Saturation levels of DO in surface water at various temperatures and salinity are shown in Table 5.2. Dissolved oxygen decreases with increasing salinity and temperature. Also, the influence of altitude can be approximated by a 7% decrease in DO saturation per 2000 ft (600 m) increase in elevation (Becker, 1924).

Dissolved oxygen concentration in streams is the result of a combination of sources and sinks. Oxygen sources include atmospheric reaeration and photosynthetic activity. Oxygen sinks include respiration, oxidation of organic wastes, consumption by inorganic matter such as ammonia, hydrogen sulfide, and ferrous salts, and oxygen demand by benthic organisms as depicted in Figure 5.5. Quantitative

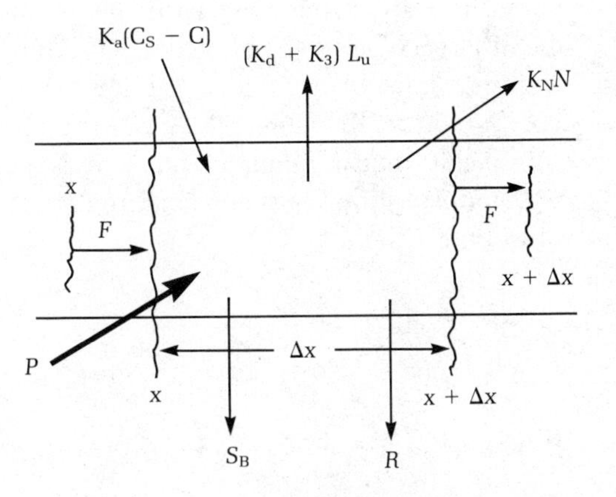

Figure 5.5 Dissolved oxygen mass balance. Quantities are defined in the text.

analysis of the oxygen balance can be made if the parameters are defined:

F = advective flowrate, liter/day

x, Δx = distance, m

C_s = saturated dissolved oxygen, mg/L

C = dissolved oxygen, mg/L
K_a = reaeration coefficient, day^{-1}
K_d = carbonaceous BOD decay coefficient, day^{-1}
K_3 = carbonaceous BOD sedimentation coefficient, day^{-1}
L_u = ultimate carbonaceous BOD, mg/L
K_N = nitrogenous BOD decay coefficient, day^{-1}
N = nitrogenous BOD, mg/L
P = photosynthetic oxygen production, mg/L-day
R = algal respiration oxygen consumption, mg/L-day
S_B = benthic oxygen demand, mg/L-day

A detailed treatment of the oxygen balance in natural systems is beyond the scope of this text. Decomposition of organic water and the photosynthesis and respiration processes are discussed in this chapter.

Reaeration

Atmospheric oxygen passes into the water through the water-atmosphere interface. This transfer of oxygen increases by increasing the surface area in contact with the atmosphere, the oxygen deficit (the difference between saturation level and the measured value), and mixing characters. Solubility of oxygen in water at various temperatures is shown in Table 5.2. A flowing stream will pick up oxygen more easily than a stagnant pond. It is postulated that the rate of oxygen absorption is proportional to its degree of undersaturation or saturation deficit D:

$$\frac{dD}{dt} = -K_a D \tag{5.13}$$

where

dD/dt = the rate of change of oxygen deficit
K_a = reaeration rate constant for existing conditions
D = oxygen deficit = $C_s - C$
C_s = saturation concentration of oxygen at a given temperature
C = existing concentration of oxygen at the same temperature

By integration between the limits D_o at time zero and D_t at time t, the solution to equation (5.13) is

$$D_t = D_o e^{-K_a t} \tag{5.14}$$

where K_a increases with temperature and the degree of mixing of the gas in water. The temperature effect follows the Van der Hoff-Arrhenius relation, but mixing effects are difficult to define. The temperature effect can be expressed by

$$K_{a_T} = (K_{a_{20°C}})1.0241^{(T-20)} \tag{5.15}$$

This formula was recommended by Churchill, Elmore, and Buckingham (1962).

Example Problem 5.7

A tank of water had been aerated in the laboratory for 3 hours at a constant temperature of 25°C. The dissolved oxygen rose from 3.0 mg/L to 6.7 mg/L. Determine K_a at 20°C.

Solution $D_t = D_o e^{-K_a t}$

C_s at 25°C is 8.4 mg/L from Table 5.2,

$$D_o = 8.4 - 3 = 5.4 \text{ mg/L}$$

and

$$D_t = 8.4 - 6.7 = 1.7 \text{ mg/L}$$

$$1.7 = 5.4e^{-K_a(3)}$$

$$(K_{a_{25°}}) = 0.385/\text{hr} = 9.25/\text{day} = 0.006/\text{min}$$

The units for K_a are generally represented by 1/day.

$$(K_{a_{20°C}}) \times 1.0241^{(25-20)} = (K_{a_{25°C}}) = 9.25/\text{day}$$

$$K_{a_{20}} = 8.21/\text{day}$$

Several factors affect the value of K_a in streams. The most important factors are turbulence and temperature. Turbulence by wave action increases the area through which diffusion of oxygen takes place. Turbulence also enhances the surface renewal of the water at the atmosphere-water interface and thus increases reaeration. It can also be concluded that in any particular stream, turbulence is a function of stream velocity and stream depth. Mathematically, various investigators developed the formula

$$K_a = C\frac{V^n}{H^m} \tag{5.16}$$

where

V = average water velocity

H = water depth

C, n, and m are constants depending on physical stream conditions that influence turbulence. Churchill, Elmore, and Buckingham (1962) recommended 5.026, 0.69, and 1.673 for C, n, and m, respectively.

5-8 Oxygen sag curve

When wastes are discharged into a body of water, the amount of dissolved oxygen will decrease as bacteria oxidize the organic matter. This drop in DO may be offset by reaeration through the air-water interface at a rate proportional to the depletion of oxygen below saturation level, that is, the oxygen deficit. The simultaneous action of oxygen consumption and oxygen production is known as the oxygen sag curve, shown in Figure 5.6.

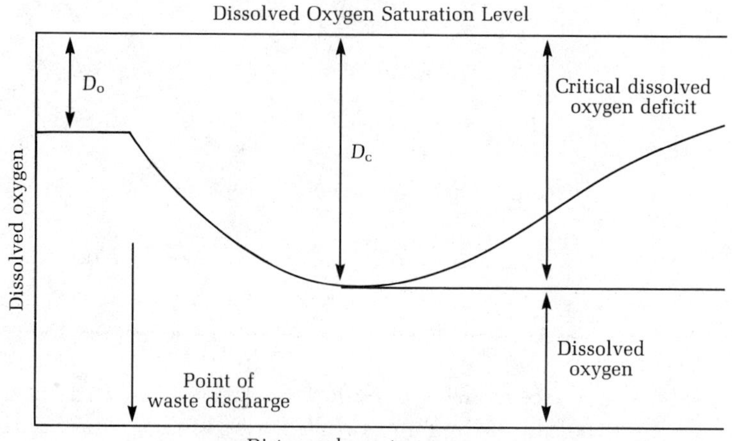

Figure 5.6 The oxygen sag curve.

Initially, the DO curve drops as the waste depletes the oxygen faster than it can be replaced. At the point D_c (critical deficit), the Do is minimum and the rate of aeration becomes equal to the rate of oxygen utilization. Beyond that point, the rate of reaeration exceeds the rate of utilization and the DO level climbs back to normal. This sequence is referred to as the natural self-purification capacity of receiving water. The Streeter-Phelps formulation states that the rate of increase in the deficit is proportional to the algebraic sum of the two terms

$$\frac{dD}{dt} = K_d L_t - K_a D_t \tag{5.17}$$

The first term ($K_d L_t$) represents the deoxygenation reaction, which increases the deficit. It is proportional to residual BOD (L) and the deoxygenation reaction rate constant K_d. The second term ($K_a D_t$) represents the reaeration reaction, which decreases the deficit, proportional to the existing deficit D and reaeration rate constant K_a. K_d and K_a are dependent upon temperature. Also, K_a depends on the depth, velocity, and physical characteristics of the stream. The solution to equation (5.17) is

$$D_t = \frac{K_d L_o}{K_a - K_d}(e^{-K_d t} - e^{-K_a t}) + D_o e^{-K_a t} \tag{5.18}$$

where L_o and D_o are initial BOD and oxygen deficits, respectively, and D_t is the deficit at time t (days). Therefore, if we know D_o, L_o, and the K values, we are able to compute the deficit at any distance downstream, provided the stream velocity is constant so that distance equals velocity times time. It is also possible to calculate the critical deficit D_c and the distance downstream from the point of discharge where D_c is reached.

Example Problem 5.8

A municipality is planning to dispose of its treated secondary effluent in a stream flowing at 50 m^3/s. The treated wastewater flow is 10 m^3/s and has a BOD of 20 mg/L. The stream water upstream from the point of wastewater discharge has a 5 mg/L BOD and 7.0 mg/L dissolved oxygen at 20°C. Calculate the DO concentration downstream from the point of discharge after 1 day in the stream, if the deoxygenation reaction rate of the mixture of stream water and wastewater is 0.5 day^{-1} and the reaeration rate is 0.8 day^{-1}. Assume that the DO in the treated wastewater is 4 mg/L.

Solution By materials balance, it is possible to calculate water flow, BOD, and DO after point of discharge:

Stream water +	Wastewater	= Mixture
$F = 50\ m^3/s$	$F = 10\ m^3/s$	$F = 60\ m^3/s$
BOD = 5.0 mg/L	BOD = 20 mg/L	BOD = 7.5 mg/L
DO = 7.0 mg/L	DO = 4 mg/L	DO = 6.5 mg/L

At 20°C, the DO saturation level is 9.2 mg/L from Table 5.2.
The initial deficit $D_o = 9.2 - 6.50 = 2.70$ mg/L
The initial BOD $L_o = 7.50$ mg/L

$$K_d = 0.5\ \text{day}^{-1}$$
$$K_a = 0.8\ \text{day}^{-1}$$
$$t = 1.0\ \text{day}$$

$$D_t = \frac{K_d L_o}{K_a - K_d}(e^{-K_d t} - e^{-K_a t}) + D_o e^{-K_a t}$$

$$\cong \frac{(0.5)(7.5)}{0.8 - 0.5}[e^{-(0.5)(1)} - e^{-(0.8)(1)}] + 2.7e^{-0.8(1)}$$

$$= 12.50(0.61 - 0.45) + 2.7(0.45)$$

$$= 2.0 + 1.21$$

$$= 3.21$$

$$\text{DO} = C_s - C_t$$
$$= 9.2 - 3.2 = 6.0\ \text{mg/L}$$

5-9 Stream pollution

Living organisms are also affected by the conditions of stream pollution. Their distribution is altered; this change is used as indicators for different types of pollution. Some organisms tolerate low DO conditions, while others cannot survive. Some organisms enhance stream recovery by their activities. For example, green algae supply dissolved oxygen needed to satisfy BOD and raise the oxygen sag curve. Also, as bacteria grow and multiply, some constituents of pollutants are concentrated into them. Therefore, sampling living organisms from the stream areas below the point of wastewater introduction will assist in defining the extent and nature of pollutional zones (Kemp, Ingram, and MacKenthun, 1970).

Some of the more important successional changes in a river when sewage is introduced, and the zones they typify, are shown in Figure 5.7. Bacteria and protozoa predominate in the zone where oxygen becomes severely limited. Further downstream, these organisms and algae become very abundant, then decline. The number of species of larger invertebrates is small at low DO levels, but tubificid worms, midge larvae, and other forms are able to tolerate conditions in this

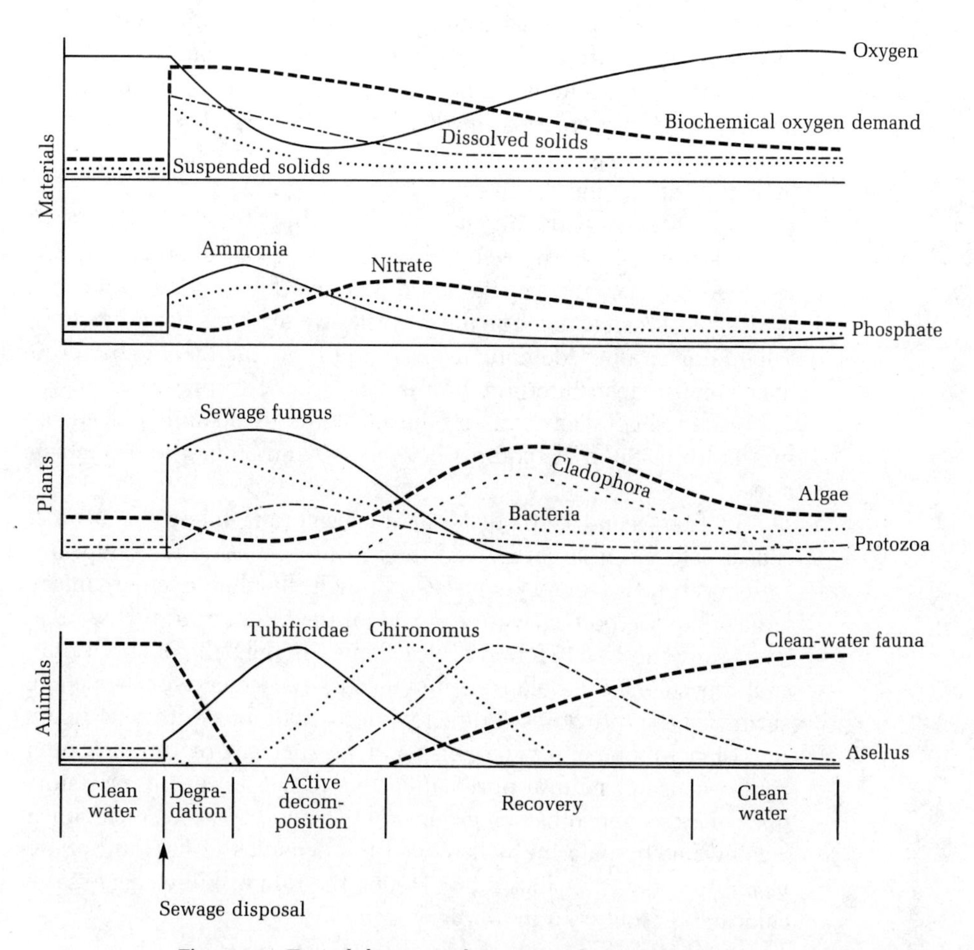

Figure 5.7 Typical changes in the water quality and the plant and animal populations of a river as it passes through various zones following the introduction of sewage (Kemp, Ingram, and Mackenthum, 1970).

zone. As the river returns to its normal conditions, the number of species able to live there increases and the predominance of lower organisms disappears.

5-10 Bacteriological examination of water

The coliform group of bacterial organisms includes all of the aerobic and facultative anaerobic, Gram-negative, non-spore-forming, rod-shaped bacteria which ferment lactose with gas formation within 48 hours at 35°C. The coliform group includes *Escherichia coli* and the *Aerobacter-Aerogenes*. Experience has established the significance of coliform group densities as criteria of the degree of pollution and thus, the sanitary quality of the sample under examination. The coliform group may not itself be pathogenic, but its presence indicates the possibility of the presence of disease-causing organisms. There are fecal and non-fecal coliforms. *Escherichia coli* organisms live in the human intestine and are found in human feces.

Only two methods are currently available for detecting coliform bacteria in water. Both are approved by the U.S. Public Health Service and are described in *Standard Methods for the Examination of Water and Wastewater*. The older of the two methods is the Multiple-Tube Fermentation Method, also referred to as the MPN (most probable number) test. Interpretation is based on the appearance of gas bubbles in small glass vials produced by the fermentation of lactose in a broth medium. It requires between 48 and 96 hours to complete a test.

In the second method, Millipore membrane filters are used to collect coliform from larger and thus more representative samples of water. When the sample is passed through the filter, microorganisms larger than the pore size are trapped on the filter surface where they are readily cultured to form visible colonies. In the Millipore procedure, coliform colonies develop a pink to dark-red color with a metallic surface sheen, providing a direct count of coliforms within 24 hours.

Disposal of coliform organisms in streams may reduce their numbers because of natural purification due to natural death, predators, toxic matter, and other environmental factors. Also, disinfection of wastewater by chlorine will reduce the densities of coliform organisms. According to Chick's law (1908), the rate of kill of bacteria by chlorination follows a first-order reaction:

$$N_t = N_o e^{-kt} \tag{5.19}$$

where

N_t = number of surviving bacteria at time t
N_o = initial number of surviving organisms
k = reaction rate constant
t = contact time with disinfectant

Example Problem 5.9

Assuming Chick's law applies:

1. How much contact time is required to kill 99% of the bacteria with a chlorine residual of 0.1 mg/L, if 80% are killed in 2 minutes contact time?
2. Calculate the percent kill after a contact time of 15 minutes.

Solution

$$\frac{N}{N_o} = e^{-kt}$$

$$0.2 = e^{-k(2)}$$

$$K = 0.805 \text{ min}^{-1}$$

1. To calculate contact time for 99% kill:

$$0.01 = e^{-0.805t}$$

$$t = 5.72 \text{ min}$$

2. To calculate percent kill after 15 minutes contact time:

$$\frac{N}{N_o} = e^{-0.805(15)}$$

$$= 0.000006$$

$$\% \text{ kill} = (1 - 0.000006)100 = 99.9994\%$$

5-11 Conclusion

Environmental regulations are enacted to protect human health and to preserve the physical, chemical, and biological integrity of our natural resources. Increased pollution loads released to the environment will upset the delicate balance of the ecosystem and ultimately will affect us. This chapter has pointed out the role of organisms in maintaining healthy environments by providing treatment of wastes and aiding the natural purification processes. Also, organisms could be a source of pollution such as pathogenic bacteria, viruses, and excessive algal growth. Pollution-control methodologies are limited by the risks we are willing to take and the economic factors involved, as well as our technological advances. Biological processes for pollution control should be fully examined and utilized whenever possible, because they are generally an extension of natural processes. This chapter also showed the importance of materials balance and energy balance in quantifying biological processes.

5-12 Problems

5.1 Define:

Photosynthesis	Reaeration
Respiration	Oxygen deficit
BOD	Nitrogenous BOD
Coliform	COD

5.2 Calculate the BOD ultimate (COD) of 20 mg/L of ethyl alcohol (C_2H_5OH). Is it assumed that C_2H_5OH is easily biodegradable?

5.3 Calculate the oxygen consumed to oxidize 1 mg/L ammonia nitrogen in a stream to its nitrate form if it follows the equation

$$NH_4^+ + 2O_2 \longrightarrow NO_3^- + H_2O + 2H^+$$

5.4 The chemical composition of gasoline is nominally C_8H_{16}. Calculate the COD and TOC in mg/L of 1 gallon of gasoline with a density of 5.5 lb/gal spilled in a pond with a water volume of 100,000 L.

5.5 Clear and black bottles were filled with stream water and incubated at 1 meter below the water surface. If the initial dissolved oxygen concentration in the bottles was 6 mg/L and the final dissolved oxygen after 2 days of incubation was 15 mg/L in the clear bottles and 1.0 mg/L in the dark bottles, calculate the gross photosynthesis, respiration, and net photosynthesis rates in mg/L-day.

5.6 Suppose an empirical analysis of algal cells is represented by $C_{106}H_{181} \cdot O_{45}N_{16}P$. Calculate (1) the relative percentage by weight of C, N, and P, and (2) the concentration of N and P required to produce 10 mg/L algal cells.

5.7 If the ultimate carbonaceous BOD at 20
is 240 mg/L, determine the reaction rate at 20°C and the 5-day BOD at 30°C.

5.8 Explain the effect of each of the following factors on dissolved oxygen in streams:

Photosynthesis	Decomposition
Respiration	Temperature
Reaeration	

5.9 The dissolved oxygen at one point of a stream is 5 mg/L. At another point 2 miles downstream, the dissolved oxygen is 8 mg/L. The stream velocity is 1 mile per hour and the oxygen is supplied through atmospheric reaeration. If the average temperature is 15°C, what is the atmospheric reaeration rate? Ignore all other possible sources and sinks of oxygen.

5.10 For a 150 mg/L solution of glucose:

1. Determine the ultimate biochemical oxygen demand (BOD_u) if the material is oxidized completely to carbon dioxide and water.

$$C_6H_{12}O_6 + 6O_2 \longrightarrow 6CO_2 + 6H_2O$$

2. If the BOD reaction is assumed to be first-order with a reaction rate constant of 0.30 per day, calculate the 5-day BOD for a 150 mg/L solution of glucose.

5.11 The 5-day biochemical oxygen demand (BOD_5) of an industrial wastewater is 300 mg/L. The chemical oxygen demand (COD) is 1000 mg/L. The ultimate BOD is equal to 80% of the COD. Determine the rate constant K_d if the BOD reaction is assumed to be first-order.

5.12 Given the following flow rates F_1 and F_2 and ultimate BOD values L_u in a river upstream from the point of discharge from a sewage treatment plant and in the effluent to the river as stated:

River	Sewage treatment plant
$F_1 = 0.4\ m^3/s$	$F_2 = 0.15\ m^3/s$
$L_{u1} = 0$	$L_{u2} = 30$ mg/L

1. Calculate BOD_u in the river immediately after mixing of river water and effluent from the treatment plant.

2. Calculate the BOD_5 at the same point if the reaction rate constant for the degradation of the organic matter is $K_d = 0.3\ day^{-1}$.

5.13 A treated effluent is discharged into a stream that has a BOD of 5 mg/L, a velocity of 3.0 km/day and a flow of 10,000 m^3/day. The effluent discharged has a BOD of 30 mg/L and a flow of 1000 m^3/day. The initial mixture of effluent and stream water has a dissolved oxygen concentration of 6 mg/L. The deoxygenation rate $K_d = 0.3\ day^{-1}$ and the reaeration rate $K_a = 0.8\ day^{-1}$. Calculate the dissolved oxygen concentration at a point 10 km downsteam from the point of effluent discharge, if the water temperature is 20°C.

5.14 Read a recent article on biological impacts of acid rain. Summarize your findings and state causes, consequences, and control methodologies. Is there a lesson to learn from the acid rain situation?

References

American Public Health Association. 1980. *Standard Methods for Examination of Water and Wastewater*, 15th ed. Washington, D.C.

Becker, H. G. 1924. Mechanism of absorption of moderately soluble gases in water. *Industrial Engineering and Chemistry*, 16, 1220.

Benefield, L. D., and Randall, C. W. 1980. *Biological Process Design for Wastewater Treatment.* Englewood Cliffs, NJ: Prentice-Hall.

Berthouex, P. M., and Rudd, D. F. 1977. *Strategies of Pollution Control.* New York: Wiley.

Chick, H. 1908. An investigation of the laws of disinfection. *Journal of Hydrology* 8, 698.

Churchill, M. A., Elmore, H. L., and Buckingham, R. A. 1962. Prediction of stream reaeration rates. *Journal of the Sanitary Engineering Division, ASCE*, 88: SA4, Proc. Paper 3199.

Keup, L. E., Ingram W. M., and Mackenthun, K. M. 1970. *Biology of Water Pollution.* Cincinnati: U.S. Department of the Interior, Federal Water Pollution Control Administration.

Lawrence, A. W., and McCarty, P. L. 1970. Unified basis for biological treatment design and operation. *Journal of the Sanitary Engineering Division, ASCE*, 96: SA3, 757.

Masters, G. M. 1974. *Introduction to Environmental Science and Technology.* New York: Wiley.

McKee, J. E. and Wolf, H. W. 1963. *Water Quality Criteria.* Resource Agency of California, State Water Quality Control Board, Publication No. 3-A.

Mitchel, R. 1974. *Introduction to Environmental Microbiology.* Englewood Cliffs, NJ: Prentice-Hall.

Monod, J. 1949. The growth of bacterial cultures. *Annual Review of Microbiology* 3, 371.

Padden, T. J. 1970. *Simulation of Stream Processes in a Model River*, Ph.D. thesis, University of Texas at Austin.

6

Water Management

The distribution of water on the earth's surface and in the ground varies in volume, form, and location. More than 97% of the water on earth is seawater, and about 90% of the rest is locked in glaciers and ice caps. Other waters become unfit for direct human consumption or industrial use when the chemical, sediment, or biological content is changed. In addition, the volume of water and flow rates are rarely constant. Thus, management of water to provide suitable quantity and usable quality is crucial for users of water. This task commonly involves the use of mass balances to determine necessary storage levels and expected flow rates.

6-1 Hydrologic data

Hydrologic data are those which reflect the various measurements of the hydrologic cycle (Figure 2.8). Precipitation measurements aid in estimating the annual and short-period pattern of water resources in a region. A certain fraction of this precipitation reaches surface holding regions such as lakes, ponds, and the ocean. The balance is returned to the atmosphere or stored in the ground. About half of the annual precipitation enters the earth, and some of this is available for water supply. For a local region, the quantity of precipitation that is available for runoff (rainfall excess) is a function of the soil infiltration

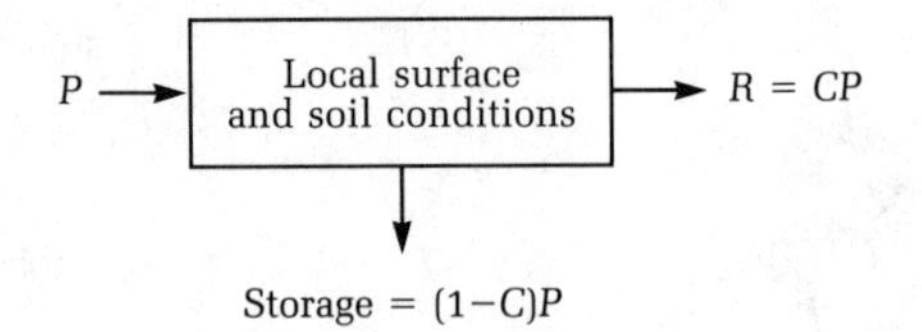

Figure 6.1 Mass balance for precipitation.

and holding capacity and the amount of paved or otherwise impervious areas. A mass balance is commonly used to express the relation of rainfall excess to precipitation and is expressed as

$$R = CP \tag{6.1}$$

where

R = rainfall excess
C = runoff coefficient
P = precipitation

This equation can be expressed in schematic form as shown in Figure 6.1.

Rainfall excess is routed by means of the sewer system to surface and possibly to underground holding systems. These holding systems are designed to serve many users and to preserve the economy of a region. Some of the common uses of the stored water are (1) water supply for irrigation, communities, and industry, (2) hydroelectric power, (3) navigation, (4) natural resource enhancement, (5) recreation, (6) food supply, (7) flood control, and (8) dilution capacity.

6-2 Sources of freshwater supply

Freshwater for any community water supply is made available from artificial and natural sources. The water is stored, treated, and distributed to users. Once used, the water is collected in sewers, treated, and discharged back to the sources. The sources of freshwater are commonly classified as (1) rainwater, (2) surface water, or (3) groundwater. Rainwater is collected from roofs and stored in cisterns (holding tanks). Surface waters are rivers or streams, lakes, and reservoir impoundments. Groundwater lies in storage areas underground, such as permeable rocks, that are fed by natural water bodies, infiltration galleries, and groundwater recharge areas.

It also should be noted that drinking water may be supplied from brackish water or seawater. These waters are made potable using a variety of treatment processes, some which include electrodialysis, reverse osmosis, distillation, and freezing. In low-rainfall areas with little or no surface or groundwater storage, and aboard ships, seawater may be the only economical source of water.

Example Problem 6.1

How much rainfall per year can be stored in a cistern next to a rural home with a roof area of 200 square meters if the annual rainfall is 90 cm?

Solution

$$90\text{ cm}\left(\frac{1\text{ m}}{100\text{ cm}}\right)(200\text{ m}^2) = 180\text{ m}^3$$

6-3 Precipitation and rainfall excess

During and immediately after a rainfall event, rainwater accumulates in surface depressions and on vegetation and infiltrates into the ground; the remaining amount is available as rainfall excess. In a simple mass balance, rainfall excess can be written as

$$R = P - I - I_A \tag{6.2}$$

where

P = precipitation volume

R = rainfall excess

I = infiltration

I_A = initial abstraction, or temporary accumulation

Equation (6.2) implies an area over which the rain has fallen. Thus, units of equation (6.2) can be simple depth (cm) or volumetric (m^3, liters). Other forces that act to reduce rainfall excess are evaporation and transpiration, which are included in equation (6.2) for long-term balances. Over long periods of time, these variables will change, thus equation (6.2) can be written as

$$R(t) = P(t) - I(t) - I_A(t) - E(t) - T(t) \tag{6.3}$$

where

$E(t)$ = evaporation as a function of time (t)
$T(t)$ = transpiration as a function of time (t)

The parameters in equation (6.3) are subject to engineering judgment. These judgments are enhanced by published data that provide information on the stochastic (time-variable) nature and geographic distribution of precipitation. The uncertainty of precipitation volume and intensity is quantified, thus engineers are able to better define their judgments. Records of precipitation, freshwater volumes, and evaporation have been kept for many geographic locations and for many years (Table 6.1).

TABLE 6.1 Hydrologic data for the United States.

Measurement	*Beginning date*	*Number of stations*
Rain and snow	1870	>100,000
Stream gauges	1890	<10,000
Groundwater	1895	—
Evaporation	1895	>500

From Miller (1982).

6-4 Infiltration

Another application of the mass balance concept is in the calculation required for estimating infiltration in the field. The test procedure consists of placing an open drum on the ground, then adding water to the drum. The water in the drum infiltrates into the ground. The storage volume of water in the drum is kept constant by adding water. Essentially the mass balance is

$$F_{in}(\Delta t) - F_{out}(\Delta t) = \Delta S \tag{6.4}$$

where

F_{in} = flow rate into the drum

F_{out} = flow rate out of the drum

ΔS = storage change

Δt = time interval

When storage is kept constant,

$$F_{in}(\Delta t) = F_{out}(\Delta t) \tag{6.5}$$

Thus, by measuring the input to the drum and keeping storage constant, the infiltration or output is equal to the input.

Example Problem 6.2

For a design storm specified for an urban area of Illinois, the depth of rainfall is 2.5 inches (6.35 cm). Initial abstraction in depressions on the pervious and impervious areas was estimated at 0.5 inch (1.27 cm). Infiltration was estimated as 1.0 inch (2.54 cm) using the double-ring infiltrometer. What is the rainfall excess in centimeters and the volume of runoff for a 10 acre (4.05 hectare) area in cubic meters?

Solution

$$R = P - I - I_A$$

$$= 6.35 - 2.54 - 1.27 = 2.54 \text{ cm}$$

$$= \frac{2.54 \text{ cm}}{100 \text{ cm/m}} (4.05 \text{ ha}) \left(\frac{1 \text{ km}^2}{100 \text{ ha}}\right)\left(\frac{10^6 \text{m}^2}{\text{km}^2}\right)$$

$$= 1.03 \times 10^3 \text{ m}^3$$

6-5 Reservoirs

Reservoirs are holding areas for excess rainfall that does not infiltrate into the ground. Built on rivers to intercept the river flows, reservoirs impound the river water providing various economic benefits. An economic loss, however, might result when the reservoir has a chance of silting, filling with river sediment, or becoming polluted. Management of sediment and pollution sources may be necessary to extend the useful life of reservoirs.

Reserviors are normally built across a river valley or a depression area suitable for the storage of water necessary for the users. But streamflow is not constant with time and frequently varies significantly from day to day, or less so from month to month. The longer the time for averaging, the less variability. Thus, if the user's demand is continuous with little time variability, the storage of water in the reservoir must be sufficient to provide for that demand when stream flow is low. On the other hand, when stream flow is high, the reservoir may function to prevent floods downstream by storing and slowly releasing water. The storage can be expressed by constructing a mass balance:

$$S_t = S_{t-\Delta t} + \bar{F}_t(\Delta t) - \bar{D}_t(\Delta t) \tag{6.6}$$

where

S_t = storage at time t

$S_{t-\Delta t}$ = storage at time $(t - \Delta t)$

$\bar{F}_t$ = average streamflow rate between time periods $(t - \Delta t)$ and t

$\bar{D}_t$ = average user demand rate between periods $(t - \Delta t)$ and t

Δt = time period

Another way of expressing the storage function of a reservior is to calculate the differential storage in each time interval, or

$$\begin{aligned} \Delta S_t &= S_t - S_{t-\Delta t} \\ &= (\bar{F} - \bar{D})\Delta t \end{aligned} \tag{6.7}$$

where

ΔS_t = incremental storage

The maximum size of a reservoir occurs when ΔS_t is the greatest, given a fixed period of streamflow records and a demand. It should be noted that demand may not remain constant but instead display seasonal effects due to community economic activities such as tourism, irrigation, or industrial demands such as the cycles of canneries.

The solution to equations (6.6) and (6.7) is relatively easy with the aid of a computer when large volumes of data are necessary; when smaller amounts of data are used, tabular or graphic forms of the equation are more popular. Greater accuracy in specifying the maximum size of a reservoir can be achieved using shorter intervals of time (say, monthly to daily data). To illustrate both the tabular and graphic solution procedure, consider the following example.

Example Problem 6.3

Calculate the required reservoir storage, assuming unlimited storage capacity, using the following streamflow volumes and projected demands. If the reservoir storage was limited to 3×10^8 m^3, how much water would be wasted?

Solution

Month	*Streamflow F* ($m^3 \times 10^6$)	*Demand D* ($m^3 \times 10^6$)	$F - D$ ($m^3 \times 10^6$)	*Cumulative* $F - D$ ($m^3 \times 10^6$)	*Wasted* ($m^3 \times 10^6$)
1	85	45	+40	40	0
2	105	45	+60	100	0
3	60	45	+25	125	0
4	110	50	+60	185	0
5	145	75	+70	255	0
6	80	80	+ 0	255	0
7	60	90	−30	225	0
8	5	90	−85	140	0
9	20	60	−40	100	0
10	40	50	−10	90	0
11	50	35	+15	105	0
12	75	45	+30	135	0
1	85	45	+40	175	0
2	100	45	+55	230	0
3	60	35	+25	255	0
4	110	50	+60	315	15
5	140	75	+65	380	65
6	80	80	+ 0	380	0
7	60	90	−30	350	0
8	5	90	−85	265	0

Assuming that the demands and streamflow are the design situations and the record for design extends over 20 months, the maximum storage would be 380×10^6 m^3, and the reservoir would not empty. Also, note that the reservoir started filling and demand was satisfied in the first month of the first year. Other operational procedures would produce other maximum reservoir sizes. If the original assumptions were used and a maximum reservoir size of 300×10^6 m^3 were built, 80×10^3 m^3 of water would be released or wasted from the standpoint of user demands. A graphical solution to this example problem is shown in Figure 6.2.

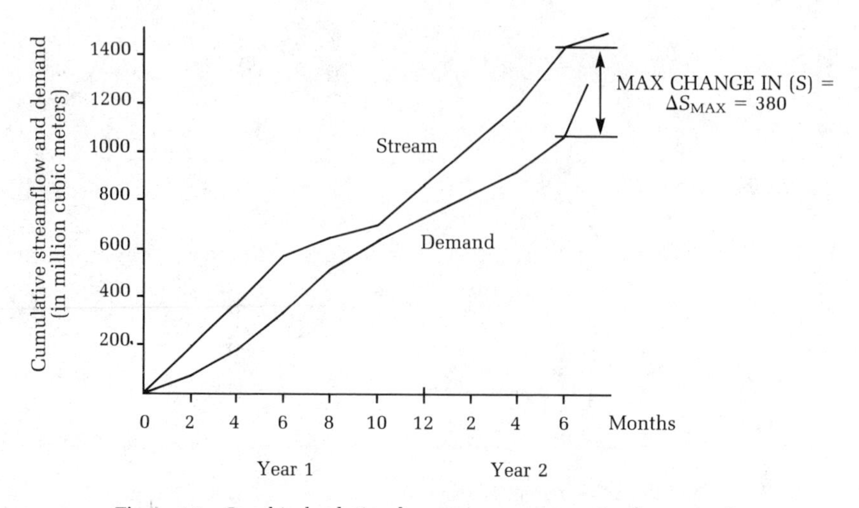

Figure 6.2 Graphical solution for storage requirements of a reservoir.

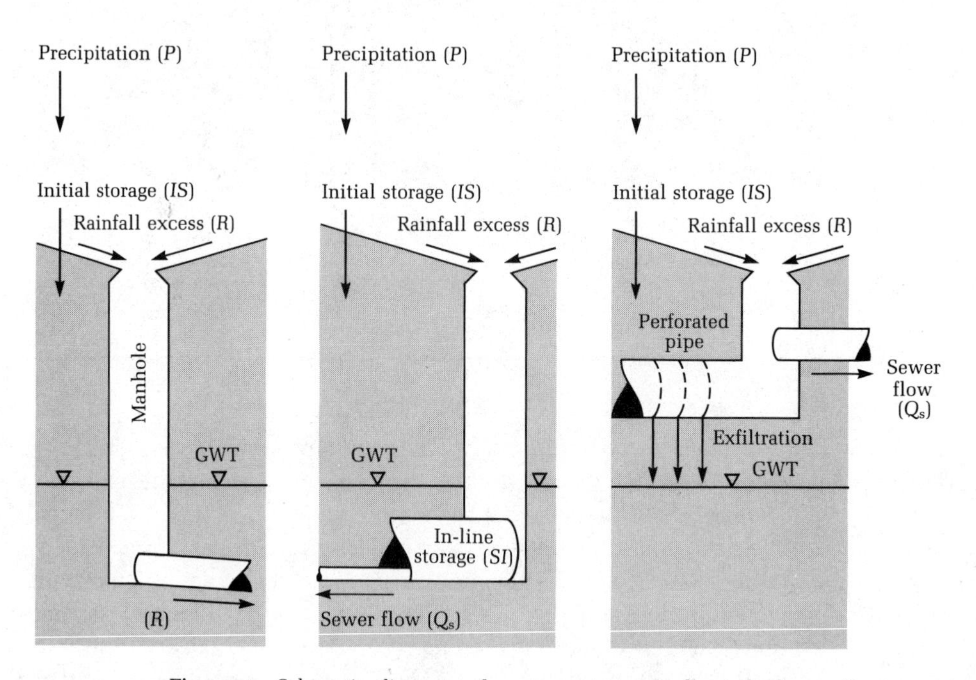

Figure 6.3 Schematic diagrams of sewer systems. A, direct discharge; B, storage-discharge; C. storage exfiltration. GWT is groundwater table.

6-6 Water storage and treatment

On a smaller scale, mass balances can be used in the design of sewer systems. Sewer systems can be classified according to the sources of the sewage: storm sewers carry only stormwater unless groundwater infiltrates into the pipes; sanitary sewers carry sanitary waste and some infiltrate. Combined sewers are designed for the flows of both sanitary and storm wastewaters.

In combined systems, stormwater is either directly discharged to the sewers or stored to reduce the flow rates into the sewer system. Slower flow rates allow smaller pipe sizes and reduce the flow rate to a treatment facility, thus the treatment facility can be smaller, or overflows, which usually have little or no treatment, can be made smaller and less frequent. Schematic representations of direct discharge systems and storage-discharge systems are shown in Figure 6.3. In direct discharge, there is little storage in the manhole or inlet structure, thus, rainfall excess is the sewer flow. However, once storage is provided, sewer flow can be reduced below that of the inlet street flow. Storage tanks must be sized to limit effluent or sewer flows. Again, a mass balance is used as an aid to design and evaluation:

$$F_I(\Delta t) - F_S(\Delta t) = \Delta S \tag{6.8}$$

where

F_I = inlet flow rate
F_S = Sewer flow rate
Δt = time interval
ΔS = storage volume

In some designs, where pollution control is needed and the groundwater table is lower than the bottom of the pipes, perforated storage pipes are used to exfiltrate the stormwater into the ground (Figures 6.3 and 6.4). An equation for these designs is

$$F_I(\Delta t) - F_S(\Delta t) - F_E(\Delta t) = \Delta S \tag{6.9}$$

where

F_E = exfiltrated water

Figure 6.4 Construction of an underground storage and treatment system. Perforated pipe allows water to filter through select material and a pervious membrane.

The exfiltration rate will likely vary with time because of the head of water in the perforated pipe and the saturation of the ground.

The designer of sewer systems is generally given a design storm of specific volume, intensity, and duration. These parameters are specified to protect property values and establish a flood frequency. Thus, the greater the storm volume expected during the design period, the greater the sizes of storage and storm sewers and the less the chance of flooding. These same mass balances also are used to evaluate existing sewer systems for a variety of street inlet flows. These flows, which vary with time, are called hydrographs and are frequently plotted or tabulated as flow rates versus time. More complete studies for stormwater designs can be found elsewhere (Wanielista, 1979; Whipple, 1983).

Example Problem 6.4

Calculate the in-line storage volume required to store the difference between the street runoff hydrograph (F_I) and sewer flow hydrograph (F_S). Use the mass balance approach with the computation time interval of 15 minutes. The physical system is similar to B in Figure 6.3. The street runoff and sewer flow as functions of time are shown below.

Time	*Street flow rate* F_I (m^3/s)	*Sewer flow rate* F_S (m^3/s)
8:00	0.06	0
8:15	0.14	0.009
8:30	0.53	0.065
8:45	1.84	0.349
9:00	1.00	0.837
9:15	0.80	1.111
9:30	0.55	1.009
9:45	0.23	0.769
10:00	0.03	0.483
10:15	0.00	0.121

Solution The problem is solved using an average flow between computation periods and the mass balance represented by equation (6.6) with $\Delta t = 15 \text{ min} \times 60 \text{ s} = 900 \text{ s}$.

Time	$S_{t-\Delta t}$, m^3	$\bar{F}_I(\Delta t)$, m^3	$\bar{F}_s(\Delta t)$, m^3	S_t, m^3
8:00	—	0	0	0
8:15	0	90.0	4.0	86.0
8:30	86.0	301.5	33.3	354.2
8:45	354.2	1066.5	186.3	1234.4
9:00	1234.4	1278.0	533.7	1978.7
9:15	1978.7	810.0	876.6	1912.1
9:30	1912.1	607.5	954.0	1565.6
9:45	1565.6	351.0	800.1	1116.5
10:00	1116.5	117.0	563.4	670.1
10:15	670.1	13.5	271.8	411.8

Thus, the maximum size storage tank is about 1979 cubic meters. A commercially available tank about equal to this volume would be specified.

6-7 Problems

6.1 Estimate the yearly input to a lake if the runoff coefficient for the lake's 150-km^2 watershed is 0.40 and the annual rainfall is 120 centimeters. At first, neglect precipitation on the lake, then add the precipitation on the lake if the lake area is 3 km^2.

6.2 What is the rainfall excess in m^3 for a single precipitation event of 70 mm, an infiltration of 16 mm, and an initial abstraction of 4 mm over a watershed area of 1 km^2?

6.3 For pollution control, you have been required to store the runoff from the first 25 mm of rainfall. The runoff coefficient for a 1-km^2 watershed is 0.5. What size (m^3) storage basin will you specify? What is the area (m^2) if the pond depth is 1 meter? Consider a rectangular pond.

6.4 The water storage pond of problem 6.3 evaporates at a rate of 3.5 cm per week, which is considered constant over a 1 month period. If there is no input (runoff) to the pond in 1 month, what is the remaining pond volume (m^3) and depth (m) after 1 month?

6.5 If the pond of problem 6.3 must be emptied by infiltration and evaporation within 3 days, what must be the infiltration rate (cm/hr) if the evaporation rate is 3.5 cm/week?

6.6 Estimate infiltration as a function of time using data from a drum infiltrometer that is placed in a soil known to have a groundwater table 6 feet below the surface. The drum diameter for infiltration estimates is 16 centimeters. The storage in the drum is held constant over the period of testing. The inflow volume as a function of time is shown below.

Inflow volume (L)	*Time (min)*
5	20
10	45
15	75
20	110
25	150
30	210
35	270
40	330
45	390

6.7 Using the street runoff hydrograph of example problem 6.4 into a rectangular pond of area 100 m^2 and 1 m deep with an infiltration rate of

0.417 cm/min, is the pond large enough to prevent overflow assuming no evaporation and transpiration?

6.8 What is the maximum size of a reservoir on a river to provide a constant user flow rate of 40×10^6 L/day if the river flow given below is considered to be the lowest yearly flow rate? Work the probelm on a monthly rate. Neglect direct rainfall on the reservoir. The reservoir bottom is impervious.

Month	*Average monthly flow rate* *(10^3 m^3/day)*	*Month*	*Average monthly flow rate* *(10^3 m^3/day)*
January	50	July	40
February	40	August	20
March	55	September	15
April	80	October	30
May	120	November	40
June	100	December	50

6.9 What is the volume of water wasted (10^3 m^3) if the maximum size of the reservoir in 6.8 is 4500×10^3 m^3? Present the solution in both graphical and tabular form.

6.10 Obtain the stream-flow records of the U.S. Geological Survey for a local stream. Size a reservoir to accumulate all the stream flow in a 1 year period. (Pick a stream that has flow in every month.)

6.11 For the same time period and stream of problem 6.10, what is the maximum size of reservoir if the demand for stored water is equal to half the smallest monthly flow? Assume that this demand is constant over the year.

6.12 For example problem 6.4, the sewer is modified to reduce the sewer flow rate. This is possible using a commercially available flow rate control device. The resulting sewer flow hydrograph is:

Time	*Sewer flow rate F_S (m^3/s)*
8:00	0.0
8:15	0.018
8:30	0.100
8:45	0.200
9:00	0.220
9:15	0.240
9:30	0.250
9:45	0.250
10:00	0.250
10:15	0.246
10:30	0.242
10:45	0.236
11:00	0.230

What size in-line storage should be used?

References

Fair, G. M., Geyer, G. C., and Okun, D. A. 1971. *Elements of Water Supply and Wastewater Disposal.* New York: Wiley.

Miller, G. T., Jr. 1982. *Living in the Environment—Concepts, Problems and Alternatives.* Belmont, CA: Wadsworth.

Wanielista, M. P. 1979. *Stormwater Management: Quantity and Quality.* Ann Arbor, MI: Ann Arbor Science Publishers.

Whipple, Grigg, N. S., Grizzard, T., Randall, C. W., Shubinski, R. P. and Tucker, L. S. 1983. *Stormwater Management in Urbanizing Areas.* Englewood Cliffs, NJ: Prentice-Hall.

7

Water Treatment

7-1 Introduction

This chapter provides a quantitative and qualitative introduction to potable (drinking) water sources, treatment, and distribution utilizing the basic concepts involving mass balances and chemistry that were presented in Chapters 3 and 4. The quantitative description is designed to enable the student to take a basic mathematical approach and determine basic quantities of materials and energy required to provide potable water to the public. This quantitative approach will be enhanced by qualitative descriptions of each facet of treatment so that the student will obtain a realistic understanding of potable water supply.

There are three main objectives of water treatment: (1) providing safe water for human consumption, (2) providing aesthetically acceptable water for human consumption, and (3) providing water for human consumption economically. Obviously there are close ties among these three objectives. Providing any potable water that is not harmful to human health requires understanding the protection of the raw water source as well as the necessary treatment technology. To a significant degree, the more the source is protected, the less treatment is required. Source protection, therefore, supports the accomplishment of all three objectives. Potable water is perhaps the most important water use, yet it is impossible to separate potable water consumption from other water uses because of the very nature of the hydrological cycle.

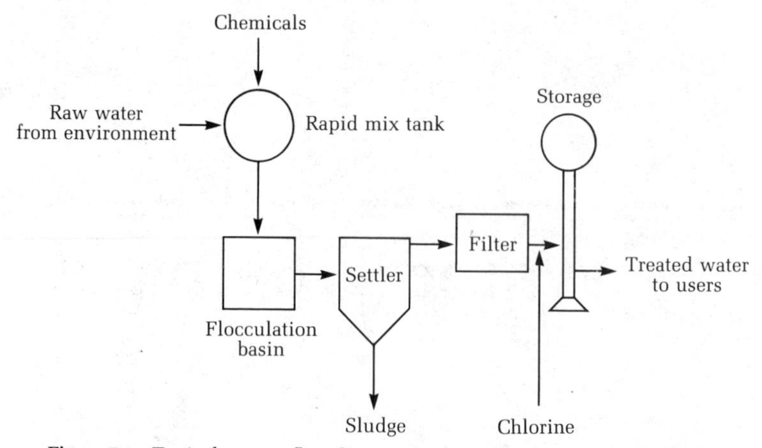

Figure 7.1 Typical process flow diagram for a potable water treatment plant.

The water-treatment plant can be designed to treat raw water from various sources in the environment. Depending on the initial quality of the raw water and the final quality desired for the treated water, one or more unit processes are needed. A typical process flow diagram is presented in Figure 7.1. Chemicals are added to remove suspended solids, color, or other materials. In order to add the chemicals properly, promote the proper reactions, and remove the precipitates, a rapid mixtank followed by a slow mix (flocculation) basin and a settler are required. Filtration next removes the solids too small to settle, and chlorine is then added to disinfect the water. All these unit processes were not always recognized as being necessary. It is interesting to briefly review the history of water treatment.

7-2 Public health

Concern over the effect of drinking water on public health was recorded as early as 2000 B.C. An early medical book contained words to the effect that impure water should be purified by being boiled over a fire, or being heated in the sun, or by dropping a hot iron into it and allowing it to cool, or by filtering it through sand and coarse gravel. The Egyptians were reported to have first utilized alum, a commonly used coagulant today, for the removal of suspended particles from water. Hippocrates (460–377 B.C.) stated that water contributes much to health and that rainwater should be boiled and strained, otherwise it would have a bad odor and cause hoarseness (American Water Works Association, 1971). The ancient Romans constructed elaborate aqueducts to transport water to Rome. It has been suggested that the Romans' extensive use of lead utensils and plumbing (the Latin *plumbum* is the source of the chemical symbol Pb) resulted in a high incidence of lead poisoning among the upper and middle classes that contributed to the eventual downfall of Rome.

In 1854, Dr. John Snow was searching to find the cause of a cholera outbreak in the St. James Parish of London that had caused hundreds of deaths in a 4-month period. His detailed work indicated that a very large number of cholera cases occurred near a public well at No. 40 Broad Street that was used for drinking water. The Broad Street well was found to be contaminated from a leaky sewer pipe. When Snow had the pump's handle removed, the cholera epidemic ended. Louis Pasteur in 1860 proved the germ theory of disease, which was supported by the earlier work of Snow.

In 1804, in Paisley, Scotland, the first filtration plant was built. The Chelsea Water Company installed filters for treatment of the Thames River water in 1829. These works were decades before the work of Pasteur and the Broad Street well incident. The technology was slow to spread. The first filtration plant in the United States was constructed in Poughkeepsie, New York in 1871 and was in service for 90 years before replacement. In the United States, during the years 1900 to 1913, the number of water plants utilizing filtration increased and the number of outbreaks of waterborne disease decreased, establishing an inverse relationship that supported filtration of drinking water taken from surface sources.

In 1835, 51 years after the discovery of chlorine, a physician noted that a small amount of chlorine added to marsh water to make it more palatable also provided a taste and odor that was disagreeable to some individuals. Chlorine was used to treat groundwaters to make the well water more palatable, although the principles of bacterial or viral disinfection were not understood. Chlorine was used in 1879 to treat waste of typhoid-stricken patients before disposal. Chlorine produced from the electrolysis of seawater was applied to sewage in 1893 prior to discharge to the Croton River in England. Professor T. M. Brown, in 1894, recognized that bacteria were not electrocuted but killed by the hypochlorite produced during seawater electrolysis. He spoke to source protection by commenting that a potable water source should not be utilized if disinfection by chlorination was required. A Leiut. Nesfield of the British Indian Medical Corps described the destruction of pathogens by chlorination and recommended the storage and transportation of chlorine gas in iron cylinders for the disinfection of water used by military personnel in the field. Thus, during the 1800s, the need to disinfect sewage as well as the need to disinfect drinking water were recognized.

A dramatic demonstration of the effectiveness of water filtration as a public health measure was recorded by Koch in 1892. Hamburg used unfiltered water; its close neighbor Altona used filtered water from the same source. In 1892, Hamburg experienced a severe cholera epidemic while Altona was hardly affected. Strangely, information for the protection and treatment of drinking water was very slow to be put to use.

Until 1920 the primary, and perhaps the only, objective of water treatment was the protection of public health. Aesthetic water quality would make water more desirable to drink, but not necessarily have any effect on human health. Examples of aesthetic water-quality parameters are color, taste, and odor. Although these parameters can be related to water quality, they are not sufficient to protect the consuming public. There is a good deal of evidence that the public adapts its tastes dependent on the water source that is available. In the area around the Broad Street well, described earlier, people apparently preferred the taste of slightly polluted water, as this well was very popular in London. Since pathogens cannot be always seen, smelled, or tasted, the primary concern of potable water should always be health, although more attention may be paid to the aesthetic quality of water.

In large part, the safe quality of drinking water in this country is taken for granted by the general public. Outbreaks of waterborne disease are uncommon, although they still occur when disinfection facilities malfunction. One such outbreak occurred in the late 1970s in a Florida migrant worker camp. Most public concern in this country takes the form of complaints concerning aesthetic parameters such as taste or odor. However, death due to waterborne disease is not uncommon to the undeveloped countries of the world. In 1960, an estimated 13,000 infants worldwide died every 24 hours due to water-borne disease. This has been decreased in the 1980s, but waterborne disease is still a major problem in the world.

The concern for the effect of drinking water on health may be on the upswing in the 1980s due to chemical carcinogens in our drinking water. These carcinogenic chemicals are much more subtle in their effect on health in that they cause not immediate (acute) disease but long-term (chronic) health effects. Sickness or death, instead of occurring in a few months, may take decades. The Environmental Protection Agency, for instance, has estimated that one excessive death per 100,000 population is caused by the consumption of 2 liters per day of water containing 0.05 $\mu g/L$ trihalomethanes over a 70-year period.

Certainly public health is directly affected by potable water. This fact needs to be considered in water-resources management. Presently, we are seeing the beginning of a major effort to control chemical carcinogens in potable water. In judging future activities to reduce these carcinogens in our water, we should recall that as late as 1929 some people still felt that disinfection of potable water was undesirable—and fluoridation is still controversial to some. Currently we can identify less than 10 to 15% of all the organics in potable water, much less state the associated acute or chronic health effects.

There is little question that we currently have drinking water that for the most part is free from biological pathogens and is aesthetically acceptable. The problem of undesirable chemical contamination in drinking water has been recognized but is not completely defined or solved.

7-3 Standards

Water-quality standards are published by various agencies and are in flux. The U.S. Public Health Service (USPHS), the World Health Organization (WHO) and the Environmental Protection Agency (EPA) are all organizations that have published drinking-water standards for both raw sources and finished potable water. In general, these standards are similar and have changed as knowledge of the health effects of various water-quality parameters has increased.

Source standards

There are no quantitative federal standards for raw water; the Public Health Service says only that the source should be of the highest quality available. (Of course, there are quantitative standards for finished water.) Many states have standards that apply to raw water sources. Raw water source standards, when used for domestic water supply, are shown in Table 7.1, and federal protection standards are shown in Table 7.2.

These standards are quite numerous but are not sufficient in themselves to protect potable water sources. Water-management districts or authorities, and county or municipal zoning boards protect potable water sources by regulating any activities that may affect the quality of the source. The protection of surface sources is easier than the protection of groundwater sources because of location. However, sometimes, because of political and economic pressure, activities are permitted that may be detrimental to potable water quality.

TABLE 7.1 Criteria ranges for raw water sources of domestic water supply in California.

Constituent	*Excellent source of water supply, requiring disinfection only, as treatment*	*Good Source of water supply, requiring usual treatment such as filtration and disinfection*	*Poor source of water supply, requiring special or auxiliary treatment and disinfection*
BOD (5-day), mg/L			
Monthly average:	0.75–1.5	1.5–2.5	Over 2.5
Maximum day, or sample:	1.0–3.0	3.0–4.0	Over 4.0
Coliform MPN per 100 mL			
Monthly average:	50–100	50–5,000	Over 5,000
Maximum day, or sample:	Less than 5% over 100	Less than 20% over 5,000	Less than 5% over 20,000
Dissolved oxygen			
mg/L average:	4.0–7.5	4.0–6.5	4.0
% saturation:	75% or better	60% or better	—
pH (average):	6.0–8.5	5.0–9.0	3.8–10.5
Chlorides, max. mg/L	50 or less	50–250	Over 250
Fluorides, mg/L	Less than 1.5	1.5–3.0	Over 3.0
Phenolic compounds, max. mg/L	None	0.005	Over 0.005
Color, units	0–20	20–150	Over 150
Turbidity, units	0–10	10–250	Over 250

From McKee and Wolf (1963).

Groundwater should meet minimum criteria such as those adopted by regulation in Florida (State of Florida, 1983). The groundwater shall be free from domestic, industrial, agricultural or other nonthermal discharges in concentrations which, alone or in combination, are:

(1) harmful to plants, animals, or organisms native to the soil and necessary for natural treatment of discharge,
(2) carcinogenic, mutagenic, teratogenic, or toxic to humans,
(3) acutely toxic to indigenous species in surface water as it is affected by groundwater,
(4) threatening to public health, welfare, or safety,
(5) creating a nuisance, or
(6) impairing reasonable and beneficial use of adjacent waters.

Florida groundwaters that are used for potable sources are further classified as G-1 where the TDS is 3000 mg/L or less or as G-2 where the TDS is 10 mg/L or less. Both of these classifications have

TABLE 7.2 Source and protection standards.

1. The water supply should be obtained from the most desirable source which is feasible, and effort should be made to prevent or control pollution of the source. If the source is not adequately protected by natural means, the supply shall be adequately protected by treatment.
2. Frequent sanitary surveys shall be made of the water supply system to locate and identify health harzards which might exist in the system. The manner and frequency of making these surveys, and the rate at which discovered health hazards are to be removed, shall be in accordance with a program approved by the Reporting Agency and the Certifying Authority.
3. Approval of water supplies shall be dependent in part upon:
   a. Enforcement of rules and regulations to prevent development of health hazards
   b. Adequate protection of the water quality throughout all parts of the system, as demonstrated by frequent surveys
   c. Proper operation of the water supply system under the responsible charge of personnel whose qualifications are acceptable to the Reporting Agency and the Certifying Authority
   d. Adequate capacity to meet peak demands without development of low pressures or other health hazards
   e. Record of laboratory examinations showing consistent compliance with the water quality requirements of these Standards.

From U.S. Public Health Service (1962).

additional federal standards imposed by rule in the form of maximum contaminant levels as defined by the Primary and Secondary Drinking Water Standards given in Tables 7.3 and 7.4.

Finished water standards

In 1914 the USPHS published standards specifying the maximum contaminant level (MCL) for bacteria in finished water. There were no USPHS standards for chemicals because the USPHS panel could not agree on them, a problem that is still with us today. These standards changed in 1925 to include more restrictive bacteriological criteria, aesthetic qualities, and requirements that the source be free from pollution, protected by the environment, or treated in plants if necessary. Major changes in 1945 and 1946 included MCLs for metals and other substances that had an adverse effect on health. These standards also spoke of the need to avoid contamination of finished water in transit such as by faculty plumbing and inadvertant connections to sewage lines. In 1962 changes to the USPHS standards established sections on radioactivity, fluorides, new bacteriological techniques, plant operation and personnel, new MCLs, and the rationale used by USPHS in establishing MCLs.

The Environmental Protection Agency has the authority by Federal Law PL 92-500 to set drinking water standards, which led to the passage of the Safe Drinking Water Act (SDWA) in 1974. This Act

TABLE 7.3 Maximum contaminant levels (MCL) for federal primary drinking water standards.

Inorganic

Chemical	*MCL (mg/L)*
Arsenic	0.05
Barium	1.
Cadmium	0.010
Fluoride	
Natural	0.05
Added T < 53.7	2.4
53.7–58.3	2.2
58.4–63.8	2.0
63.9–70.6	1.8
70.7–79.2	1.6
79.3–90.3	1.4
Lead	0.05
Mercury	0.05
Nitrate	0.002
Selenium	10.
Silver	0.01

Radioactivity

Source	*MCL*
Natural	
Gross alpha	15 pCi/L
Ra-226 and 228	5 pCi/L
Cultural	Dose to any part of body: 4 millirems per year

Turbidity

Monthly average	1 NTU or 5 NTU if data indicate no interference with chlorine residual, disinfection, or bacterial testing
2-day average	5 NTU

Biological

Method	*Monthly average*	*Individual samples*	
		N < 20	*N > 20*
Membrane filter for coliforms	1/100 mL	4/100 mL in 2 or more	4/100 mL in 5%
Multiple tube	10%	3 in 2 or more	3 in 5%

Organics

Chemical	*MCL (mg/L)*
Pesticides (chlorinated hydrocarbons)	
Endrin	0.0002
Lindane	0.004
Methoxychlor	0.01
Toxaphene	0.005
Chlorophenoxys	
2,4-D	0.1
2,4,5-TP	0.01
Trihalomethanes	
Chloroform	0.10
Dichlorobromoform	0.10
Dibromochloroform	0.10
Bromoform	0.10

From USEPA, "National Primary Drinking Water Regulations," Federal Register 40:59566-59588.

TABLE 7.4 Maximum contaminant levels for federal secondary drinking water standards.

Water quality parameter	*MCL (mg/L unless otherwise noted)*
Chlorides	250
Color	15 cpu*
Copper	1
Corrosivity	Not corrosive
Foaming agents	0.5
Hydrogen sulfide	0.05
Iron	0.3
Maganese	0.05
Odor	3 TON**
pH	6.5 to 8.5 pH
Sulfate	250
Total dissolved solids	500
Zinc	5

*Cobalt platinum unit.
**Taste and odor number.
From USEPA, "National Secondary Drinking Water Regulations," Federal Register 42: 17144–17147.

envisioned that the states would enforce the SDWA by passing the appropriate state laws; however, some states did not. About 80% of the states have elected to obtain primacy, that is, becoming the prime enforcing agency for the SWDA. The states that do not choose primacy, approximately 20%, have an in-state EPA group that takes this responsibility. That is why sometimes a group will be identified as U.S. EPA, the normal EPA group, or, for example, Illinois EPA, an EPA group administering federal laws for the state. The current EPA standards for finished water are listed in Tables 7.3. and 7.4. As previously stated, the primary standards are for the protection of health and the secondary standards are for aesthetic acceptability.

These standards are not inclusive of all elements that can be hazardous to human health. They were originally created to apply to contaminants that were common to natural sources, as potable water was to be taken from the best available—assumedly pristine—source. Very few pristine sources are presently available, and the present standards therefore are inadequate. Over 700 compounds have been identified in drinking water that have adverse health effects. The standards will undoubtedly change in the future: perhaps more surrogate measures like the bacteria standard for coliform bacteria will be used. Coliform bacteria are not pathogenic but are common to the intestines of warm-blooded mammals and have successfully been used as an indicator of the presence of pathogenic organisms. Total organic halogens (TOX), total organic carbon (TOC), and total organic nitrogen

(TON) are good potential surrogate parameters for drinking-water standards.

These standards will continue to change in all respects—biological, organics, inorganics, and aesthetics. The largest area of change will involve the organics standards, but this is not to imply that other standards should be taken for granted. There have been outbreaks of waterborne diseases in the 1970s. High lead concentrations in finished waters have been recorded recently. Engineers and scientists involved with potable water must continually be aware of changes in technology and seek to deliver the safest water that is economically feasible to the consumer.

7-4 Potable water sources

There are two major sources of potable water, surface water and groundwater. (Recently, brackish water or seawater has come into use where surface and groundwater sources are scarce.) Each has different advantages and disadvantages. The supply of drinking water is not simply what source may be available, but the quantity and quality of the sources that may be available. In some areas of the world and in this country, wastewater is receiving serious consideration as a potable water source. Many municipalities are extending the present supply by installing water-saving devices in toilets and showers. Such devices have been estimated to reduce potable water requirements by 10% to 20%.

Surface water sources include rivers, lakes, and surficial aquifers that are essentially unconfined. An aquifer is an underground water supply that is intermixed in a geological formation or stratum. If the water-bearing formation is confined by an impervious formation, an aquiclude, it is a confined aquifer. If the aquifer is restrained under pressure such that the elevation of the water in the aquifer will rise if it is opened (drilled) to the atmosphere, it is an artesian aquifer; such aquifers are not uncommon and may need only to be tapped without pumping to supply water for a small area. Such wells are common among the ranches of northeastern Oklahoma.

Some obvious advantages of a surface water supply are availability and visibility. It is easily reached for supply, and major pollution can be spotted easily. Surface sources typically have softer water and do not require as much treatment for hardness removal. Surface waters, being open to the atmosphere, are high in oxygen, which oxidizes and removes the iron and manganese in raw water. Surface waters are typically free of hydrogen sulfide, which produces an objectionable odor similar to rotten eggs. Surface waters are cleanable if contaminated.

On the other hand, surface waters are variable in quality and easily polluted by natural discharge or cultural waste. Surface waters have high biological activity, which can produce taste or odor in finished water. Surface waters can be high in color and turbidity, which requires additional treatment. Surface waters are generally higher in organics, which form trihalomethanes (known carcinogens) when chlorine is used for disinfection.

Groundwater sources are generally better protected than surface supplies. The quality is more uniform, thus treatment is consistent and therefore easier. Natural color and organics are lower in groundwater than surface waters, therefore treatment for color removal is not typically required. This in turn means that trihalomethanes are lower in finished water produced from groundwater. This reduces the treatment cost because treatment for trihalomethanes is not required. Groundwater is less likely to have taste and odor contamination produced by biological activity. Groundwater is not as apt to be corrosive because the low dissolved oxygen content reduces the likelihood of oxygen participating in a half cell reaction necessary for corrosion.

Disadvantages of groundwater include the comparative inaccessibility of groundwater supplies. Hydrogen sulfide concentrations are produced in low-oxygen environments and are typically found in groundwaters. The reducing conditions of groundwaters usually solubilize iron and manganese which, when exposed to oxygen in potable use, stain surfaces such as walls or laundry. Once groundwater aquifers are contaminated, there is no known way to cleanse them. Groundwaters are often so hard that they must be softened to minimize scale formation during potable use. The advantages and disadvantages of ground and surface potable water sources are summarized in Table 7.5.

Some areas of the world and the United States have grown to the extent that a water deficit exists: the potable water supply is exceeded by the demand. In these areas, growth seems still to be desirable, so wastewater is being considered as a source of potable water. Such an area is Denver, Colorado, which will spend $28 million over a 15-year period to determine if treated domestic sewage can safely be consumed by the public. Augmentation of the present supply by 1995 will be necessary if Denver is to keep growing.

There are as of 1983 no water quality standards for potable water reuse of treated sewage. The concern with this practice is the potential

TABLE 7.5 Advantages and disadvantages of surface and ground potable water supplies.

Surface		*Ground*	
Advantages	*Disadvantages*	*Advantages*	*Disadvantages*
Availability	Easily polluted	Protection	High H_2S
Visible	Variable quality	Low color	High hardness
Cleanable	High color	Low turbidity	Inaccessible
Low Fe and Mn	High turbidity	Constant quality	Not cleanable
Low H_2S	Biological taste and odor	Low corrosivity	
Low hardness	High organics	Low organics	
	High trihalomethane Formation potential	Low trihalomethane Formation potential	

adverse long-term (chronic) health effects. Because of the need to supplement potable water sources, the National Research Council (NRC) brought together the most knowledgeable professionals in this country and, as requested by the EPA, published their comments in 1982 in the *Quality Criteria for Water Reuse*. These criteria recommend a three-tiered program of testing before any treated water is consumed. First, a pilot plant like the proposed treatment plant is constructed. Then, treated wastewater from the plant is produced and tested at three levels against a conventional source, if available. The first level of testing is for specifics like the EPA's defined 129 priority pollutants, the Drinking Water Standards, TOX, TOC, and any others. The second level of testing utilizes concentrated organics at a 1000:1 ratio with tests for mutagenicity. The third and final level of testing consists of mammalian cells and animals for acute and chronic toxicity testing. Only after the reused water passes all three levels of testing and outperforms the conventional source on the same tests is it deemed safe for consumption. Obviously, such testing is expensive and may not be required in the future, but presently it is the only method of determining if the water is safe to drink.

7-5 Treatment processes

The treatment processes will be discussed as unit processes that are commonly used for conventional treatment. The water produced must meet the primary standards introduced earlier; most waters meet the secondary standards. Indeed, city officials are much more likely to receive immediate complaints if the final product looks, tastes, or smells than if the lead or THM standards are exceeded.

Many treatment processes exit for the removal of selected pollutants. Each treatment process can be categorized by chemical changes or pollutant removals. The treatment processes considered herein are those in most common use today. Schematic flow diagrams showing the incorporation of these processes are illustrated in Figure 7.2.

In most water treatment processes, the contaminants are removed by a phase separation process (sedimentation or filtration). In preparation for the phase separation, various processes are employed to convert soluble contaminants to an insoluble form, such as softening, coagulation, and flocculation, prior to sedimentation. Selected chemicals used in treatment processes are shown in Appendix C.

Pretreatment processes are screening, pre-sedimentation, chemical addition, and aeration (Eliassen, 1969). Aeration of groundwaters could be a stand-alone process for the removal of hydrogen sulfide and carbon dioxide from water. It is also the first step in the removal of iron and manganese. However, oxidation is usually required for final removal.

Aeration-oxidation

Aeration is used in water treatment plants for the removal of H_2S and CO_2. Any volatile gas may be removed to some degree by aeration. The theory for gas transfer between an air-liquid interface may be accurately described by the first-order equation

$$dC/dt = K_g(C_s - C_t) \tag{7.1}$$

where

dC/dt = concentration change with time
K_g = a rate constant dependent on operation conditions
C_s = equilibrium concentration
C_t = concentration at time t

Integration of equation (7.1) yields equation (7.2):

$$C_t - C_o = C_s - C_o(1 - \exp(-K_g t)) \tag{7.2}$$

where

C_o = initial concentration

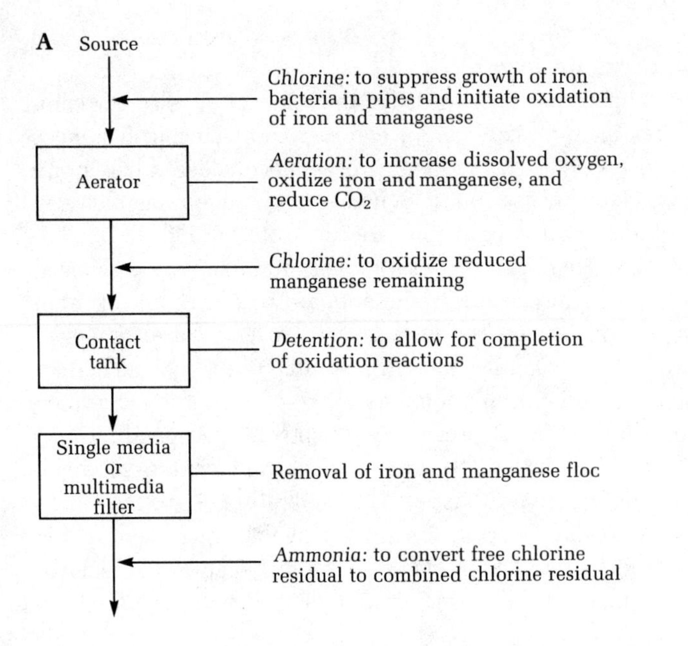

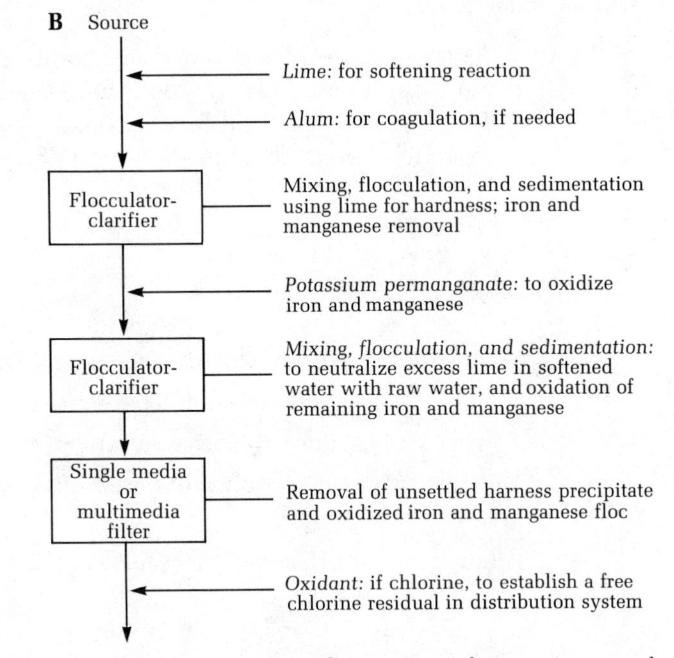

Figure 7.2 Schematics of water-treatment plant. A, iron and manganese removal plant using aeration and chlorine for oxidation. B, treatment for partial softening and iron and manganese removal. C, chemical-coagulation treatment plant with provisions for handling high turbidity. D, chemical-coagulation treatment plant with special provisions for taste and odor.

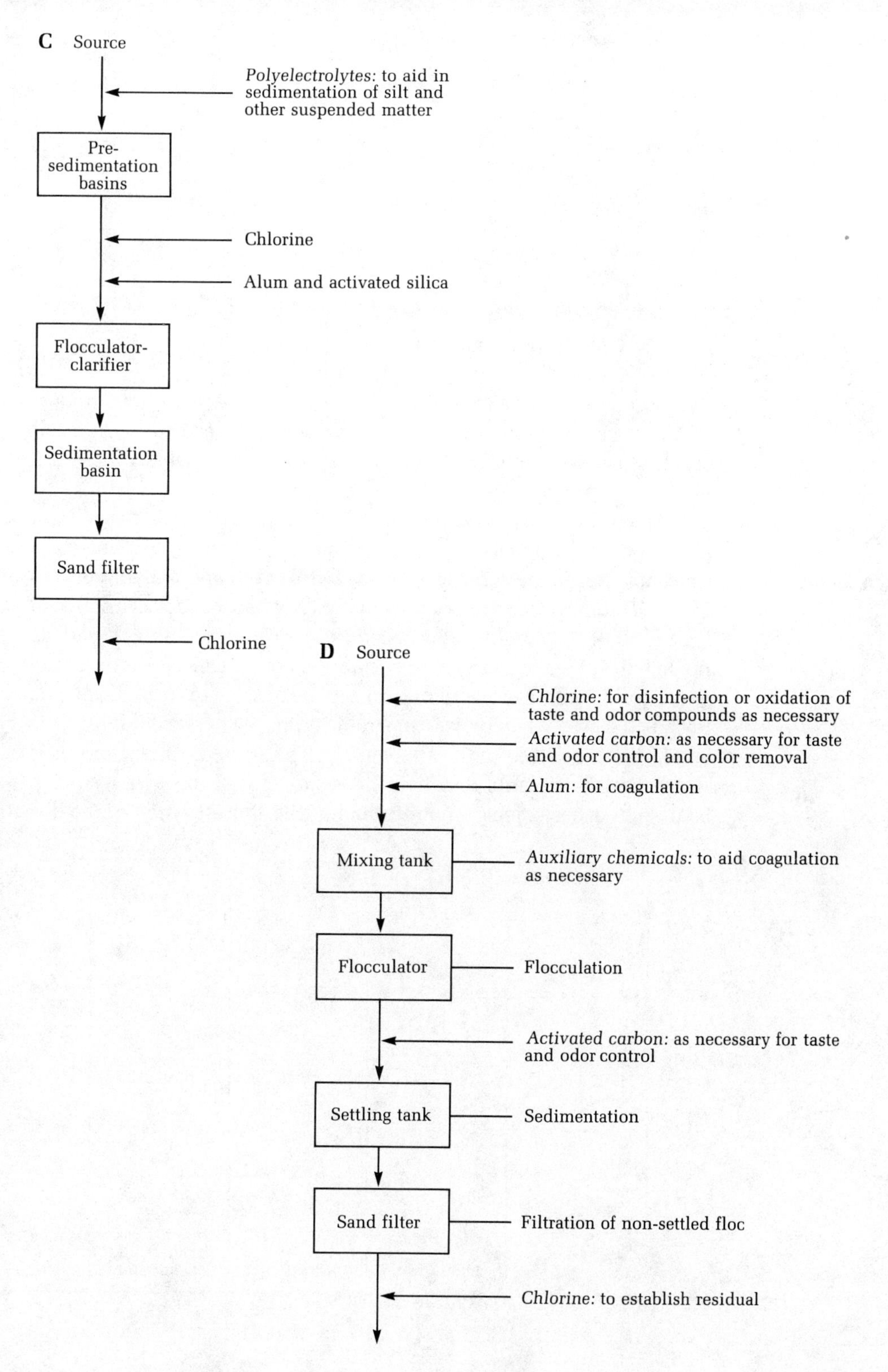
C Source
Polyelectrolytes: to aid in sedimentation of silt and other suspended matter
Pre-sedimentation basins
Chlorine
Alum and activated silica
Flocculator-clarifier
Sedimentation basin
Sand filter
Chlorine
D Source
Chlorine: for disinfection or oxidation of taste and odor compounds as necessary
Activated carbon: as necessary for taste and odor control and color removal
Alum: for coagulation
Mixing tank
Auxiliary chemicals: to aid coagulation as necessary
Flocculator
Flocculation
Activated carbon: as necessary for taste and odor control
Settling tank
Sedimentation
Sand filter
Filtration of non-settled floc
Chlorine: to establish residual

and

$$K_g = k_g(A/V) \tag{7.3}$$

where

k_g = gas transfer rate constant
A = area of liquid surface in contact with gas
V = volume of liquid surface in contact with gas

K_g is a function of the area of the gas-liquid interface per unit volume of liquid being aerated, which is commonly assumed to be in small spheres. The gas transfer rate constant k_g varies and usually increases with temperature.

This process can be visualized as a mass transfer between (1) a liquid phase with bulk contaminant concentration L_B and interface concentration L_i and (2) a gas phase with bulk contaminant concentration G_B and interface concentration G_i. A schematic is shown in Figure 7.3. If the equilibrium concentration of the gas in the liquid, as predicted by Henry's law, is less than the actual concentration of the gas in the liquid, then the dissolved gas will escape to the atmosphere. Such is the case when typical anaerobic groundwaters containing CO_2 and H_2S are exposed to the atmosphere. The smell of rotten eggs near water-treatment plants results from escape of H_2S to the atmosphere as it approaches chemical equilibrium. If the concentration of the dis-

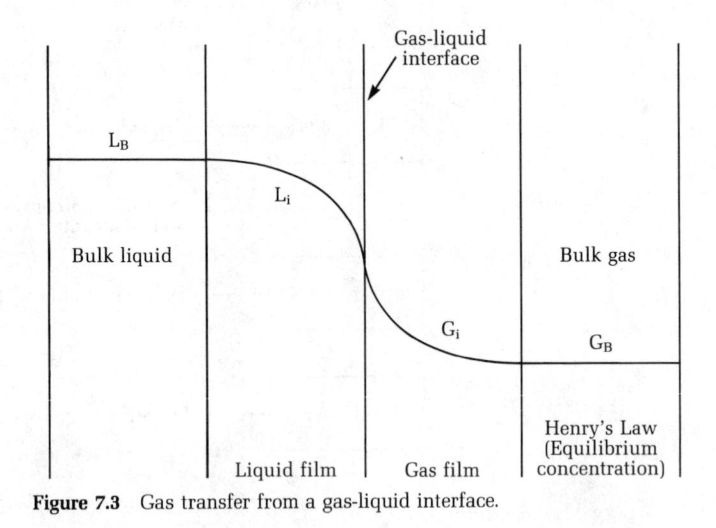

Figure 7.3 Gas transfer from a gas-liquid interface.

solved gas in the liquid is lower than that predicted by Henry's law, then absorption will occur.

CO_2 has a relatively low solubility in water, whereas H_2S has a moderate solubility, both of which can be determined by Henry's law. However, both of the gases can be removed chemically, and it is not economical to try to remove them all by aeration. Typically, aeration is only used for CO_2 or H_2S removal from groundwaters. If the CO_2 concentration is less than 10 mg/L, it is removed by lime precipitation. H_2S removal sometimes accompanies CO_2 removal and produces odors that affect the surrounding neighborhood, so oxidation is typically preferred. The typical aeration equation is given in equation (7.4), which combines the gas transfer rate constant into the mass transfer equation:

$$C_t - C_o = C_s - C_o\left(1 - \exp\left(k_g \frac{At}{V}\right)\right) \quad (7.4)$$

Example Problem 7.1

Determine the maximum allowable concentration of CO_2 liquid at 25°C and 1 atm pressure. State if groundwater containing 32 mg/L CO_2 at 25°C would absorb or desorb CO_2.

Solution

$$K_H(CO_2)_g = (CO_2)_L$$

$$K_H = \frac{(CO_2)_L}{(CO_2)_g} = 10^{-1.5} \text{ at } 25°C$$

Partial pressure CO_2 in the atmosphere $= 10^{-3.5}$ atm

Equilibrium CO_2 in water open to atmosphere:

$$\frac{(CO_2)_L}{(CO_2)_g} = \frac{(CO_2)_L}{10^{-3.5}} = 10^{-1.5}$$

$$(CO_2)_L = 10^{-5.0} \text{ mol/L}$$

$$= 10^{-5} \text{ mol/L}\left(\frac{44 \text{ g}}{\text{mol}}\right)\left(\frac{10^3 \text{ mg}}{\text{g}}\right) = 0.44 \text{ mg/L}$$

The equilibrium concentration of dissolved CO_2 is 0.44 mg/L, which is less than 32 mg/L, so the CO_2 will escape from the water to the atmosphere.

Example Problem 7.2

A water-treatment plant sprays this same water into the air for 2.5 seconds. The water takes the shape of spheres with a diameter of 0.50 cm. The initial concentration of CO_2 is reduced from 32 mg/L to 10 mg/L. Find the gas transfer coefficient (neglecting the trace amount of CO_2 remaining (0.44 mg/L)).

Solution For CO_2:

$$C_s - C_o = -32 \text{ mg/L}$$

$$C_t - C_s = 10 \text{ mg/L}$$

$$C_t - C_o = -22 \text{ mg/L}$$

Using equation (7.3):

$$-22 \text{ mg/L} = -32(1 - \exp(-K_g\, 2.5))$$

$$K_g = 1.16 \text{ s}^{-1}$$

$$K_g = k_g \frac{A}{V} = \frac{k_g\, 6}{D} \text{ (spheres)}$$

$$k_g = \frac{K_g\, D}{6} = \frac{1.16(0.50)}{6} = \frac{0.10 \text{ cm}}{\text{s}}$$

H_2S can be effectively removed by oxidation, after chlorination, or by the addition of potassium permanganate. This method of removing hydrogen sulfide by chemical oxidation can be preferred because it is able to remove all H_2S odors and it does not create an odor problem. However, hydrogen sulfide removal by oxidation is more expensive than aeration.

$$H_2S + 2Cl_2 + 2H_2O + O_2 \longrightarrow H_2SO_4 + 4HCl \quad (7.5)$$

$$5H_2S + 4KMnO_4 + 7O_2 \longrightarrow 4MnO_2 + 3H_2SO_4 + 2K_2SO_4 + 2H_2O \quad (7.6)$$

Precipitation processes

Precipitation processes common in conventional water treatment are lime softening and chemical coagulation. Each of these processes has different purposes, but each produces a solid precipitate that forms sludge. Sludge, as it is discussed here, is a mixture of solids and water that is a problem to dispose of because there is so much of it.

Softening. Excessive calcium and magnesium ions in potable water cause undesirable scaling on home fixtures. Water hardness is defined as the sum of the polyvalent metal ions in solution; however, it may be approximated by the sum of the calcium plus the magnesium ions:

$$\text{Total hardness} = Ca^{+2} + Mg^{+2} \tag{7.7}$$

The calcium and magnesium ions are positively charged (cations). Since the sum of all dissolved ions in water is electrically neutral, the hardness must be balanced by equivalent anions or negatively charged ions. The majority of these are bicarbonate ions, HCO_3^-, which are produced from the CO_2 in the atmosphere and the dissolution of calcium carbonate rock in the aquicludes. The major reactions, given in Chapter 4, include the solubilization of CO_2 gas (equation 7.8), the ionization of aqueous carbonic acid (equation 7.9), the ionization of bicarbonates (equation 7.10), and the solubility product of calcium carbonate (equation 7.11):

$$(CO_2)g + H_2O = H_2CO_3 \qquad pK = 1.5 \tag{7.8}$$

$$H_2CO_3 = H^+ + HCO_3^- \qquad pK = 6.3 \tag{7.9}$$

$$HCO_3^- = H^+ + CO_3^{-2} \qquad pK = 10.3 \tag{7.10}$$

$$CaCO_3 = Ca^{+2} + CO_3^{-2} \qquad pK = 8.3 \tag{7.11}$$

Alkalinity is defined as the sum of the titratable bases and is dominated by the carbonate system in natural waters. The equation for alkalinity is given in equation (7.12) where all species are in moles per liter:

$$\text{Alkalinity} = 2[CO_3^{-2}] + [HCO_3^-] + [OH^-] - [H^+] \tag{7.12}$$

The $MgCO_3$ or $CaCO_3$ that is dissolved into natural water contributes carbonate hardness in that the dissolved magnesium or calcium is electrically neutralized by species from the carbonate system. Hardness contributed from sulfates, chlorides, or other compounds of magnesium or calcium besides carbonate is non-carbonate hardness. This classification is shown schematically in Figure 7.4.

Non-carbonate and carbonate hardness and alkalinity can now be associated with water softening. The natural hardness in water is reduced by precipitating calcium and magnesium from solution as $CaCO_3$ or $Mg(OH)_2$ only. These reactions occur at high pH, and lime, $Ca(OH)_2$, is generally used to raise the pH. The lime-soda softening

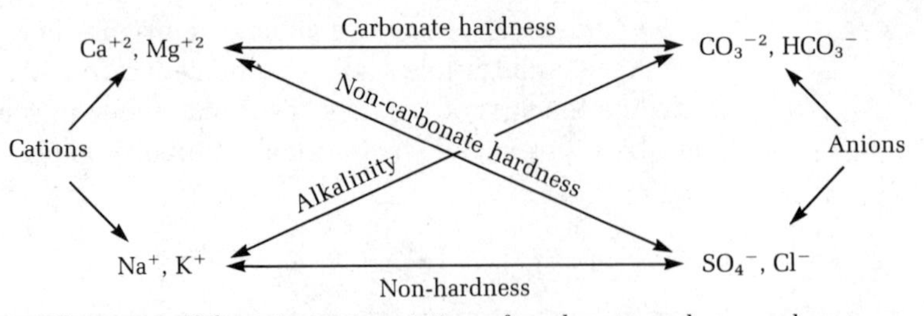

Figure 7.4 Schematic representation of carbonate and non-carbonate hardness equilibrium in natural waters.

reactions are given in equations (7.12) through (7.17). The equations will be used to determine how much chemical is required to soften water.

$$CO_2 + Ca(OH)_2 = CaCO_3\downarrow + H_2O \tag{7.12}$$

$$Ca(HCO_3) + Ca(OH)_2 = 2CaCO_3\downarrow + 2H_2O \tag{7.13}$$

$$Mg(HCO_3)_2 + 2Ca(OH)_2 = 2CaCO_3\downarrow + Mg(OH)_2 + 2H_2O \tag{7.14}$$

$$Ca(SO_4) + Na_2CO_3 = CaCO_3\downarrow + Na_2SO_4 \tag{7.15}$$

$$Mg(SO_4) + Ca(OH)_2 + Na_2CO_3 = Mg(OH)_2\downarrow + CaCO_3\downarrow + NaSO_4 \tag{7.16}$$

$$Na(HCO_3) + Ca(OH)_2 = CaCO_3\downarrow + NaOH + H_2O \tag{7.17}$$

CO_2 is always present in trace amounts in natural waters. Equation (7.12) does not represent any hardness removal, but CO_2 does exert a lime demand. Equations (7.13) and (7.14) represent removal of carbonate hardness. Note that two equivalents of lime are required per equivalent of magnesium carbonate hardness removed whereas only 1 equivalent is required for calcium carbonate. Equations (7.15) and (7.16) are for removal of non-carbonate hardness. An equivalent of alkalinity must be added for every equivalent of calcium non-carbonate hardness removed. Magnesium non-carbonate hardness, equation (7.16), is expensive to remove because $CaSO_4$ is produced and alkalinity must then be added to remove the $CaSO_4$ (see equation 7.15).

$NaHCO_3$ (equation 7.17) is called negative hardness (or excess alkalinity) and must be removed when magnesium hardness is to be removed. This is because effective magnesium removal does not occur

TABLE 7.6 Stoichiometric estimate for lime-soda softening, in milliequivalents.

	Lime		*Soda*		
Lime-soda demand	*meq/meq*	*Dose*	*meq/meq*	*Dose*	*Eq. no.*
CO_2	1	x*	0	0	7.12
Carbonate hardness					
Calcium	1	x	0	0	7.13
Magnesium	2	x	0	0	7.14
Non-carbonate hardness					
Calcium	0	0	1	x	7.15
Magnesium	1	x	1	x	7.16
Negative hardness	1	x	0	0	7.17
Excess					
$CaCO_3$ solubility	0.5	0.5	0	0	—
$Mg(OH)_2$ precipitation	0.5	0.5	0	0	—
Carbonate alkalinity					
Balance	0	0	x	x	—

*x = to be determined by water quality.

until pH 11, and HCO_3^- must be neutralized to CO_3^{-2} before pH 11 can be reached.

Estimates of the necessary amounts of lime and soda ash to soften any water can be made by using Table 7.6. This table is derived from the reactions presented above, plus provisions for excess doses of 0.5 milliequivalents each and for ensuring the alkalinity balance. The excess, denoted as $CaCO_3$ solubility, is simply an excess lime dose to push the reaction to precipitation as the normal detention time in the plant reactor is 2 to 3 hours. The excess for $Mg(OH)_2$ is simply an estimate for the lime required to reach pH 11, where $Mg(OH)_2$ will be removed. The alkalinity balance is an approximation to ensure that enough carbonate alkalinity is present to precipitate the desired $CaCO_3$ as shown in equation (7.18):

$$(Ca)_R + (Ca)_A \leq (CO_2)_P + (C.ALK)_P + (C.ALK)_A \tag{7.18}$$

where

$(Ca)_R$ = meq/L calcium removed

$(Ca)_A$ = meq/L calcium added for calcium removal

$(CO_2)_P$ = meq/L CO_2 present

$(C.ALK)_P$ = meq/L carbonate alkalinity present

$(C.ALK)_A$ = meq/L carbonate alkalinity added

In Chapter 4, an equation was introduced that enables any ion to be reported in units of another by multiplying by that milliequivalent ratio. In the water-treatment industry, most units are reported in mg/L as $CaCO_3$. The milliequivalent weight of $CaCO_3$ is 50 mg. To calculate the lime or soda demand, (1) determine the calcium and magnesium carbonate and non-carbonate hardnesses to be removed, (2) allocate the lime-soda doses by category in Table 7.6, (3) check the alkalinity balance, and (4) sum the necessary lime and soda doses. The lowest hardness that can be achieved by lime-soda softening is approximately 35 mg/L as $CaCO_3$. This hardness is all calcium hardness as all of the magnesium hardness can be removed if necessary.

Example Problem 7.3

Determine the lime and soda doses required to soften the following water to a total hardness (TH) of 100 mg/L as $CaCO_3$.

$$\begin{aligned} CO_2 &= 4.4 \text{ mg/L as } CO_2 \\ TH &= 200 \text{ mg/L as } CaCO_3 \\ CaH &= 200 \text{ mg/L as } CaCO_3 \\ AlK &= 200 \text{ mg/L as } CaCO_3 \end{aligned}$$

Solution Since TH = CaH, MgH = 0:

$$\begin{aligned} TH &= CaH + MgH = 200 \\ \text{Ca hardness removed} &= 200 - 100 = 100 \text{ mg/L} \\ CH &\leq ALK = 200 \end{aligned}$$

This means that all of the TH is CH.

$$\text{CH removal} = 100 \text{ mg/L}\left(\frac{\text{meq}}{50 \text{ mq}}\right) = 2.0 \text{ meq/L}$$

$$CO_2 \text{ removal} = 4.4 \text{ mg/L}\left(\frac{\text{meg}}{22 \text{ mg}}\right) = 0.2 \text{ meq}$$

Lime-soda calculations:

	Lime		Soda	
	meq/meq	Dose (meq/L)	meq/meq	Dose (meq/L)
CO_2	1	0.2	0	0
CH-Ca	1	2.0	0	0
CH-Mg	2	0	0	0
NCH-Ca	0	0	1	0
NCH-Mg	—	0	—	0
Neq H	1	0	0	0
Excess $CaCO_3$	0.5	0.5	0	0
Excess $Mg(OH)_2$	0.5	0	0	0
Total	—	2.7	—	0

The total doses are

$$Ca(OH)_2 = 2.7 \text{ meq/L}\left(\frac{37 \text{ mg}}{\text{meq}}\right) = 99.9 \text{ mg/L}$$

or

$$Ca(OH)_2 = 99.9(8.345) = 833.7 \text{ lbs/mgal}$$

Recarbonation. Recarbonation is a two-fold process in potable water treatment. It is normally done only after lime softening. Primary recarbonation is used to precipitate the excess calcium that is always provided as lime to raise the pH when magnesium is removed by softening. Magnesium removal does not effectively begin until pH 11, and $CaCO_3$ precipitation is essentially complete at pH 10.3 to 10.6. All of the calcium from the lime necessary to raise the pH from 10.3 to 11 stays in solution. However, when CO_2 is introduced into a water at pH 11, it is transformed to ${CO_3}^{-2}$:

$$CO_2 + 2OH^- \longrightarrow {CO_3}^{-2} + H_2O \tag{7.19}$$
$${CO_3}^{-2} + Ca^{+2} \rightleftarrows CaCO_3\downarrow$$

These processes for removal of excess calcium are necessary because the excess calcium added for magnesium removal is not accounted for in equation (7.18). The point could be made that the non-carbonate calcium hardness produced from the removal of magnesium non-carbonate hardness could also be removed this way. However, most engineers like to nudge the chemical dose upward a bit, so the soda allocation for its removal is left in Table 7.1

Secondary recarbonation is used for adjusting the pH. This is necessary because supersaturation of $CaCO_3$ may cause precipitation onto sand filters, causing excessive head loss. The reaction is

$$CO_3^{-2} + CO_2 + H_2O \longrightarrow 2HCO_3^{-} \tag{7.20}$$

Secondary recarbonation is a common unit process that is used at every lime softening plant. Unfortunately not all of the $CaCO_3$ will precipitate in the reactors. If a large amount of CO_3^{-2} is present, then $CaCO_3$ will form and may seal existing filters and pipes. Such precipitates are commonly seen on bathroom walls and kitchen fixtures.

Coagulation. Coagulation is a process for destabilization of colloids in which aluminum or iron salts are added for the removal of color or turbidity from raw water. Alum, $Al_2(SO_4)_3 \cdot 14H_2O$, is the most common coagulant although $FeCl_3$, $FeSO_4 \cdot XH_2O$, and other coagulants can be used. The discussion here will center on alum, although any iron or aluminum salt reacts in the same way. The important reactions are

$$Al_2(SO_4) + 6H_2O \longrightarrow 2Al(OH)_3 + 3H_2SO_4 \tag{7.21}$$

$$H_2SO_4 \longrightarrow 2H^+ + SO_4^{-2} \tag{7.22}$$

$$H^+ + HCO_3^- \rightleftharpoons H_2CO_3 \tag{7.23}$$

Both aluminum and iron salts undergo hydrolysis (7.21) to form the metal hydroxide. Typically, it is the in-place formation of the metal hyrdoxide that captures the color or turbidity which is removed during sedimentation. Three protons (or H^+) are produced for each mole of aluminum changed to aluminum hydroxide. These protons will lower the pH and soluble species $Al(OH)^{+2}$, $Al(OH)_2^{+}$, and Al^{+3} will form, which may reduce the effectiveness of coagulation. If the pH is too high, $Al(OH)_4^{-}$ can form, which may also reduce the effectiveness of coagulation. The optimum range for alum coagulation is approximately from pH 5 to pH 7. Since there is natural alkalinity in most waters, few plants add alkalinity to offset the pH drop due to alum addition because the pH is seldom driven below 5; nevertheless equation (7.24) shows the soda alkalinity required to offset pH drop due to alum addition.

$$Al_2(SO_4)_3 \cdot 14H_2O + 6H_2O + 3Na_2CO_3 \longrightarrow 2Al(OH)_3\downarrow + 3H_2CO_3 + 3Na_2SO_4 \quad (7.24)$$

Reaction vessels. There are many different types of reaction vessels used for precipitation reactions in water treatment. The common choice of water plant designers is a solids upflow contact unit. These reactors accomplish three important engineering processes: (1) rapid mixing of all chemicals to provide immediate dispersion for optimum treatment, (2) the growth of the nucleated solids after coagulation to a large floc or crystal, and (3) the separation of the floc from the treated water. Such a reaction vessel is shown in Figure 7.5.

When a chemical like alum is first introduced into a process, it should be dispersed as quickly as possible. This is done by a rapid mix in which small nuclei form. The actual forming of color or turbidity particles occurs in two mixing processes: the larger coagulated nuclei first form a large floc particle for settling and treatment, then slow mixing is required to promote growth. This part of the reaction is referred to as flocculation. The water is then treated and separated from the resulting suspension by sedimentation. The sludge is discharged from the bottom of the reactor and the treated water overflows the top to be carried to filtration or another treatment process.

The required mixing or energy input for adequate rapid mixing can be determined from

$$G = \sqrt{\frac{550P}{V\mu}} \quad (7.25)$$

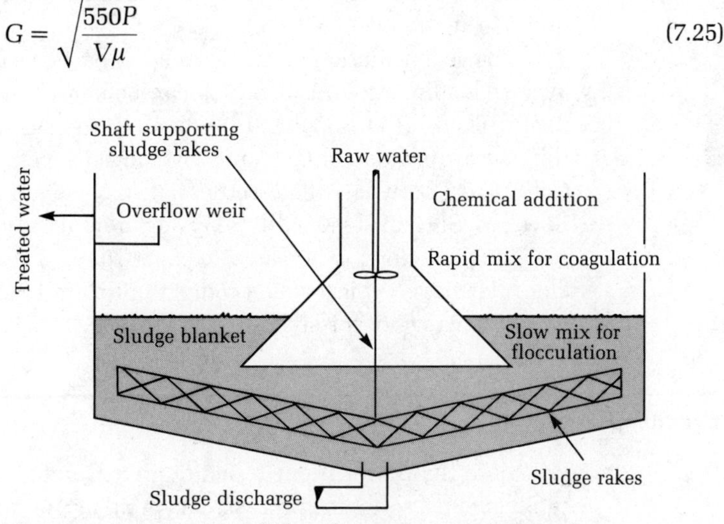

Figure 7.5 Functions of a solids upflow contact reaction vessel.

where

G = velocity gradient, s^{-1}
P = power, HP
V = volume, ft^3
μ = absolute viscosity of water, lbf-s/ft^2
550 = conversion factor, ft-lbf/s-HP

or, in metric units,

$$G = \sqrt{\frac{P}{V\mu}}$$

where

P = power, W
V = volume, m^3
μ = viscosity, cp

The velocity gradient G is a means of measuring relative velocities between particles. Desirable G values for rapid mix and flocculation during softening are 800 s^{-1} and 200 s^{-1}, respectively. The G values for alum coagulation can be estimated at 800 and 100 s^{-1} for rapid mix and flocculation.

The sedimentation of the flocculated particles is controlled by the surface loading rate (SLR), which can be shown to be independent of depth. The SLR is calculated by dividing the surface area of the settling basin (solids contact unit) by the flow per unit time. Because $CaCO_3$ or $Mg(OH)_2$ floc settles at a faster rate than $Al(OH)_3$ flow, an SLR of 1440 gal/ft^2/day (4 cm/min) is used for design of softening reactors, whereas 720 gal/ft^2/day (2 cm/min) is used for alum clarifiers. Typical solids contact units are 15 feet deep to provide adequate capacity for sludge at the bottom.

Example Problem 7.4

Calculate the power input to maintain a G value of 1000 s^{-1} during rapid mix in a softening process. The rapid mix is conducted in a 5 ft × 5 ft × 5 ft chamber.

$$G = 1000/s = \sqrt{\frac{550\,P}{V\mu}}$$

$$\mu = 0.273 \times 10^{-4}\ \text{lbf-s/ft}^2 \text{ at } 50^\circ\text{F}$$

$$1000/s = \sqrt{\frac{550P}{125 \times 0.273 \times 10^{-4}}}$$

$$P = \frac{1 \times 10^6}{1.6117 \times 10^5}$$

$$= 6.20\ \text{HP}$$

Example Problem 7.5

Calculate the surface area required for settling alum floc in a 10 mgd coagulation plant. Estimate the detention time. Assume a depth of 15 feet.

$$\text{SLR} = 720\ \text{gal/ft}^2\text{/day}$$

$$\text{Area} = \text{Flow/SLR}$$

$$\text{Area} = \frac{10 \times 10^6\ \text{gal/day}}{720\ \text{gal/ft}^2\text{/day}} = 13{,}889\ \text{ft}^2$$

$$\text{Area} = \frac{\pi D^2}{4} = 13{,}889$$

$$D = \sqrt{\frac{13{,}889 \times 4}{\pi}} = 133\ \text{ft}$$

A solids contact unit with a 133-ft diameter would meet design criteria.

$$T = \frac{V}{Q} = \frac{AD}{Q} = \frac{13{,}889\ \text{ft}^2 \times 15\ \text{ft}}{10 \times 10^6\ \text{gal/day}\ (1/7.48)\ (\text{ft}^3\text{/gal})}$$

$$= 0.16\ \text{day} = 3.74\ \text{hours}$$

Filtration

Filtration is the final process for removing suspended solids in the water-treatment plant. Turbidity, the amount of light scattered relative to a chemical standard, and length of filter operation before cleaning are used as a means of gauging the effectiveness of filtration. The turbidity standard for finished water is 1 turbidity unit (NTU), with 5 units being acceptable if specific criteria are met. Water-plant operators generally keep the turbidity of the finished water below 1 NTU. It is difficult to explain to the public why their drinking water is cloudy or otherwise objectionable. Consequently, the operators try hard to avoid any visible, odor, or taste problems in their product.

As the filter is operated, solids carried over from settling are accumulating on the filter surface. These solids have the effect of either decreasing throughput, as is the case in declining-rate filters, or increasing the head required for constant throughput. The latter tendency is characterized by a rising water level in the filters. A diagram of a multi-media gravity filter is shown in Figure 7.6. Settled water is simply routed to the top of the filters and passed downward through the media en route to disinfection.

Mixed media are used for filter media in order that layers of media with different effective sizes can be used. The filter then can remove large particles on the surface and allow the smaller particles to pass into the layered media before removal. This has the effect of increasing the filter run life and improving quality. The specific gravity of the top medium is lowest, which increases with the order of the media used so that they will settle in the same order after backwashing.

During backwash, the deposited solids must be freed from the media, flushed to the backwash troughs, and carried away for disposal

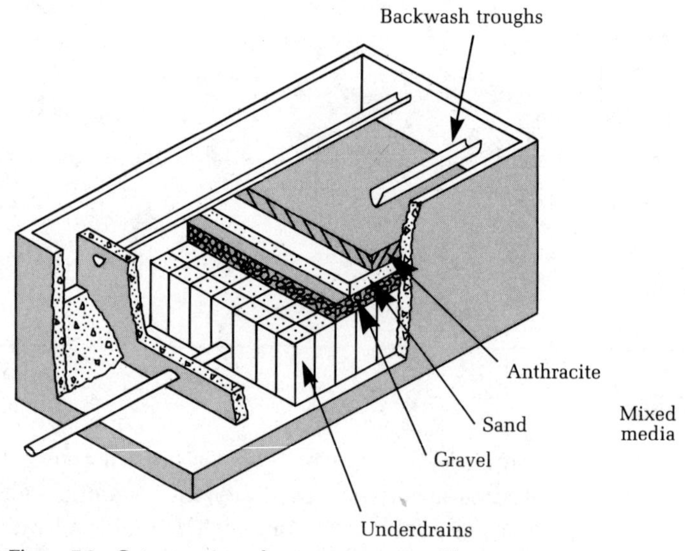

Figure 7.6 Cutaway view of a water-treatment filter.

or reclamation. The backwash rate must expand or fluidize the bed by 20% to 50%, but not so much as to wash away the filter media. Typical design rates for backwash are 20 gal/ft^2/min for 5 to 15 minutes, whereas normal filtering rates are 2 to 5 gal/ft^2/min. Filtration runs can vary from 20 to 140 hours, with 30 to 40 hours between washings being considered optimum.

Example Problem 7.6

Determine the filtration surface area for a 10 mgd potable water treatment plant if the filters are taken out of service for 1 hour each day for backwashing. Calculate what percentage of the treated water is used for backwashing if actual filter washing is done for 15 minutes.

Solution Required filtration rate = 4 gal/ft^2-min

$$\text{Minimum filter area} = \frac{\text{Flow}}{\text{Rate}} = \frac{(10 \times 10^6 \text{ g/day})(1 \text{ day}/1440 \text{ min})}{4 \text{ gal/ft}^2\text{-min}}$$

$$\text{Actual area} = 1736 \text{ ft}^2 \times \frac{24 \text{ hr/day}}{23 \text{ hr/operation}} = 1812 \text{ ft}^2$$

$$\text{Backwash rate} = 20 \text{ g/ft}^2\text{/min} \times 10 \text{ min} \times 1812 \text{ ft}^2$$
$$= 362{,}400 \text{ gal}$$

$$\text{Percent backwash} = \frac{362{,}400}{10{,}000{,}000} \times 100 = 3.6\%$$

Disinfection

Disinfection is the removal of all pathogenic organisms, whereas sterilization is the removal of all organisms. Several different unit processes contribute to overall disinfection of potable water: softening or coagulation, filtration, and oxidation.

Because of the higher pH involved, softening disinfects water better than coagulation. The measure of disinfection efficiency is removal of coliform bacteria. Softening at pH 11 kills most coliform bacteria, whereas coagulation typically removes 30 to 70% of all bacteria present. Filtration can effectively remove pathogenic bacteria, and both processes in series will produce a high-quality water. However, disinfection by oxidation is necessary to kill all pathogenic organisms.

The test for disinfection uses coliform bacteria. These bacteria have a subgroup, fecal coliform, that is present only in the intestines of warm-blooded animals. These bacteria, although not-pathogenic, are considered more resistant to disinfection than the pathogenic organisms. Therefore, the absence of coliforms after chemical disinfection indicates that pathogenic organisms are also absent. This reasoning underlies the practice of disinfection in this country. Disinfection is practiced not only at the water plant but in the distribution system too. The coliform standard is checked throughout the distribution system at frequent intervals.

The most common disinfectant in this country is chlorine. It is a powerful oxidizing agent that is readily reduced, as are most disinfecting agents, and readily soluble in water. Chlorine is commonly purchased in 2000-lb quantities as a liquid under pressure, which is then metered into a dosing stream so as not to exceed the solubility of dissolved chlorine as Cl_2. The gas reacts with water to form HCl and HOCl, hypochlorous acid. The HOCl is the disinfecting agent. The hypochlorite ion, OCl^-, has a chlorine atom with a positive valence which will oxidize nearly any species to reach the chloride state Cl^-, which is stable. The pertinent reactions are

$$Cl_2 + H_2O \rightleftharpoons HCl + HOCl \quad pK = 3.35 \text{ at } 25°C \tag{7.26}$$

$$HOCl \rightleftharpoons H^+ + OCl^- \quad pK = 7.6 \text{ at } 25°C \tag{7.27}$$

Many states require a 0.2 mg/L free chlorine residual or a 0.6 mg/L total chlorine residual at the tap. Such a residual ensures that any pathogens present would have been destroyed in a reaction with the disinfectant. Design criteria usually call for a 30-minute detention time in the disinfection process at the water-treatment plant. These requirements have eliminated most outbreaks of waterborne disease in this country.

Hypochlorous acid will react with ammonia to form chloramines, which have the ability to disinfect. The disinfection ability of chloramines is not as great as chlorine; however, they dissipate much more slowly than chlorine and are desirable in some waters.

$$NH_3 + HOCL \rightleftharpoons NH_2Cl + H_2O \tag{7.28}$$

$$NH_2Cl + HOCl \rightleftharpoons NHCl_2 + H_2O \tag{7.29}$$

$$NHCl_2 + HOCl \rightleftharpoons NCl_3 + H_2O \tag{7.30}$$

$$2NH_3 + 3Cl_2 \rightleftharpoons N_2 + 6HCl \tag{7.31}$$

Monochloramine (NH_2Cl) is the desirable form of chloramines for disinfection which is produced by a weight ratio of 1 part NH_3 to 4 or 5 parts Cl_2 in practice. Other forms are not as effective disinfectants

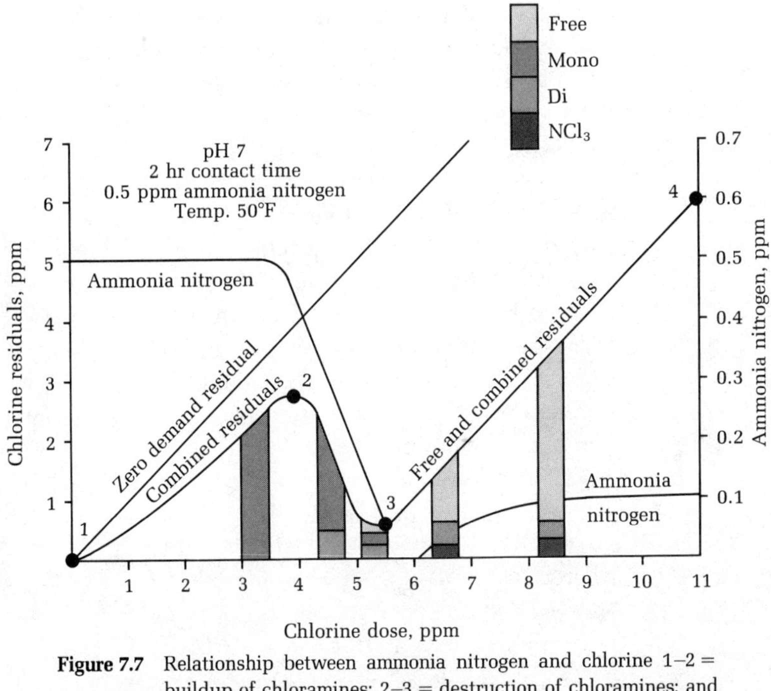

Figure 7.7 Relationship between ammonia nitrogen and chlorine 1–2 = buildup of chloramines; 2–3 = destruction of chloramines; and 3–4 = buildup of a free chlorine residual.

and cause taste problems. The sequence of a typical breakpoint chlorination curve is shown in Figure 7.7. There are three distinct phases on this curve. First, from point 1 to 2, the build-up of chloramines; second from 2 to 3, the destruction of chloramines; and from 3 to 4, the build-up of a free chlorine residual. The complete destruction of chloramines occurs when the chlorine dose to ammonia present ratio is 2 to 1.

7-6 Corrosion

Corrosion is a common problem in water-distribution systems and homes. Galvanic corrosion was discussed earlier in Chapter 4. There are many other mechanisms of corrosion other than galvanic corrosion; however, all forms of corrosion involve the transfer of electrons or oxidation. The main methods to control corrosive tendencies in water produced for drinking are stabilization and addition of a complexing agent. Both of these methods essentially disrupt the electrical connection between the water and the metal that can be oxidized or corroded.

The most common complexing agents are silicates and polyphosphates. These materials are added to about 1 mg/L as phosphate or silicate. They can combine chemically with exposed metal and form a barrier between the metal and the water. Sodium hexametaphosphate, $(NaPO_3)_6$, and sodium silicate, $Na_2 \cdot SiO_2$, are the common forms.

Almost every water plant attempts to produce a stable water; that is, a water that will neither precipitate nor dissolve $CaCO_3$. Usually such a water is difficult to produce, so a water that is slightly encrusting (will form a slight film of $CaCO_3$) is produced. This has the effect of blocking corrosion; however water pipes can be plugged by encrusting water. This tendency is controlled by the pH of the treated water: if it is higher than pH_s (stability pH), an encrusting water is produced and if it is less than pH_s, a corrosive water is produced. The pH_s is determined by

$$pH_s = pK_2 - pK_{sp} + pCa + pALK \tag{7.32}$$

where

$$\begin{aligned} pK_2 &= 10.33 \text{ at } 25^\circ\text{C} \\ pK_{sp} &= 8.33 \text{ at } 25^\circ\text{C} \\ pCa &= \text{negative log Ca, mol/L} \\ pALK &= \text{negative log ALK, mol/L} \end{aligned}$$

Example Problem 7.7

Determine the stabilization pH of the following water at 25°C.

$$\begin{aligned} Ca^{+2} &= 100 \text{ mg/L as } CaCO_3 \\ ALK &= 50 \text{ mg/L as } CaCO_3 \\ pH &= 8.1 \end{aligned}$$

Solution

$$100 \text{ mg/L Ca}\left(\frac{\text{meq}}{50 \text{ mg/L}}\right)\left(\frac{\text{mmol}}{2 \text{ meq}}\right)\left(\frac{10^{-3} \text{ mol}}{\text{mmol}}\right) = 10^{-3} \text{mol/L } Ca^{+2}$$

$$50 \text{ mg/L ALK}\left(\frac{\text{meq}}{50 \text{ mg/L}}\right)\left(\frac{\text{mmol}}{\text{meq}}\right)\left(\frac{10^{-3} \text{ mol}}{\text{mmol}}\right) = 10^{-3} \text{ mol/L } HCO_3^-$$

$$\begin{aligned} pH_s &= pK_2 - pK_{sp} + pCa + pALK \\ &= 10.33 - 8.33 + 3 + 3 = 8.0 \end{aligned}$$

The water is slightly encrusting as the actual pH exceeds pH_s and $CaCO_3$ will precipitate.

7-7 Removal of organics and trace metals

There are many metals listed in the drinking water standards. Removal of these metals depends on (1) their source, (2) their chemical state (dissolved or suspended), and (3) their particular chemistry. The source of metals in drinking is usually the raw water, although corrosion can cause undesirable concentrations of iron, lead, copper and zinc. If the stabilization of water and complexing agents cannot stop corrosion, it may be necessary to lower the dissolved oxygen content of the finished water as oxygen can act as a half cell supporting corrosion. If the metals are in the suspended state in the raw water, then coagulation would remove them. If they are dissolved, most of them could be removed by a high-pH process that would precipitate the hydroxides. The chemistry of the metal controls the removal processes. Lead dissolving from old pipes can be reduced by increasing the alkalinity and causing a lead carbonate precipitate to form.

Organics are present in all potable waters. The only organic standard that is presently a problem is the trihalomethane (THM) standard. Trihalomethanes are formed during chlorination when organics in settled water are oxidized to chloroform and bromoform. The current standard is 0.10 mg/L. Ten to twenty percent of all Americans are drinking water with trihalomethanes in excess of standard. This should be corrected in the 1980s, but the problem of undesirable trace organics in drinking water will not be solved in this century. The general THM reaction is

$$\text{Organics} + \text{Chlorine or Bromine} = \text{THMs} + \text{Other TOX} \qquad (7.34)$$

This reaction is general because the exact structure of organics in water is unknown, and most of the reaction products are also unknown. Organic halogens other than trihalomethanes constitute 65% to 95% of the total. The health effects of these are presently unknown, but there is a probability that the undefined TOX may have an adverse health effect.

Many different processes have been recommended for controlling THMs, including removal with granular activated carbon and ion exchange, but these methods are far too costly for application. The only method that has been successfully used nationally is to use a disinfectant other than chlorine. Chloramines have commonly been used for this purpose.

7-8 Water transmission and distribution

In the technical literature transmission usually refers to networks used to transport water before treatment and distribution refers to networks used to transport water after treatment. Water for many of its users is seldom found where it is needed. It is unequally distributed, or if treated for subsequent use, it must still be distributed among users. Users are not at the same place, nor do they need water always at the same time or at constant rates. Thus, the engineer is faced with problems of variable-rate transmission and distribution of water. Water budgets (material balances) are in popular use to account for the whereabouts of water.

Unequal space distribution of water is illustrated by the California State Water Plans and Projects. Most rainfall and water occur in the northen part of the state, while about 80% of the water demand for the state is found in relatively dry areas. Over 10,000 million liters of water per day (4 million acre-feet/year) is transmitted and distributed to the urban and agricultural areas around San Francisco Bay, Sacramento Valley, San Joaquin Valley, southern California, and other smaller regions. There are 16 reservoirs, 1065 km (662 mi) of transmission, and 8 power plants. The transmission system consists of aqueducts, which are large trapezoidal channels. These are open channels because the water surface is open to the atmosphere (see Figure 7.8). A closed conduit is one, such as a pipe, which is not open to the air. Flow in pipes and conduits can be by gravity or pressure, assisted

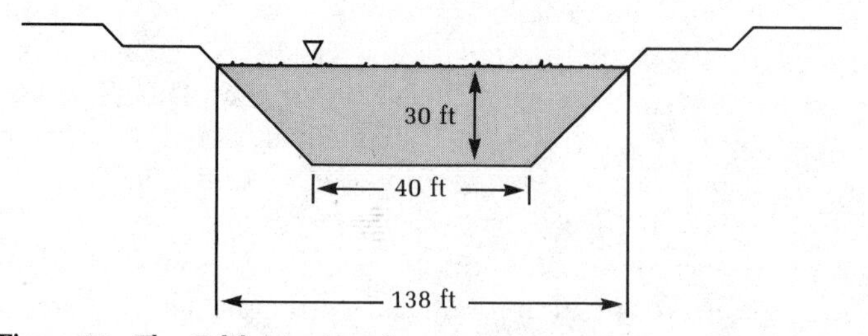

Figure 7.8 The California Aqueduct channel cross-section near Gustine, illustrating an open channel.

usually by pumps. In the California system, pumping is required to cross the Tehachapi Mountains and the water must be raised about 600 meters (2000 feet).

We have learned in earlier chapters (Chapter 3 in particular) that an energy balance can be used to describe fundamental relationships for water flow. Hydraulics, which is briefly described as a study of the transport, distribution, and position of water, relies on energy balances. The total energy of water at a point is the sum of its position energy, pressure energy, and velocity energy. When fluid flows from one point to another, an energy loss takes place. A mathematical expression of this energy principle is known as Bernoulli's theorem. In equation form, between two points in a system (such as that in Figure 7.9), one can write

$$Z_1 + \frac{P_1}{w} + \frac{v_1^2}{2g} = Z_2 + \frac{P_2}{w} + \frac{v_2^2}{2g} + h_L \tag{7.34}$$

where

Z = position energy or elevation
P = pressure energy
w = specific weight of the fluid
v = velocity
g = acceleration due to gravity
h_L = energy lost between points 1 and 2

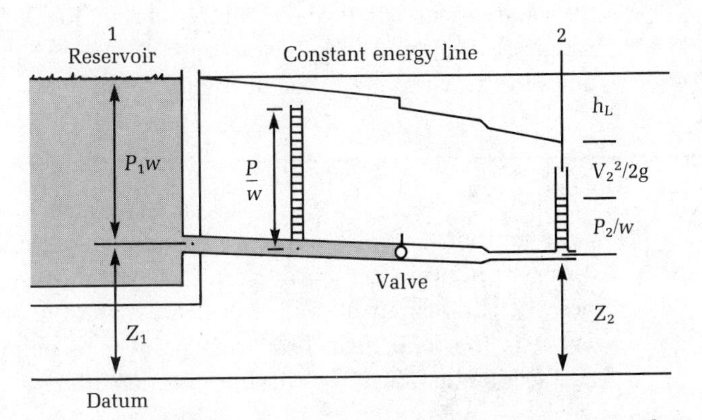

Figure 7.9 Bernoulli's energy balance between an upstream (zero velocity reservoir) and a downstream discharge (see equation 7.34).

TABLE 7.7 Some minor losses of head coefficient to illustrate head loss in water transport system.*

Name	Coefficient K	Name	Coefficient K
Check value	8	Exit loss	1.0
Glove valve	10	Entrance loss:	
Gate valve	0.2	Projecting pipe	1.0
90° elbow	0.3	Rounded	0.25
45° elbow	0.2	$\frac{1}{2}$ contraction	0.33
$22\frac{1}{2}$° elbow	0.15	2X expansion	0.56

* All valves wide open. Partial closure will be higher. Exact values are provided by manufacturers.

Energy loss illustrated in Figure 7.9 is noted as major losses, or those due to pipe friction, and minor losses, or those due to valves, pipe bends, changes in pipe diameters, and others. Minor head losses are usually expressed in terms of K velocity heads or

$$h_L = K\frac{V^2}{2g} \tag{7.35}$$

where

K = coefficient of head loss (see Table 7.7)

The designer of a water transport system must know the flow rate and energy requirements, among other important design considerations. However, energy is one of the most important. If friction losses are greater than the available energy for flow, pumping may be required. Pumping can be incorporated into Equation (7.34). It is added to the left-hand side of the equation.

Example Problem 7.8

Assume that the water storage discharge center line in Figure 7.9 is 4 meters above a datum at point 1 and 3 meters above a datum at point 2. The velocity at point 2 is 4.43m/s and the pressure head is 2 m. What is the total head loss from point 1 to point 2 if the reservoir surface is 6 m above the discharge center line?

Solution Rewriting the Bernoulli equation:

$$Z_1 + \frac{P_1}{w} + \frac{V_1{}^2}{2g} = Z_2 + \frac{P_2}{w} + \frac{V_2{}^2}{2g} + h_L$$

$$4 + 6 + 0 = 3 + 2 + \frac{(4.43)^2}{2(9.81)} + h_L$$

$$h_L = 4 \text{ meters}$$

Flow rate and velocity equations

Some design considerations will be presented as an introduction to flow rate and velocity equations. To size the pipe or open channel, information on the demand for water, required pressure, time variability, and geographic variability are but a few necessary and very important measures. Local records will sometimes produce usable data. An assumed daily demand is shown in Figure 7.10. The daily fluctuations themselves will vary and depend to a great extent on major users such as industry and tourism. Some industries and tourism may be seasonal. Thus, demand is changed when major users change. Another example is climate effects, such as increased lawn watering during hot dry weather. Peak flow rates may be as high as 5–10 times the average or as low as 2 times the average. Once flow rates have been estimated, the specific points of use on the distribution or transmission system are determined. These may be major users and affect the peak demands so that fluctuations may be less beyond the specific point of major use. For a potable water distribution system, pressure becomes important, and 30 psi (200 kPa) is usually considered to be the minimum desirable in any area. Demand for fire fighting also can affect flow rate and volume estimates. Fire demand may reduce pressure for short periods of time.

For closed conduits (pipes) in water-distribution systems, the

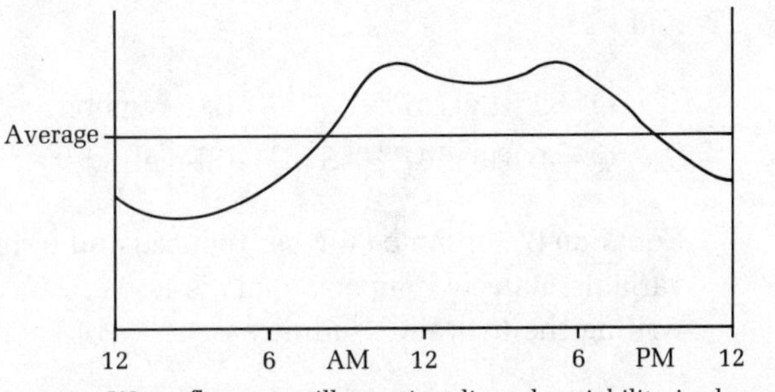

Figure 7.10 Water flow rates illustrating diurnal variability in demand.

TABLE 7.8 Hazen-Williams coefficients.

Type of pipe	C
PVC (plastic)	140
Finished wood and mansory	120
Vitrified clay	110
Old cast iron	100
Old riveted steel	95

From Vennard (1975).

Hazen-Williams formula has been used to determine pipe size and friction loss for a given flow rate. The Hazen-Williams formula is written as

$$v = 1.318\ CR^{0.63}S^{0.54} \quad \text{(U.S. customary)} \tag{7.36}$$

$$v = 0.849\ CR^{0.63}S^{0.54} \quad \text{(SI-metric)}$$

where

v = average velocity

C = friction coefficient (see Table 7.8)

R = hydraulic radius = A/P where A = cross-section area of flow and P = wetted perimeter

S = hydraulic slope = h_L/L where h_L = friction loss and L = length of pipe

Flow rate is added to the Hazen-Williams formula when the continuity equation is substituted for velocity. The continuity equation states that flow rate is the product of flow area and velocity:

$$Q = Av \tag{7.37}$$

and

$$Q = 1.318\ CAR^{0.63}S^{0.54} \quad \text{(U.S. customary)} \tag{7.38}$$

$$Q = 0.849\ CAR^{0.63}S^{0.54} \quad \text{(SI-metric)}$$

Equation (7.38) can be further reduced and written in terms of a design variable, namely diameter. This is done for a pipe flowing full by rewriting the hydraulic radius as

$$R = \frac{A}{P} = \frac{\pi D^2}{4\pi D} = \frac{D}{4} \tag{7.39}$$

Substituting into the Hazen-Williams formula;

$$Q = 0.432\ CD^{2.63}S^{0.54} \quad \text{(U.S. customary)} \tag{7.40}$$
$$Q = 0.278\ CD^{2.63}S^{0.54} \quad \text{(SI-metric)}$$

Example Problem 7.9

An old 460-mm (18-in.) cast-iron pipe delivers a flow rate of 0.25 m^3/s (8.8 ft^3/s). The pipe is 1200 m (3940 ft) long. What is the head loss over the total pipe length?

Solution Assume that the Hazen-Williams equation can be used to compute the head loss.

$$Q = 0.278\ CD^{2.63}(h_L/L)^{0.54}$$

or

$$h_L = \frac{10.7\ Q^{1.85}L}{C^{1.85}\ D^{4.87}}$$

Substituting:

$$h_L = \frac{10.7(0.25)^{1.85}(1200)}{100^{1.85}(0.46)^{4.87}}$$
$$= 8.65\ \text{m}(28.4\ \text{ft})$$

Diagram D-3 in Appendix D can also be used to solve this problem.

For an open channel, the Manning equation is used to aid in the design and operation. In the terms of the Hazen-Williams equations, the Manning equation is

$$v = \frac{1.486}{n} R^{2/3}S^{1/2} \quad \text{(U.S. customary)} \tag{7.41}$$

$$v = \frac{1}{n} R^{2/3}S^{1/2} \quad \text{(SI-metric)}$$

where

n = coefficient of roughness (see Table 7.9)

and, in terms of flow rate, it is

$$Q = \frac{1.486}{n} R^{2/3} A S^{1/2} \quad \text{(U.S. customary)} \tag{7.42}$$

$$Q = \frac{1}{n} R^{2/3} A S^{1/2} \quad \text{(SI-metric)}$$

TABLE 7.9 Values of *n* to be used with the Manning equation.

Surface	*Best*	*Good*[a]	*Bad*
Uncoated cast-iron pipe	0.012	0.013	0.015
Coated cast-iron pipe	0.011	0.012	0.013
Commercial wrought-iron pipe, galvanized	0.013	0.014	0.017
Vitrified sewer pipe	0.010	0.013	0.017
Cement mortar surfaces	0.011	0.012	0.015
Concrete-lined channels	0.012	0.014	0.018
Cement-rubble surface	0.017	0.020	0.030
Canals and ditches			
Earth, straight and uniform	0.017	0.020	0.025
Rock cuts, smooth and uniform	0.025	0.030	0.035
Rock cuts, jagged and irregular	0.035	0.040	0.045
Winding sluggish canals	0.0225	0.025	0.030
Dredged-earth channels	0.025	0.0275	0.033
Canals with rough stony beds, weeds on earth banks	0.025	0.030	0.040
Earth bottom, rubble sides	0.028	0.030	0.035
Natural-stream channels			
1. Clean, straight bank, full stage, no rifts or deep pools	0.025	0.0275	0.033
2. Same as (1), but some weeds and stones	0.030	0.033	0.040
3. Winding, some pools and shoals, clean	0.033	0.035	0.045
4. Same as (3), lower stages, more ineffective slope and sections	0.040	0.045	0.055
5. Same as (3), some weeds and stones	0.035	0.040	0.050
6. Same as (4), stony sections	0.045	0.050	0.060
7. Sluggish river reaches, rather weedy or with very deep pools	0.050	0.060	0.080
8. Very weedy reaches	0.075	0.100	0.150

From Branter (1976).
[a] Frequently used for design.

Example Problem 7.10

What is the discharge in a concrete channel that is trapezoidal in cross section with side slopes of 1 to 1, bottom width = 3 meters, channel slope of 0.005, and a depth of flow at steady-state conditions equal to 1 meter?

Solution Using the Manning equation, first solve for the hydraulic radius:

$$R = \frac{A}{P} = \frac{(3 \times 1) + 2/2(1 \times 1)}{3 + 2\sqrt{1^2 + 1^2}} = 0.69$$

then for flow rate:

$$Q = \frac{1}{0.014} 0.69^{2/3}(4)(0.005)^{1/2} = 15.78 \text{ m}^3/\text{s}$$

Hardy-Cross method

The Hardy-Cross method is a computational procedure used to accurately design a water distribution network that has many crossover points such as shown in Figure 7.11. Crossover points or nodes illustrate that pipes can operate in more than one loop with many draw-off points. Essentially, the method uses a series of mass balances at each node, using at first assumed flows until an acceptable energy balance (head loss) is obtained. The Hardy-Cross method is an excellent example of the use of the mass and energy balance approach. In any distribution system, continuity must be preserved and the pressure at any node must be the same.

The basic Hazen-Williams formula is used for balancing heads by correcting assumed flows. At all points in the computation procedure, inflows must equal outflows ($Q_{in} = Q_{out}$ for all nodes and for the network). For a fixed diameter of pipe, equation (7.40) reduces to

$$Q = K_a S^{0.54} \tag{7.43}$$

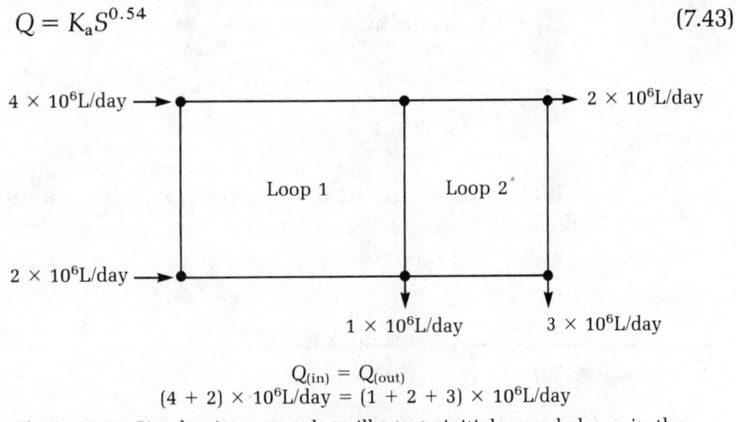

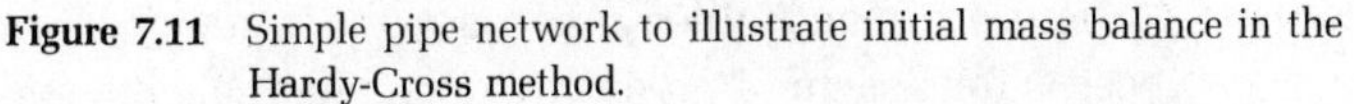

$$Q_{(in)} = Q_{(out)}$$
$$(4 + 2) \times 10^6 \text{L/day} = (1 + 2 + 3) \times 10^6 \text{L/day}$$

Figure 7.11 Simple pipe network to illustrate initial mass balance in the Hardy-Cross method.

where

$$K_a = 0.278\ CD^{2.63}$$

and since

$$S = h_L/L$$

thus

$$h_L = KQ^{1.85} \tag{7.44}$$

where

$$K = L^{0.54}/K_a$$

Since head loss must be balanced at each node,

$$\sum h_L = 0 \tag{7.45}$$

For each loop in a network,

$$h_{L_1} - h_{L_2} = 0 \tag{7.46}$$

and

$$0 = K_1(Q_1 - \Delta Q)^{1.85} - K_2(Q_2 - \Delta Q)^{1.85} \tag{7.47}$$

After expansion by binomials,

$$\Delta Q = \frac{\sum H}{1.85 \sum (H/Q)} \tag{7.48}$$

Thus, using a mass balance and energy balance approach, the adjusted flows to achieve a head loss balance can be calculated. Computer programs (Flint, 1981) are available to do this type of work.

7-9 Conclusion

There are many other processes for water treatment and other concerns for design of pipes and open channels beyond what is discussed in this text. This chapter has presented a discussion of some of the more common treatment processes. One of the problems in the water-treatment industry today is the lack of application of basic chemistry and mass balance to the design and operation of water plants. Some knowledgeable engineers and operators do not utilize these principles and consequently experience problems. This chapter was intended to be an introduction to water treatment utilizing those principles with some application. There exist today some very detailed standards founded on a concern for public health, aesthetics, and cost. The public, professional, and regulatory demands on engineers will be constantly increasing as technology advances. Successful engineering is becoming more and more dependent on a thorough understanding of chemical and material balance fundamentals.

7-10 Problems

7.1 Assume that chlorination may be described by a first-order kinetic relation with a rate constant of 0.5 min^{-1}.

1. Determine the volume of a plug flow reactor required to reduce the concentration of coliform bacteria from 5000 organisms per 100 mL to 1 organism per 100 mL. The water production rate is 1 mgd.
2. Repeat the calculation for a complete mix reactor.
3. Which reactor would be the best choice?

7.2 Express the following ion concentrations in mg/L as $CaCO_3$.

1. 100 mg/L Ca^{+2}
2. 400 mg/L HCO_3^-
3. 60 mg/L Na^+
4. 50 mg/L CO_2

7.3 Determine the quantity of sludge produced during removal of 100 mg/L hardness expressed as $CaCO_3$ for each of the following types of hardness:

1. Calcium carbonate
2. Calcium non-carbonate
3. Magnesium carbonate
4. Magnesium non-carbonate

Express your answer in mg/L of sludge.

7.4 Based on current chemical costs for lime and soda (assume that soda costs twice as much as lime per unit mass), which type of hardness is most expensive to remove by chemical precipitation?

1. Calcium or magnesium
2. Carbonate or non-carbonate

7.5 Chemical analysis of a groundwater produced these results (ion concentrations except as noted):

$$\begin{aligned}
Mg &= 36.5 \text{ mg/L} \\
Ca &= 72 \text{ mg/L} \\
Na &= 23 \text{ mg/L} \\
Cl &= 49.7 \text{ mg/L} \\
CO_2 &= 8.8 \text{ mg/L} \\
\text{Alkalinity} &= 230 \text{ mg/L as } CaCO_3 \\
pH &= 7.8
\end{aligned}$$

Determine the following hardness in mg/L as $CaCO_3$:

Total hardness
Calcium carbonate hardness
Calcium non-carbonate hardness
Magnesium carbonate hardness
Magnesium non-carbonate hardness
Negative hardness

7.6 Repeat problem 7.5 if all concentrations remain the same except alkalinity, which is raised to 350 mg/L as $CaCO_3$.

7.7 HOCl is a much more effective disinfectant than OCl^-. The pK_a for hypochlorous acid is 7.6. Which pH, 7 or 8, would provide better disinfection efficiency?

7.8 Using half reactions for the oxidation of NH_3 to N_2 and the reduction of Cl_2 to Cl^-, write a balanced equation for nitrogen removal by breakpoint chlorination.

7.9 Determine the chlorine dose (in mg/L Cl_2) and consumption of alkalinity (in mg/L as $CaCO_3$) associated with breakpoint chlorination of a water source containing 1.0 mg/L of ammonia.

7.10 Estimate the Manning roughness coefficient in an open rectangular channel with a bottom width of 6 feet, with a flow depth of 3 feet, and an average velocity of flow over the cross-section of 4 feet per second. The slope of the channel is 6 feet per 1000 feet.

7.11 What is the velocity of flow in a rectangular channel 10 meters wide and 2 meters deep if the n factor is 0.017 and channel slope is 1 meter per 500 meters? Also, what is the discharge?

7.12 What is the head loss of a 300-mm PVC pipe with a flow rate of 0.20 m^3/s? The pipe is 500 meters long with an open gate valve and 90°elbows. Neglect entrance and exit losses

7.13 Using a conservation of energy approach (Bernoulli's equation) and Figure 7.9 with a velocity at point 2 equal to 2.4 m/s and other parameters as defined in example problem 7.8, calculate the head loss.

References

American Water Works Association. 1971. *Water Quality and Treatment, A Handbook of Public Water Supplies,* 3rd ed. New York: McGraw-Hill.

Branter, E. F., and King, H. W. 1976. *Handbook of Hydraulics,* 6th ed. New York: McGraw-Hill.

Eliassen, R. 1963. *Water Quality Criteria,* California State Water Quality Board Publication 3A.

Fair, G. M., Geyer, J. C., and Okum, D. A. 1971. *Elements of Wastewater Supply and Wastewater Disposal,* 2nd ed. New York: Wiley.

Flint, H. M. 1981. Hydraulic network analysis using the microcomputer. *Civil Engineering,* October 1981, 63.

McKee, J. E., and Wolf, H. W. 1963. *Water Quality Criteria.* California State Water Quality Board Publication 3-A.

Sanks, R. L. 1980. *Water Treatment Plant Design.* Ann Arbor, MI: Ann Arbor Science.

State of Florida, Chapter 17-3.403, F.A.C. 1983. Classification of groundwaters. Tallahassee: State Department of Environmental Regulation.

U.S. Environmental Protection Agency. 1978. *Safe Drinking Water Criteria.*

U.S. Public Health Service. 1962. *Public Health Service Drinking Water Standards.* Public Health Service Publication No. 956.

Vennard, J. K. and Street, R. L. 1975. *Elementary Fluid Mechanics,* 5th ed. New York: Wiley.

Weber, W. J., Jr. 1972. *Physicochemical Processes for Water Quality Control.* New York: Wiley-Interscience.

8

Wastewater Treatment

8-1 Introduction

Municipal and industrial wastewaters are treated to meet a specified quality based on advances in treatment technology, economic considerations, and effluent limitations. Treated effluents should comply with local, state, and federal legislation enacted to eliminate or control potential health hazards. Examples are the Water Quality, Air Quality, and Toxic and Hazardous Wastes Acts. Compliance with these laws requires not only proper environmental engineering design and operation of pollution-control facilities, but careful analysis and accurate measurements of pollutants and environmental quality parameters.

The Federal Water Pollution Control Administration (FWPCA) in 1969 issued a set of criteria to be used by the states as guidelines in setting their own standards. Streams were classified according to water usage: (a) recreation and aesthetics, (b) public water supplies, (c) fish, other aquatic life, and wildlife, (d) agriculture, and (e) industry. The FWPCA was incorporated into the Environmental Protection Agency in 1970, and comprehensive federal water-pollution legislation was enacted under Public Law 92-500 on October 18, 1972. This law and its amendments provide for the restoration and maintenance of the chemical, physical, and biological integrity of the nation's water.

Section 101 sets three goals: (1) to eliminate the discharge of pollutants, (2) wherever possible, to have water quality suitable for sustaining fish, shellfish, and wildlife and for recreational purposes, and (3) to prohibit the discharge of toxic pollutants.

The law directs the EPA administrator to establish guidelines within which individual states must operate their permit programs. The best control technology economically available will be necessary and a non-discharge requirement can be imposed if it is both technologically and economically achievable. Therefore, the degree of treatment is governed by the fate and impact of discharged pollutants on the water quality of the receiving stream.

Treatment processes apply physical, chemical, and biological principles to remove pollutants from the liquid phase and concentrate them in the solid phase or convert them into gases. For example, organic pollutants can be converted into carbon dioxide, methane, and other substances. Separation between solids and liquid fractions is accomplished by gravity sedimentation, centrifugation, filtration, or other physical means. This chapter discusses the major concepts used in biological treatment and emphasizes the application of materials balances to wastewater treatment processes such as keeping an account of pollutants in the liquid, solid, and gaseous phases, or in defining the degree of treatment efficiency to meet effluent standards before selecting a treatment process.

8-2 Treatment processes

There are many wastewater treatment processes in use tailored to both the characteristics of the waste and the desired degree of treatment; some of these are listed in Figure 8.1. Pre-treatment or primary treatment is used to remove settleable suspended solids and to prepare the wastewater for subsequent treatment. Secondary treatment generally includes the decomposition of soluble organic matter and separation of suspended solids by sedimentation. In modern secondary treatment plants for domestic wastewater, the removal efficiency of BOD_5 and suspended solids (SS) is better than 90%. Generally, the standard effluent concentration of BOD_5 or SS is lower than 30 mg/L. The primary treatment may produce an effluent with 50–70% SS removal and 30–40% BOD_5 removal. Advanced or tertiary waste treatment can achieve the desired degree of BOD_5, SS, N, and P removal by physical, chemical, and biological processes. Allowable concentrations of pollutants released to the environment are based on the waste load that can be assimilated.

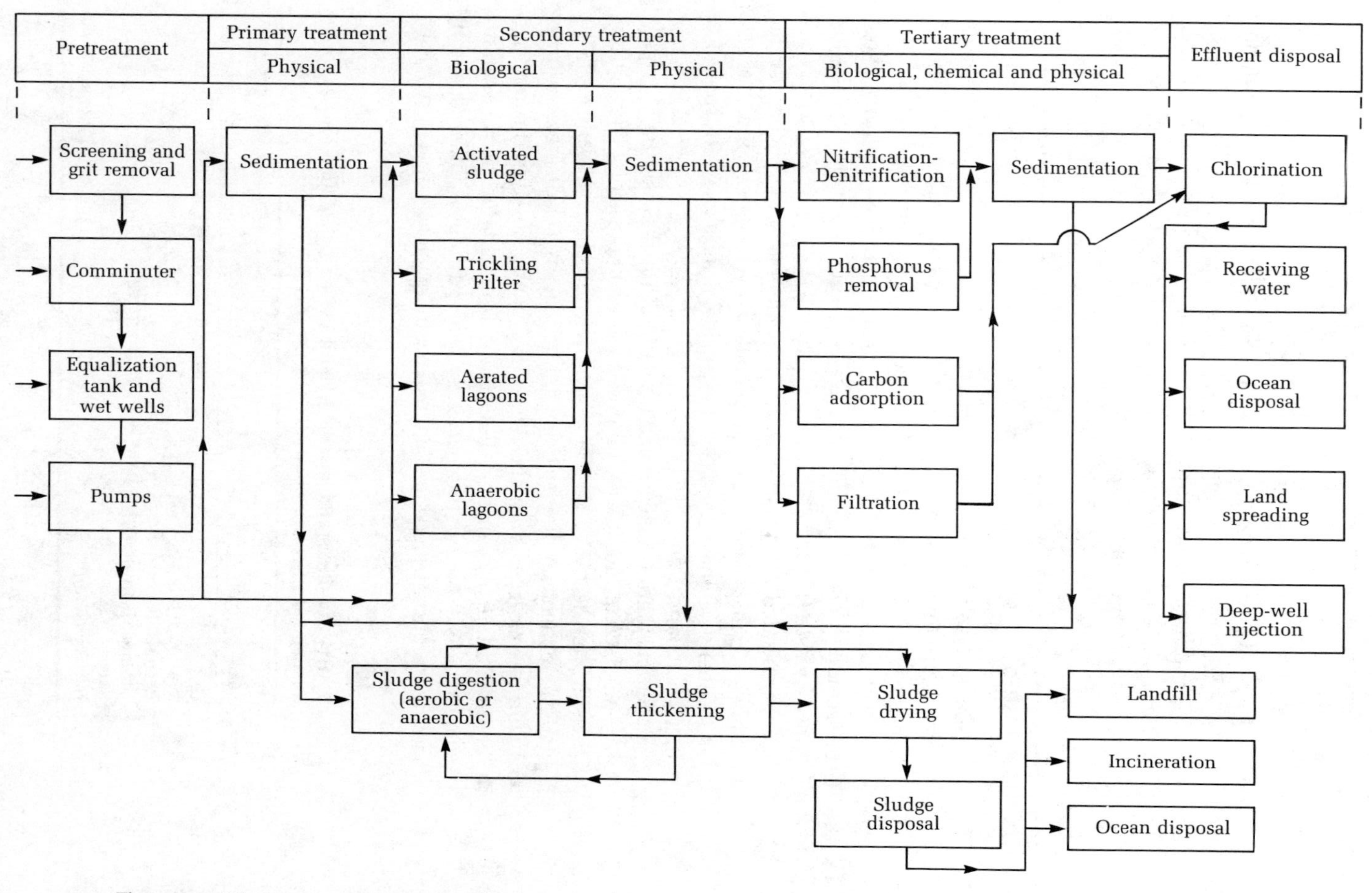

Figure 8.1 Wastewater treatment processes.

The basic reaction for the conversion of organic matter is:

Bacteria + Oxygen + Organic Matter $\longrightarrow$
New Bacterial Cells + Carbon Dioxide + Water + Energy.

It can readily be seen that this process requires a food source (organic matter), bacterial cells (which accumulate in wastewater solids), and an oxygen source (aeration).

8-3 Sedimentation

Suspended solids are removed from wastewater by gravity in sedimentation tanks, which are used at all stages of treatment as shown in Figure 8.1. They are used to remove grit or sand and silt during pretreatment, suspended solids in primary clarifiers, biological flocs in secondary clarifiers, and chemically coagulated solids during advanced waste treatment.

Suspended solid particles in wastewater may be bacterial cells, colloids, chemical precipitates, clay, silt, or even coarse sand. The size and density of these particles vary widely, as shown in Table 8.1. (Berthouex and Rudd, 1977).

Sedimentation takes place by various types of settling, namely, free or discrete, flocculant, hindered, and compression. This settling classification is based on the concentration of the suspended solids and the tendency of the particles to interact. Free or discrete sedimentation is concerned with the removal of nonflocculating discrete particles in dilute suspensions. Flocculant sedimentation removes dilute suspensions of flocculating particles. Zone settling or hindered settling occurs in concentrated suspensions due to the closely spaced particles.

TABLE 8.1 Sizes and specific gravities of particles often found suspended in wastewater.

Solid material	*Specific gravity (kg/L)*	*Size*
Grit	2.0–3.0	1 mm$^+$
Wastewater organic solids	1.2–1.4	1 mm$^+$
Bacterial flocs	1.01–1.1	10–100 μm
Granular activated carbon	1.4–1.6	0.5–4 mm
Sand	2.65	1 mm$^+$
Silt and clay	2.6–3.5	Colloidal (1 μm) to 100 μm

Finally, compression or consolidation of sediments takes place at the bottom of the basin. The following discussion will be limited to free settling of discrete nonflocculant particles in dilute suspensions.

Particles are assumed to be insoluble, free to move in the suspending medium, and large enough to sink at a reasonable rate. In general, particles will settle out of a fluid when the gravitational force that causes settling dominates the buoyant, electrostatic, turbulent, and molecular forces that tend to keep them suspended. Particles will accelerate until the drag force equals the impelling force; then settling occurs at a constant velocity V_s. The settling velocity of spherical particles through gas or liquid under different flow conditions can be described by

$$V_s = \sqrt{\frac{4g}{3C_D}\left(\frac{\rho_s - \rho_l}{\rho_l}\right)d} = \sqrt{\frac{4g}{3C_D}(S_s - 1)d} \qquad (8.1)$$

where

V_s = terminal settling velocity (cm/s)
g = gravitational constant (cm/s^2)
d = particle diameter (cm)
ρ_s = particle density (g/cm^3)
ρ_l = liquid density (g/cm^3)
μ = fluid viscosity (g/cm-s)
S_s = specific gravity of solid particles
C_D = coefficient of drag; varies with flow regime around the particle

The flow regime around the particle may vary from laminar (slow) to turbulent (rapid). C_D for laminar flow is substituted in equation (8.1) and Stoke's law was developed to describe the slow settling of spherical particles as follows:

$$V_s = \frac{gd^2(\rho_s - \rho_l)}{18\mu} \qquad (8.2)$$

μ = fluid viscosity (g/cm-s) and values for water viscosity are shown in Table D-1 in Appendix D.

Example Problem 8.1

The following table shows the relative distribution of a mixture of particles left to settle by gravity in a quiescent column of water.

Diameter (μm)	*Specific gravity*	*Percentage by weight*
10	2.65	5
20	1.50	15
100	1.20	65
200	1.10	15

Determine the settling velocities for each particle size if water viscosity is 1.31×10^{-2} g/cm-s (poise) at 10°C.

Solution The size and specific gravity values given in the above table can be used to calculate settling velocities in water using Eq. (8.2) as shown below:

Diameter (μm)	Settling velocity (cm/s)
10	0.007
20	0.008
100	0.083
200	0.166

As 80% of the particles have a diameter of 100 μm or greater, 80% of the particles will settle out at velocities equal to or greater than 0.083 cm/s.

Camp (1946) developed relations to describe removal of discrete particles in an ideal settling tank. He suggested that the terminal velocity of a particle which settles a distance equal to the effective depth of the tank in a detention time can be thought of as an overflow rate. The flow of wastewater divided by the surface area is known as the overflow rate, V_o, and is equivalent to the settling velocity for the desired size particles to be removed:

$$V_o = \frac{F_o}{A_s} \tag{8.3}$$

where

A_s = surface area; the plan area of the tank (m^2)
F_o = wastewater flow (m^3/day)

From example problem 8.1, the overflow rate equivalent to settling velocity of 0.083 cm/s can be calculated as follows:

$$\text{Overflow rate} = \frac{0.083\ \text{cm}}{\text{s}} \times \frac{86{,}400\ \text{s}}{\text{day}} \times \frac{\text{m}}{100\ \text{cm}} \times \frac{\text{m}^2}{\text{m}^2}$$
$$= 71.7\ \text{m}^3/\text{day-m}^2$$

Because settling tanks are not ideal, allowance should be made for turbulence, dead zones, entrance and exit areas, and short circuits. A factor of 2 or 3 is required; thus clarifiers at sewage treatment plants have overflow rates of 20–50 m^3/m^2-day. However, more accurate overflow rates can be determined from experimentation and pilot studies.

The top surface area of a rectangular tank is equivalent to the length of the tank L multiplied by its width W. A circular tank has a top surface area equal to πr^2 where r is the tank radius. Another important design parameter is the detention time DT, which is equal to the tank volume V divided by the wastewater flow F_o. Also, the tank volume can be represented by the top surface area A_s times the water depth h. Therefore, the detention time is equal to the side water depth divided by the overflow rate as follows:

$$DT = \frac{V}{F_o} = \frac{A_s h}{F_o} = \frac{h}{V_o} \tag{8.4}$$

Example Problem 8.2

A primary clarifier for a municipal wastewater treatment plant is to be designed for an average flow of 10,000 m^3/day. The state's regulatory agency criteria are: average overflow rate = 20 m^3/m^2-day and detention time = 3.0 hours. Determine the clarifier diameter and the side water depth in the clarifier.

Solution

$$V_o = \frac{F_o}{A_s}$$

or

$$A_s = \frac{F_o}{V_o} = \frac{10{,}000 \text{ m}^3/\text{day}}{20 \text{ m}^3/\text{m}^2\text{-day}} = 500 \text{ m}^2$$

Using a circular tank with diameter d:

$$d = 2r = \sqrt{\frac{4A_s}{\pi}}$$

$$= \sqrt{\frac{4 \times 500 \text{ m}^2}{3.14}} = 25 \text{ m}$$

If detention time = 3 $h = V/F_o$,

$$V = \left(\frac{3 \text{ hr}}{24 \text{ hr/day}}\right)(10{,}000 \text{ m}^3/\text{day}) = 1250 \text{ m}^3$$

$$= \pi r^2 h = \pi \left(\frac{25}{2}\right)^2 h$$

$$h = \frac{1250}{3.14 \times \left(\frac{25}{2}\right)^2} = 2.5 \text{ m}$$

Sedimentation tanks usually take the schematic form shown in Figure 8.2. This figure shows the incoming flow F_o with concentration X_o and the underflow F_u with concentration X_u. Solids are withdrawn ("wasted") continuously or intermittently from the underflow. From the materials balance, the incoming flow is equal to the effluent flow F_e plus the underflow:

$$F_o = F_e + F_u \tag{8.5}$$

Similarly, the incoming solids should be equal to the solids discharged in the effluent plus solids in the underflow:

$$F_oX_o = F_eX_e + F_uX_u \tag{8.6}$$

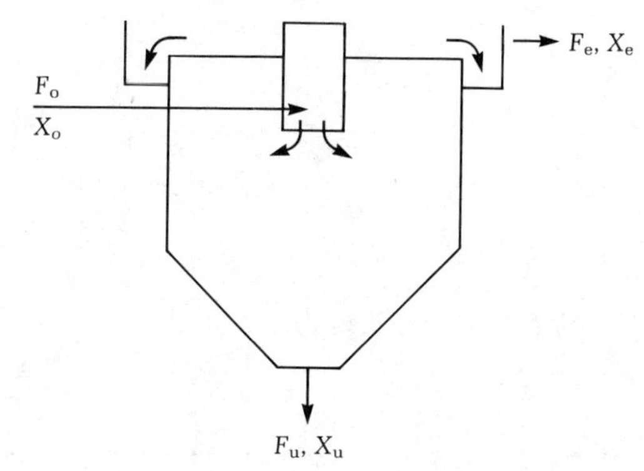

Figure 8.2 Schematic of sedimentation tank.

X_u can be represented in units of mass-volume^{-1} or percentage by weight. For example, 1 percent solids approximates 10,000 mg/L concentration of SS.

Example Problem 8.3

From example problem 8.2, the incoming flow is 10,000 m^3/day and influent concentration is 200 mg/L SS. The efficiency of sedimentation is 70% and the solids concentration in the underflow is 0.5%. Calculate the underflow rate.

Solution From equation (8.5),

$$F_o = F_e + F_u$$

or

$$F_e = F_o - F_u$$

$$X_o = 200 \text{ mg/L}$$

$$X_e = 200\left(\frac{100 - 70}{100}\right) = 60 \text{ mg/L}$$

$$X_u = 0.5\% = \frac{0.5 \text{ g}}{100 \text{ g}} = \frac{5 \text{ g}}{\text{L}} = 5{,}000 \text{ mg/L}$$

$$F_o = 10{,}000 \text{ m}^3/\text{day}$$

From the mass balance equation (8.6),

$$F_oX_o = F_eX_e + F_uX_u$$

$$10{,}000\ \text{m}^3/\text{day} \times 200\ \text{mg/L} = (10{,}000 - F_u)\ \text{m}^3/\text{day} \times 60\ \text{mg/L} + F_u\ \text{m}^3/\text{day} \times 5000\ \text{mg/L}$$

$$1{,}400{,}000 = 4940\ F_u$$

$$F_u = 283.4\ \text{m}^3/\text{day}$$

Notice that F_u is less than 3% of F_o. Often F_e can be considered equal to F_o.

8-4 Activated sludge process

Activated sludge is a heterogeneous microbial culture composed mostly of bacteria, protozoans, rotifers, and fungi. Bacteria are mostly responsible for assimilating the organic matter in waste water, whereas the protozoa and rotifers are helpful in removing the dispersed bacteria that otherwise would not settle out. During this process, much of the soluble and colloidal organic material remaining after primary sedimentation is metabolized by the microorganisms to carbon dioxide, ammonia, and water. A sizable fraction is converted to cellular material, which can be separated by gravity sedimentation.

The layout for a typical activated sludge plant is presented in Figure 8.3. Effluent from the primary clarifier is aerated in the presence

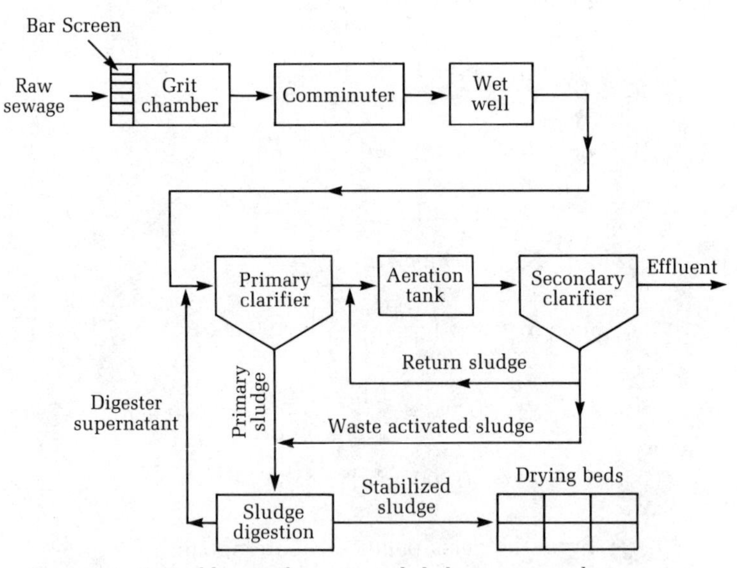

Figure 8.3 Typical layout of an activated sludge treatment plant.

Figure 8.4 Aeration tanks (background) are followed by secondary clarifiers (foreground).

of microbial suspension in the aeration tank. Hydraulic detention time in the aeration tank depends on the strength of the wastewater influent and the removal rate by existing biomass. The aeration tank is followed by a gravity sedimentation tank (Figure 8.4) for separation of solids from the liquid. A fraction of the settled solids is cycled back to the aeration tank as return sludge to keep a constant concentration of mixed liquor suspended solids (MLSS) in the tank. Excess solids resulting from the growth of bacterial cells in the aeration tank are wasted and removed from the system to the sludge digestion tank. Solids settled in the primary clarifier tank are also directed to the sludge digestion tank for further treatment and stabilization. Digestion of sludge can be aerobic or anaerobic, and well-digested sludge is dried in sludge beds for ultimate disposal. The supernatant from sludge digestion tanks is diverted to the primary clarifier for additional treatment. Secondary treated effluent is disinfected and disposed of in adjacent water bodies or by land spreading or other effluent disposal methods as listed in Figure 8.1.

Design of aeration tank

Kinetic models based on steady-state conditions and using mass balance of material flow within the system are used to describe the activated sludge process (Benefield and Randall, 1980). A typical flow

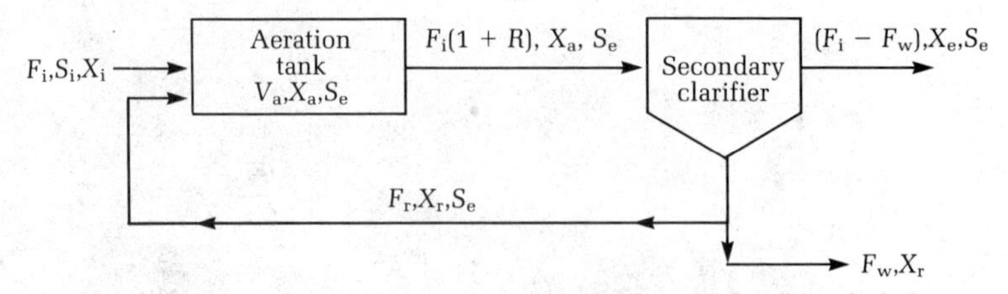

Figure 8.5 Typical flow scheme for a completely mixed activated sludge plant. Terms are defined in the text.

scheme for a completely mixed activated sludge process is shown in Figure 8.5. Raw wastewater enters the aeration tank and is mixed with recycled sludge from the clarifier. Biological processes inside the tank reduce the BOD concentration from S_i to S_e, resulting in an increase in bacterial cells or MLVSS, X_a, in the aeration tank. The solids are concentrated in the bottom of the secondary clarifier. A fraction of the solids is recycled to the aeration tank; another fraction is wasted and removed permanently from the system. Also, some solids X_e are removed with the secondary clarifier effluent. A materials balance for the organic matter or BOD loading and the suspended solids can be made using the terms of Figure 8.5 where

F_i = rate of fresh wastewater inflow to the aeration tank (m^3/day)

S_i = substrate concentration of BOD in raw sewage (mg/L)

V_a = volume of aeration tank (m^3)

X_a = mixed liquor volatile suspended solids in the aeration tank (mg/L)

X_e = suspended solids in the effluent after treatment (mg/L)

S_e = steady-state substrate concentration after treatment (mg/L)

F_r = rate of sludge recycle = RF_i

R = sludge recycle ratio = F_r/F_i

F_w = rate of sludge wasting (m^3/day)

X_r = mixed liquor volatile suspended solids in the recycled sludge from the secondary clarifier (mg/L)

Mass balances for the microorganisms and for the substrate flowing in the aeration tank, as shown in Figure 8.5, can be described as follows:

Accumulation of microorganisms within the system boundary	=	Inflow of microorganisms into the system boundary	−	Outflow of microorganisms from the system
		+ Net growth of microorganisms within the system boundary		

or in a simplified form,

Accumulation = Inflow − Outflow + Net growth

or

$$\frac{dX}{dt} V_a = F_i X_i + F_r X_r - (F_i + F_r) X_a + V_a X_a (\mu - K_d) \qquad (8.7)$$

where

$\frac{dX}{dt}$ = rate of microorganisms growth measured in terms of mass of mixed liquor volatile suspended solids (MLVSS) per unit volume-time

V_a = reactor volume (aeration tank)

F_i = flow rate

X_i = concentration of microorganisms in influent

X_a = concentration of microorganisms in reactor

μ = specific growth rate

K_d = decay rate

Specific growth rate μ can be substituted by equation (5.4) from Chapter 5 as:

$$\mu = \frac{\mu_{max}(S_e)}{K_s + S_e}$$

Also, the substrate balance can be written similar to the microorganism mass balance as follows:

$$\frac{dS}{dt} V_a = F_i S_i + F_r S_e - (F_i + F_r) S_e - V_a X_a \left(\frac{q_{max} S_e}{K_e + S_e} \right) \tag{8.8}$$

where

$q_{max} = \mu_{max}/Y$ as shown by equation (5.2) in Chapter 5.

Equations (8.7) and (8.8) can be solved if biokinetic constants μ_{max}, Y, K_d, and K_s were determined. Typical values for treatment of domestic wastewater and some industrial wastewater effluents treated biologically under aerobic or anaerobic environment are available (Sykes, 1975). Detailed analysis of equations (8.7) and (8.8) and derivation of design parameters are beyond the scope of this text.

The most important parameters for the design and operation of activated sludge plants are: (1) the biological solids retention time (BSRT) θ_c, which is defined as the average number of days a unit of biomass remains in the treatment system, and (2) the food-to-biomass ratio F/M which is the ratio of the ultimate BOD inflow per day in the aeration tank to the total biomass represented by the mixed liquor volatile suspended solids (MLVSS) in the aeration tank. θ_c may also be known as the mean cell residence time, and the MLVSS may be replaced by MLSS. From the mass balance of material flow shown in Figure 8.5,

$$\theta_c = \frac{X}{\Delta X/\Delta t} = \frac{\text{Total mass of cells in reactor}}{\text{Mass rate of cells removed per day}} \tag{8.9}$$

$$= \frac{X_a V_a}{F_w X_r + (F_i - F_w) X_e}$$

where

$\Delta X/\Delta t$ = total active biomass withdrawn from the system (kg of MLVSS/day)

X = total active biomass in aeration tank (kg of MLVSS)

The mass rate of cells removed daily from the plant should be equal to the growth rate of microorganisms in order to maintain a constant and uniform MLVSS in the aeration tank. Another important parameter in design and operation of the plant is the food-to-microorganism ratio

$$\frac{F}{M} = \frac{F_i S_i}{X_a V_a} \tag{8.10}$$

The food-to-microorganism ratio (F/M) may be known as organic loading. Also, the specific substrate utilization rate q, as previously defined in Chapter 5, is

$$q = \frac{F_i (S_i - S_e)}{X_a V_a} \tag{8.11}$$

The recycled solids X_r to the aeration tank is a function of the settling characteristics of the solids in the secondary clarifier. When secondary clarifiers are operating properly, solids captured should approximate 100%, and the maximum solids concentration in the sludge return line $(X_r)_{max}$ can be estimated by

$$(X_r)_{max} = \frac{10^6}{SVI} \tag{8.12}$$

where

SVI = sludge volume index (mL/g), or volume of sludge occupied by 1 g of solids when 1 L of the activated sludge from the aeration tank is kept under quiescent conditions in a graduated cylinder for 30 minutes

---

Example Problem 8.4

An aeration tank of 30 × 10 × 5 meters treats 5000 cubic meters per day of primary wastewater effluent containing 150 mg/L of ultimate BOD. Under steady-state conditions, the aeration tank contains 2000 mg/L of MLVSS. The sludge recycle ratio is 0.3 and wasted sludge is 500 kg/day. Calculate

1. Hydraulic detention time in aeration tank.
2. Concentration of return sludge.

3. Biological solids retention time if 90% substrate removal efficiency is desired, and the effluent suspended solids are not to exceed 20 mg/L.
4. Specific substrate utilization rate.

Solution Draw a schematic to describe the problem statement as shown below. The removal efficiency is 90%, therefore most of the variables can be quantified:

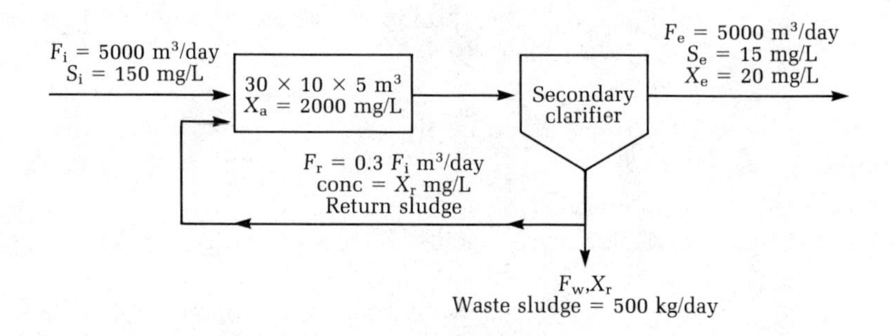

1. Hydraulic detention time = Tank volume/Flow = θ_h. The flow is considered the incoming fresh feed flow F_i plus the return sludge from F_r. However, some investigators use only the incoming fresh feed flow to determine the fresh feed hydraulic detention time.

$$F_i = 5000 \text{ m}^3/\text{day}$$

$$F_r = (0.3)(5000) = 1500 \text{ m}^3/\text{day}$$

$$\text{Tank volume} = V = 30 \times 10 \times 5 = 1500 \text{ m}^3$$

$$\theta_h = \frac{1500 \text{ m}^3}{(5000 + 1500) \text{ m}^3/\text{day}} = 0.23 \text{ day} = 5.54 \text{ hours}$$

2. Concentration of return sludge: For this steady-state system, a mixed liquor suspended solids (MLVSS) concentration of 2000 mg/L is maintained in the aeration tank. These solids are supplied by return sludge and biological growth as shown from a material balance:

$$\begin{matrix}\text{Solids inflow} \\ \text{in fresh feed}\end{matrix} + \begin{matrix}\text{Solids inflow} \\ \text{in return sludge}\end{matrix} + \begin{matrix}\text{Biological} \\ \text{growth}\end{matrix} = \begin{matrix}\text{Solids outflow} \\ \text{from tank}\end{matrix}$$

The concentration of solids inflow in fresh feed is considered insignificant or zero, and the biological growth is equivalent to solids lost from the system.

Biological growth = Solids in waste sludge + Effluent solids

$$\frac{\Delta X_g}{\Delta t} = \frac{\Delta X_w}{\Delta t} + \frac{\Delta X_e}{\Delta t}$$

$$= 500\ \text{kg/day} + (5000\ \text{m}^3/\text{day})(20\ \text{mg/L}) \times \left(\frac{1000\ \text{L/m}^3}{10^6\ \text{mg/kg}}\right)$$

$$= 500 + 100$$

$$= 600\ \text{kg/day}$$

For materials balance of MLVSS:

$$F_i X_i + F_r X_r + \text{Growth} = (F_i + F_r) X_a$$

$$0 + (1500\ \text{m}^3/\text{day}) X_r\ \text{mg/L}\left(\frac{1000\ \text{L/m}^3}{10^6\ \text{mg/kg}}\right) + 600\ \text{kg/day}$$

$$= (5000 + 1500)\ \text{m}^3/\text{day}\left(2000\ \text{mg/L} \times \frac{1000\ \text{L/m}^3}{10^6\ \text{mg/kg}}\right)$$

$$1.5\ X_r = 13{,}000 - 600 = 12{,}400$$

$$X_r = 8267\ \text{mg/L}$$

3. Biological solids retention time, θ_c:

$$\theta_c = \frac{X}{\Delta X/\Delta t} = \frac{\text{Total MLVSS in aeration tank}}{\text{MLVSS removed from the system}}$$

$$= \frac{(1500\ \text{m}^3)(2000\ \text{mg/L})(1000\ \text{L/m}^3)(\text{kg}/10^6\ \text{mg})}{600\ \text{kg/day}}$$

$$= \frac{3000\ \text{kg}}{600\ \text{kg/day}} = 5\ \text{days}$$

Notice from the flow balance of the whole system:

$$F_i + F_r = F_e + F_r + F_w$$

F_w is generally very small, and F_i approximates F_e.

4. Specific substrate utilization q:

$$q = \frac{F_i(S_i - S_e)}{X_a V_a} \qquad \text{(as shown in equation 8.11)}$$

$$= \frac{(5000\ \text{m}^3/\text{day})(150 - 15)\text{mg/L}}{(2000\ \text{mg/L})(1500\ \text{m}^3)}$$

$$= 0.225/\text{day}$$

q represents the mass of BOD or substrate utilized per unit mass of microorganisms per unit time, as stated earlier. From the previous example, it is shown that some of the biokinetic parameters discussed in Chapter 5 can be estimated by monitoring the operational parameters of a full-scale plant.

Air supply

The minimum allowable dissolved oxygen in activated sludge processes is 0.5 to 1.0 mg/L, and the air-flow rates vary between 90–100 m^3/kg BOD removed for conventional aeration, 30–45 for high-rate aeration, and 125 for extended aeration. Air can be supplied by compressors through a diffuser system or by mechanical aerators that entrain atmospheric air directly. A simplified approach to estimate the oxygen required for an activated sludge system is to calculate the mass of ultimate BOD (BOD_u) removed from the wastewater by treatment. For example, if the influent flow is F_i (m^3/day), the influent BOD_u is S_i, and the effluent BOD_u is S_e, then the oxygen required (in kg/day) approximates:

$$(F_i\ \text{m}^3/\text{day})(S_i - S_e)\ \text{mg/L}\ (1000\ \text{L/m}^3)(1\ \text{kg}/10^6\ \text{mg})$$

8-5 Activated sludge process modifications

The activated sludge process is designed to remove soluble and insoluble organics from a wastewater stream and to convert this material into gases, water, and a microbial suspension. Classically, this is accomplished by the conventional activated sludge process where wastewater is mixed with a biological culture in a long, narrow aeration basin with a volume sufficient to provide about 6 hours of contact time. Operating experience has revealed several problems related to unequal distribution of sudden or shock loads, toxic or high-strength waste, and oxygen requirements. Therefore, numerous modifications to the conventional process are used. For the high-rate activated sludge process a low MLSS concentration is maintained, which results in a high specific growth rate and specific substrate utilization rate. Tapered aeration is similar to the conventional process but differs in the air diffuser arrangement: more air is injected at the head of the tank, where the oxygen demand is greatest, and then decreased along the tank length as the demand decreases.

Figure 8.6 shows various activated sludge modifications. In the step-aeration process, wastewater is fed into the tank at different

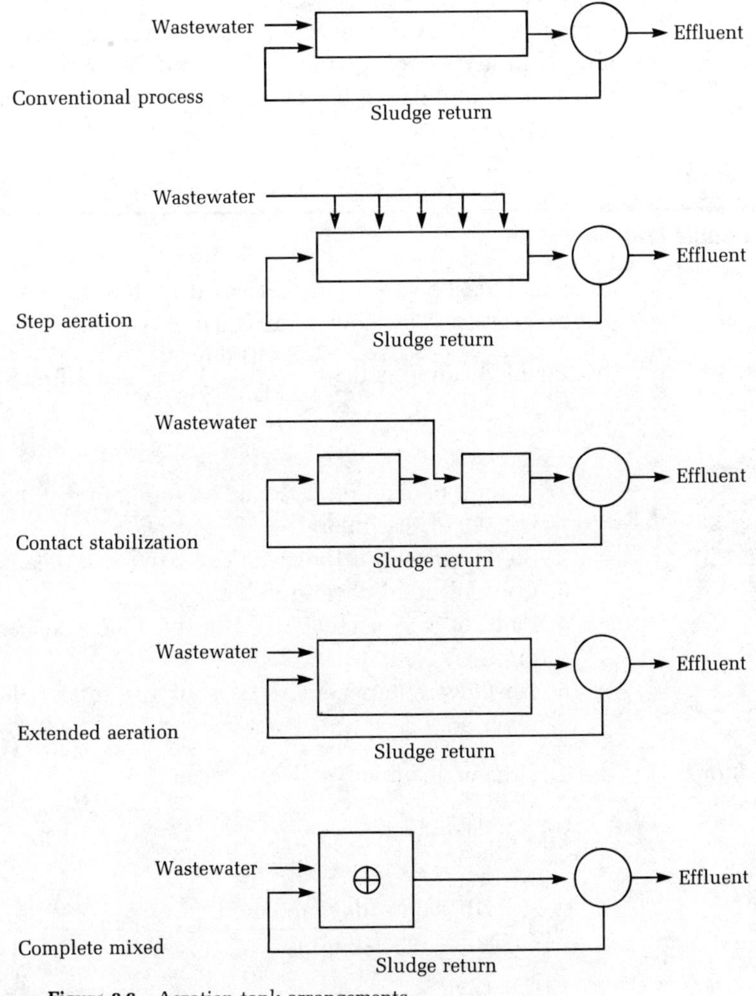

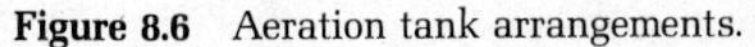

Figure 8.6 Aeration tank arrangements.

points along its length, and return sludge is mixed with a portion of the wastewater and enters at the head of the aeration tank.

Extended aeration plants are generally small (less than 3785 m^3/day), because of the large aeration tank volumes required. They employ complete mixing, and primary clarification is generally omitted.

The contact stabilization process requires two tanks, a contact and a stabilization or reaeration tank. The first tank provides contact between the biomass and the wastewater for a short retention time in order to adsorb soluble and colloidal organics. The biomass is then reaerated in a secondary clarifier, and the sludge is channeled to the second aeration tank where the organic material adsorbed onto the biomass surface is stabilized. Design criteria for common activated sludge processes are presented in Table 8.2. The mean cell residence time θ_c varies between 5 and 15 days and the organic loading q varies between 0.2 and 0.6 kg BOD removed per kg MLVSS-day for most of the aeration tanks.

Example Problem 8.5

Municipal wastewater is to be treated by the conventional activated sludge process. The flow is 10,000 m^3/day with 250 mg/L BOD. If the organic loading is $0.3\,\frac{\text{kg BOD/day}}{\text{kg MLVSS}}$ and the MLVSS is 3000 mg/L, calculate:

1. Volume of aeration tank and retention time if the return sludge is 25% of the inflow.
2. Sludge wasted if the mean cell residence time is 6 days.
3. Concentration of return sludge.
4. Flow of waste sludge if effluent solids concentration is 20 mg/L.
5. Air-flow requirement if 90% of the inflow BOD should be removed.

Solution

1. Organic loading:

$$F/M = \frac{F_i S_i}{X_a V_a}$$

$$0.3 = \frac{(10{,}000\ \text{m}^3/\text{day})(250\ \text{mg/L})}{(3000\ \text{mg/L})V_a}$$

$$V_a = 2778\ \text{m}^3$$

TABLE 8.2 Design criteria for activated sludge processes.

Process modification	θ_c *(days)*	q *(kg BOD removed/kg MLVSS – day)*	*MLVSS (mg/L)*	$\frac{V}{F_i}$ (h)	$\frac{F_r}{F_i}$	*BOD removal efficiency (%)*
Conventional	5–15	0.2–0.4	1500–3000	4–8	0.25–0.5	85–95
Completely mixed	5–15	0.2–0.6	3000–6000	3–5	0.25–1.0	85–95
Step aeration	5–15	0.2–0.4	2000–3500	3–5	0.25–0.75	85–95
High rate	0.2–0.5	1.5–5.0	600–1000	1.5–3	0.05–0.15	60–75
Extended aeration	Not applicable	0.05–0.15	3000–6000	18–36	0.75–1.5	75–95
Contact stabilization	5–15	0.2–0.60	1000–3000[a] 4000–10,000[b]	0.5–1.0 3–6	0.25–1.0	80–90

[a] Contact tank.
[b] Stabilization tank.

The hydraulic retention time is based on the incoming flow and return sludge:

$$h = \frac{2778 \text{ m}^3}{(10{,}000 + 2500) \text{ m}^3/\text{day}} = 0.22 \text{ days} \times 24 \text{ hour/day}$$

$$= 5.33 \text{ hours}$$

2. Total MLVSS in the aeration tank:

$$(2778 \text{ m}^3)(3000 \text{ mg/L})(1000 \text{ L/m}^3)(\text{kg}/10^6 \text{ mg}) = 8334 \text{ kg}$$

$$\theta_c = \frac{X}{\Delta X/\Delta t} = \frac{8334 \text{ kg}}{\Delta X/\Delta t} = 6 \text{ days}$$

$$\Delta X/\Delta t = 1389 \text{ kg/day}$$

3. To maintain steady operation, 1389 kg of MLVSS per day is removed from the system either as waste sludge or solids in the wastewater effluent after treatment. The sludge wasting can be estimated by the materials balance shown in Figure 8.7.

$$(12{,}500 \text{ m}^3/\text{day})(3000 \text{ mg/L})(1000 \text{ L/m}^3)(\text{kg}/10^6 \text{ mg})$$
$$= 37{,}500 \text{ kg/day}$$

$$37{,}500 \text{ kg/day} = (2500 \text{ m}^3/\text{day})(X_r \text{ kg/m}^3) + 1389 \text{ kg/day}$$

$$X_r = \frac{37{,}500 - 1389}{2500} = 14.44 \text{ kg/m}^3 = 14{,}440 \text{ mg/L}$$

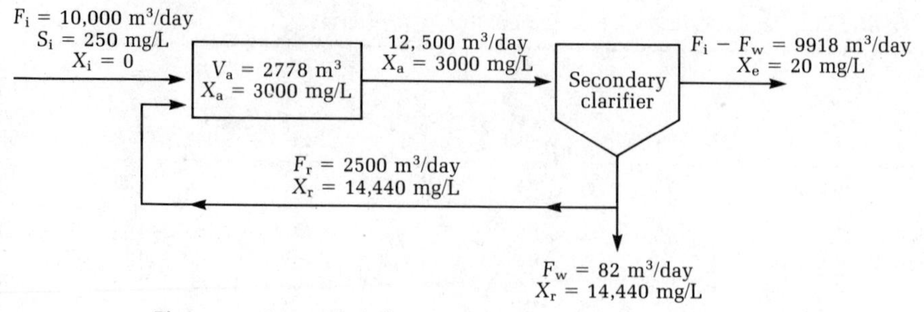

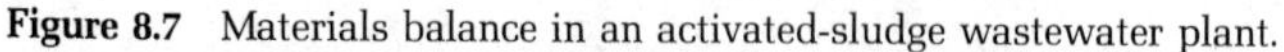

Figure 8.7 Materials balance in an activated-sludge wastewater plant.

4. Waste sludge $= 1389 \text{ kg/day} = F_w X_r + (F_i - F_w)X_e$

$$= F_w \text{ m}^3/\text{day} \times 14.44 \text{ kg/m}^3 + (10{,}000 - F_w) \text{ m}^3/\text{day} \times 20 \text{ mg/L} \times 1000 \text{ L/m}^3 \times \text{kg}/10^6 \text{ mg}$$

$$1389 = 14.44\ F_w + 200 - 0.02\ F_w$$

$$1189 \text{ kg/m}^3 = 14.42\ F_w$$

$$F_w = 82 \text{ m}^3/\text{day}$$

5. The air flow required for 90% BOD removal efficiency:

$$\text{Mass of oxygen/day} = (10{,}000 \text{ m}^3/\text{day}) \times (250 \text{ mg/L} \times 0.9)(1000 \text{ L/m}^3)(\text{kg}/10^6 \text{ mg})$$

$$O_2/\text{day} = 2250 \text{ kg } O_2/\text{day}$$

The oxygen transfer efficiency is 4–12%. Assume an efficiency of 8%. Therefore, O_2/day required = 2250 kg O_2/day × 1/0.08 = 28,125 kg O_2/day. Air is 23.2% oxygen by weight, therefore, air requirements = 28,125/0.232 = 121,228 kg air/day. If the specific weight of air at 0°C and 1.0 atmosphere is 1.201 kg/m^3, the air requirement is

$$\frac{121{,}228 \text{ kg air/day}}{1.201 \text{ kg/m}^3} = 100{,}940 \text{ m}^3/\text{day}$$

or

$$\frac{\text{Air requirement}}{\text{kg of BOD removed}} = 100{,}940/2250 = 45 \text{ m}^3/\text{kg BOD}$$

8-6 Anaerobic treatment

Anaerobic waste treatment is primarily used for stabilization and volume reduction of municipal and industrial sludges. It is also effective in treating complex organic wastes from packing houses, breweries, distilleries, fatty acids, and wood-fiber and synthetic milk processing. Anaerobic treatment involves hydrolysis of complex waste components including fats, proteins, and polysaccharides to their component subunits by a heterogeneous group of facultative and anaerobic bacteria. This is known as "acid fermentation" and results in simpler organic compounds, alcohols, and new bacterial cells. These end products—acetic acid, propionic acid and other intermediates—are converted to gases, mainly methane and carbon dioxide, by the anaerobic bacteria during the methane fermentation process as follows:

$$\underset{\text{(acetic acid)}}{CH_3COOH} \xrightarrow[\text{bacteria}]{} CH_4 + CO_2$$

and

$$\underset{\text{(propionic acid)}}{CH_3CH_2COOH} + 0.5\,H_2O \xrightarrow[\text{bacteria}]{} 1.25\,CO_2 + 1.75\,CH_4$$

Stabilization of organic matter occurs mainly during methane fermentation. It is generally assumed to be the rate-controlling step in anaerobic waste treatment processes. The quantity of methane released can be estimated from the equation above.

The volume of methane formed per pound of ultimate BOD oxidized is 5.63 ft^3 at 0°C and 1.0 atmosphere (McCarty, 1968) or 0.347 m^3 per kilogram. The total volume of gas produced is about two-thirds methane and one-third carbon dioxide.

Example Problem 8.6

From the fermentation of acetic acid, calculate the volume of methane gas produced at 35°C and 1.0 atmosphere per gram of acetic acid. Also, calculate the methane volume per gram of ultimate BOD or COD.

Solution

$$CH_3COOH \longrightarrow CH_4 + CO_2$$

Mol. wt.: 60 16 44

Methane produced per gram of acetic acid = 16/60 = 0.267 g. One gram-mole of gas at standard temperature and pressure has a volume of 22.4 liters. Hence, the volume of methane produced at 0°C and 1.0 atmosphere is

$$\frac{0.267 \text{ g } CH_4}{16 \text{ g } CH_4/\text{g-mole}} \times \frac{22.4 \text{ L}}{\text{g-mole}} = 0.373 \text{ L}$$

or 0.373 liter of CH_4 is produced by fermentation of 1 gram of CH_3COOH at STP. At 35°C and 1.0 atmosphere, CH_4 produced is

$$\frac{P_1V_1}{T_1} = \frac{P_2V_2}{T_2} \text{ or } \frac{(1.0)(V_2)}{35 + 273} = \frac{(1.0) \times (0.373 \text{ L})}{273}$$

$$V_2 = \frac{(0.373 \text{ L})(308)}{273} = 0.421 \text{ L}$$

Also,

$$CH_3COOH + 2O_2 \longrightarrow 2CO_2 + H_2O$$

Mol. wt.: 60 64

One gram of acetic acid requires 1.067 grams of oxygen, the ultimate BOD which is equal to COD.

$$\frac{1.067 \text{ g } BOD_u}{\text{g Acetic acid}}$$

Therefore, methane produced at 35°C and 1.0 atmosphere is

$$\frac{0.422 \text{ L}}{1.067} = \frac{0.4 \text{ L } CH_4}{\text{g COD or } BOD_u}$$

8-7 Sludge digestion

Sludge digestion converts the bulky, odorous, and putrescible sludge material from primary and secondary settling tanks to a well-digested sludge that can be dewatered easily and made relatively free of noxious odors. Volatile suspended solids (VSS) are broken down and a reduction in sludge volume will result from aerobic or anaerobic sludge digestion. The end products of aerobic digestion are cells, CO_2,

TABLE 8.3 Quantities and qualities of sludge produced by different treatment processes.

	Sludge produced per million liters of sewage		*Sludge characteristics*		
Treatment process	*Liquid sludge (L)*	*Dry solids (kg)*	*Moisture (percent)*	*Sp. gravity of liquid*	*Sp. gravity of solids*
Primary sedimentation					
Undigested	2,950	150	95.0	1.02	1.40
Digested in separate tanks	1,450	90	94.0	1.03	—
Digested and dewatered on sand beds	—	90	60.0	—	—
Activated sludge	19,400	290	98.5	1.005	1.25
Primary sedimentation and activated sludge					
Undigested	6,900	280	96.0	1.02	—
Digested	2,700	170	94.0	1.03	—
Digested and dewatered on sand beds	—	170	60.0	—	—

and water. The end products of anaerobic digestion are methane gas, CO_2, unused intermediate organics, and a relatively small amount of cells. During digestion the amount of fixed suspended solids remains relatively constant, the volatile suspended solids is lowered, and the volume of sludge is considerably reduced. Table 8.3 presents quantities and qualities of sludge produced by different treatment processes. The calculations in this table are based on the assumptions that sewage flow is 378 liters per person and raw sewage contains 300 mg/L suspended solids. It is obvious from the table that the volume of sludge and the dry solids content are reduced during sludge digestion.

Methane bacteria are strict anaerobes and very sensitive to conditions of their environment. Therefore, a decrease in gas production and lowering in the percentage of methane gas are signs of failure in the anaerobic digestion process. The gas production is generally 65–69% by volume methane and 31–35% carbon dioxide under favorable operational conditions. It is estimated that the gas produced is 1.0–1.1 m^3/kg of VSS destroyed. This gas contains approximately 0.70 m^3 methane/kg VSS destroyed. In other terms, methane produced at 35°C and 1 atmosphere amounts to 0.40 m^3/kg of COD or ultimate BOD. As the heating value for methane gas is about 36,000 kJ/m^3, many municipalities around the country have investigated the possibility of burning methane produced from anaerobic digestion to help power their wastewater facilities.

Digested sludge is generally thickened and dewatered to reduce its moisture content by natural evaporation, percolation, filtration,

Figure 8.8 Typical sludge drying beds.

and mechanical means. The available dewatering processes include vacuum filters, centrifuges, filter presses, drying beds, and lagoons. Drying beds and lagoons are most commonly used for small plants. Digested sludge is placed on beds over sand and gravel layers for several days or weeks in a layer 20–30 cm thick and allowed to dry as shown in Figure 8.8. After drying, the sludge is removed and either disposed of in a landfill or used as a soil fertilizer.

8-8 Advanced wastewater treatment

Advanced or tertiary treatment of wastewater produces effluent quality beyond the secondary level. For example, secondary treatment removes 25–55% of the total nitrogen and 10–30% of the total phosphorus, while 90% or better removal efficiencies may be required. These nutrients may accelerate the eutrophication of streams and lakes that receive the effluent and lead to excessive growth of algae and other aquatic plants. Nutrients can be removed by many processes listed in Figure 8.1. Physical processes include filtration, distillation, and reverse osmosis; chemical processes include electrodialysis, chemical precipitation, carbon adsorption, ammonia stripping, and ion exchange; and biological processes include bioaccumulation and harvesting of algae and aquatic plants grown on the nutrients, bacterial assimilation, and nitrification/denitrification.

Before an engineer can consider tertiary treatment needs, it is necessary to establish water-quality requirements for specific water uses. Increasing emphasis is being placed on the removal of phosphorus, nitrogen, refractory organics which are difficult to decompose biologically, and total inorganic solids when extensive water reuse is necessary for industrial expansion. Phosphorus is precipitated by the addition of coagulants such as calcium, iron, and aluminum salts to the aeration tank or secondary effluent. Examples of the chemical reactions are:

$$Al_2(SO_4)_3 \cdot 14H_2O + 2PO_4^{-3} \longrightarrow 2AlPO_4\downarrow + 3SO_4^{-2} + 14H_2O$$
$$5Ca(OH)_2 + 3HPO_4^{-2} \longrightarrow Ca_5OH(PO_4)_3\downarrow + 3H_2O + 6OH^-$$

The precipitate in the first reaction is aluminum phosphate and in the second reaction is calcium hydroxyphosphate or, to use its mineral name, hydroxylapatite.

Phosphorus can also be removed biologically in bacterial and plant cells by luxury uptake. Similarly, nitrogen can be removed biologically by nitrification/denitrification.

Nitrifying bacteria, *Nitrosomonas,* oxidize ammonium to nitrites and nitrifying bacteria, *Nitrobacter,* oxidize nitrites to nitrates. The overall reaction for ammonium conversion to nitrates has been proposed by McCarty (1970) on the basis of both laboratory and theoretical calculation:

$$NH_4^+ + 1.83\ O_2 + 1.98\ HCO_3^- \longrightarrow 0.021\ C_5H_7O_2N + 1.04\ H_2O + 0.98\ NO_3^- + 1.88\ H_2CO_3$$

Nitrifying bacteria utilize inorganic carbon sources, as shown from the previous equation, and they are classified as autotrophic organisms. They produce approximately 1 mole of NO_3^- for each mole of NH_4^+ converted. The nitrates produced can be reduced to nitrogen gas by denitrifying bacteria. These exist under anaerobic conditions, and McCarty (1969) developed the following empirical equation to describe the overall nitrate removal:

$$NO_3^- + 1.08\ CH_3OH + H^+ \longrightarrow 0.065\ C_5H_7O_2N + 0.47\ N_2 + 0.76\ CO_2 + 2.44\ H_2O$$

From this equation, it can be seen that the denitrifying bacteria utilize organic carbon and they are classified as heterotrophic organisms. It appears that approximately one-half mole of nitrogen gas is produced for each mole of nitrates reduced. Denitrifying organisms require an organic carbon source which is satisfied by the addition of methanol (CH_3OH), anaerobic environment, and the presence of nitrates.

Advanced treatment is expensive and costs 2 to 3 times that of secondary treatment of wastewater. Therefore, a cost-benefit analysis should be investigated thoroughly during the decision-making process. It may be possible to treat wastewater to drinking water standards, if so desired.

8-9 On-site wastewater treatment

Approximately 18 million housing units, or 25% of all housing units in the United States during 1980, disposed of their wastewater using on-site wastewater treatment and disposal systems (U.S. Environmental Protection Agency, 1980). The number of on-site systems is increasing with an estimated one-half million new systems to be installed every year. These systems may include septic tanks, intermittent sand filters, aerobic treatment units, and others. The effluent from these systems may be safely disposed of onto the land, into surface water, or evaporated into the atmosphere by a variety of methods.

Septic tanks are the most widely used on-site wastewater treatment option in the United States. They are watertight concrete, fiberglass, or steel boxes designed and constructed to receive wastewater from a home or cluster of homes. They are primarily sedimentation basins, although a minor degree of organic matter digestion and solids destruction occurs due to anaerobic digestion. Units are ordinarily sized to provide a 24-hour detention time at average daily flow. Settleable solids and partially decomposed sludge settle to the bottom of the tank and accumulate. A scum of fats and greases rises to the top. The partially clarified liquid is allowed to flow just below the floating scum layer, through an outlet as shown in Figure 8.9. The effluent of a septic tank is offensive and further treatment is required, either in an additional process or by soil bacteria.

Effluent from septic tanks can be satisfactorily discharged to a soil absorption field on many natural soils. Soils should be capable of long-term transmission of water as evident by standard percolation tests. The flow which can be applied per unit area per day through a drainage field can be estimated as a function of percolation rate (Steel and McGhee, 1979) as follows:

$$F = 204\sqrt{t}$$

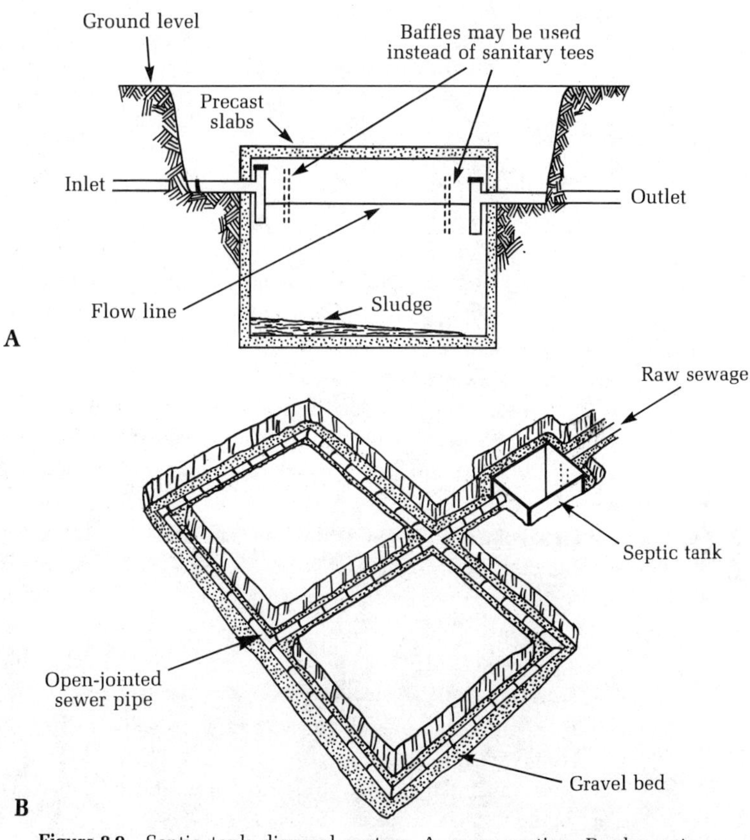

Figure 8.9 Septic tank disposal system. A, cross section, B, absorption field.

where

F = the flow in L/m^2-day

t = time (min) required for water surface to fall 25 mm

If the percolation rate t is 4 minutes or less, which means a subsidence rate over 6 mm/min, a tile disposal field will generally prove to be an effective method of treatment. Local regulatory agencies may require that the septic tank be located at specified distances from the home, water wells, and water lines to reduce any risk of contaminating the potable water supply. Regulations aside, careful consideration should be given to design, construction, and maintenance of septic tanks and disposal fields in order to avoid pollution of subsurface water or nearby surface water.

8-10 Problems

8.1 An aeration tank has a capacity of 3 million gallons and is treating 8 million gallons per day. The inflow suspended solids is neglected, and the outflow suspended solids from the clarifier effluent is 25 mg/L. The recycled sludge is 2.0 million gallons per day and the return suspended solids concentration is 10,000 mg/L. Calculate the BSRT, θ_c, if MLSS in the aeration tank is 2200 mg/L.

8.2 An activated sludge process removes 10,000 kg BOD per day. The aeration tank volume is 2000 m^3. The mixed liquor volatile suspended solids concentration is 2000 mg/L. Estimate how much oxygen per day is consumed, and what volume of air per minute must be supplied at STP.

8.3 A wastewater plant flow is 1 cubic meter per minute. The aeration tank is to operate with 2500 mg/L MLVSS and a mean cell retention time of 6 days for 90% removal of BOD and suspended solids. The influent BOD and SS are 250 mg/L. Calculate the size of the aeration tank, if organic loading is 0.3/day, and the amount of wasted sludge.

8.4 Calculate the volume (liters) of methane gas produced by degradation of 1 gram of propionic acid at 50°C and 1.0 atmosphere. Also, determine the methane produced per gram of COD.

8.5 Sludge fed to an anaerobic digester is 80 percent volatile solids by weight. Digested sludge is 55% volatile solids. What fraction of the volatile solids put into the digester have been destroyed?

8.6 An anaerobic digester that converts organic matter into gas is used to digest 1500 kg of wet sewage sludge. The sludge input is 4% solids by weight, and the solids are 70% organic and 30% inert. The digested sludge taken out is 6 percent by weight solids which are 50% organic. Assume that the supernatant contains small amounts of solids that can be ignored when making material balance. Calculate how much water is removed from the digester as supernatant (kg), how many kilograms of organic solids is converted to gas, how many cubic feet of methane can be produced if reduction of 1 pound of organic matter will result in gas production of 18 cubic feet of gas of which 70% is methane and 30% CO_2.

8.7 Calculate the settling rates of calcium carbonate crystals formed in a clarifier if the temperature is 10°C. The crystal can be represented by a sphere of 0.12 mm diameter, and its density is 1.25 g/cm^3. Also, express the settling velocity in terms of m^3/day-m^2.

8.8 Design the secondary clarifier for the plant in problem 8.3 if the desired detention time is 2 hours and the overflow rate is 20 m^3/day-m^2.

8.9 Define:

activated sludge	endogenous respiration
facultative bacteria	grit chamber
acid fermentation	advanced wastewater treatment
oxygen deficit	Stoke's law
lag phase	step aeration

References

Benefield, L. D. and Randall, C. W. 1980. *Biological Process Design for Wastewater Treatment.* Englewood Cliffs, NJ: Prentice-Hall.

Berthouex, P. M. and Rudd, D. F. 1977. *Strategy of Pollution Control.* New York: Wiley.

Camp, T. R. 1946. Sedimentation and design of settling tanks. *Transactions of the American Society of Chemical Engineers.* 111, 895.

McCarty, P. L. 1968. Anaerobic treatment of soluble wastes. In *Advances in Water Quality Improvement*, ed. E. F. Gloyna and W. W. Eckenfelder. Austin: University of Texas Press.

McCarty, P. L. 1970. Phosphorus and nitrogen removal by biological systems. Presented at Wastewater Reclamation and Reuse Workshop, Lake Tahoe, CA, June 25–27, 1970.

McCarty, P. L., Beck, L. and St. Amants, P. 1969. Biological denitrification of wastewaters by addition of organic materials. In *Proceedings of the 24th Purdue Industrial Waste Conference.* Lafayette, IN.

Metcalf and Eddy, Inc. 1979. *Wastewater Engineering: Treatment, Disposal,* Reuse, 2nd ed. New York: McGraw-Hill.

Steel, E. W. and McGhee, T. J. 1979. *Water Supply and Sewerage*, 5th ed. New York: McGraw-Hill.

Sykes, R. M. 1975. Theoretical heterotrophic yields. *Journal of the Water Pollution Control Federation* 47, 2019.

U.S. Environmental Protection Agency. 1980 *Design Manual—On-site Wastewater Treatment and Disposal Systems.* EPA 625/1-80-012. 1980.

9

Energy Supplies and Power Generation

9-1 Introduction

About 200 years ago, an era began that may well be the most significant in history—the industrial revolution. Beginning in the mid-1700s, technological innovations individually and together produced great improvements in industry, transportation, health care, communications, and standard of living.

The United States in particular was a leader in the technological or industrial revolution for two reasons. First, it was blessed with an abundance of natural resources; a large and virtually untapped country lay before the pioneers, explorers, and industrialists of that time. Second, it had an abundance of human resources. The American society and system of government seemed to breed a person with a spirit of innovation and free enterprise, which led to rapid development of the natural resources and rapid improvements in technology. The use of energy was crucial to the industrial and transportation systems that were developing so rapidly. Thus it happened in the United States that energy resources were developed that were cheap and abundant.

Today, the United States is no longer in a rapid growth period; however, energy consumption still is important to U.S. industrial output and is intricately woven into our economy. It is important to realize that the present energy situation did not spring up overnight.

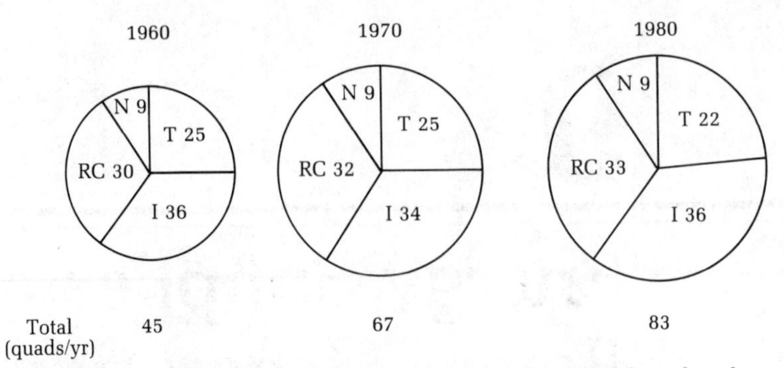

Figure 9.1 U.S. energy usage, total and percent by sector for selected years. From Exxon (1977) and U.S. Department of Energy (1981). T, transportation; I, industrial; RC, residential-commercial; N, non-energy use.

As we have pointed out, the development of energy resources paralleled the development of industry and technology in this country. In the early years, the principal fuel was wood, followed by a switch to coal, then to the use of natural gas and oil as our principal energy resources. As a nation, we are dependent on energy in every facet of our lives: agriculture, transportation, medicine, industry, and personal homes. In addition, we use energy resources (such as oil, coal, and natural gas) to produce non-energy materials (plastics, lubricants, solvents, chemicals).

9-2 The U.S. energy picture

It is interesting to note energy usage by sector of the economy in the U.S. and how it has changed in recent years. In Figure 9.1 we present energy use by sector and by year, in the form of pie charts. The pies get bigger with the years, indicating the growth in total energy usage from 45 quadrillion BTU (45 quads) in 1960 to 82 quadrillion in 1980. Within each pie, energy use is distributed among the three end-use sectors industrial, residential-commercial, transportation, and the non-energy use sector.

In this figure we do not display electricity generation as an end use. Electricity is ultimately consumed in one of the economic sectors previously mentioned. However, conversion of fossil fuel energy to electrical energy is inefficient, and demand for electricity in this country is large, so in order to generate usable electricity, large amounts of raw energy resources are consumed.

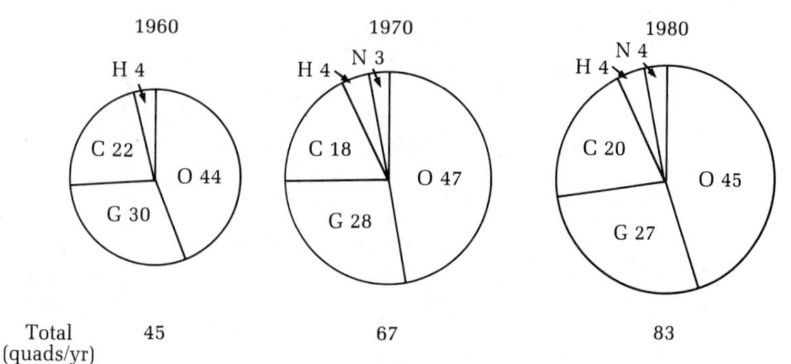

Figure 9.2 U.S. energy supply, total and percent by type of supply for selected years. From Exxon (1977) and U.S. Department of Energy (1981). O, oil; C, coal; H, hydroelectric; G, gas; N, nuclear.

Typically, generating electricity accounts for about one-fourth of all U.S. energy consumption. Distributed electricity is much less than this one-fourth of the total supply. In Figure 9.1 we have allocated the "energy overhead" of electricity generation to the users in proportion to their electrical use. About two-thirds of the electrical energy generated goes to residential or commercial use, with the remaining one-third going to industrial consumption. In Figure 9.2, we chose the same years as Figure 9.1, but now the pies represent energy supplies by type. Supply equals demand in these charts because excess supplies are stored or not produced. Note how, during this period of time, the energy supplies have changed their relative positions.

In Figure 9.3 we have combined energy supply and demand for the year 1980 to show how different sectors of the economy depend on different forms of energy to varying degrees. In particular, note that the transportation sector is essentially totally dependent on liquid petroleum fuels. This is not surprising when we consider the large number of automobiles, trucks, buses, railroads, and airplanes that we have in this country, all of which use liquid fuels. Also in Figure 9.3, we have broken out electricity use from direct use of the primary energy sources and shown energy losses (heat) attributable to electricity generation.

In 1980, the major U.S. supplies of energy included oil, coal, gas, nuclear reactors, and hydroelectric generators. These sources varied from supplying a large part of the total to a fairly small percentage. In Table 9.1 we depict U.S. energy usage in 1980 by its source, using typical units for that source as well as a common set of units, so that the total contributions of each type can be compared more readily.

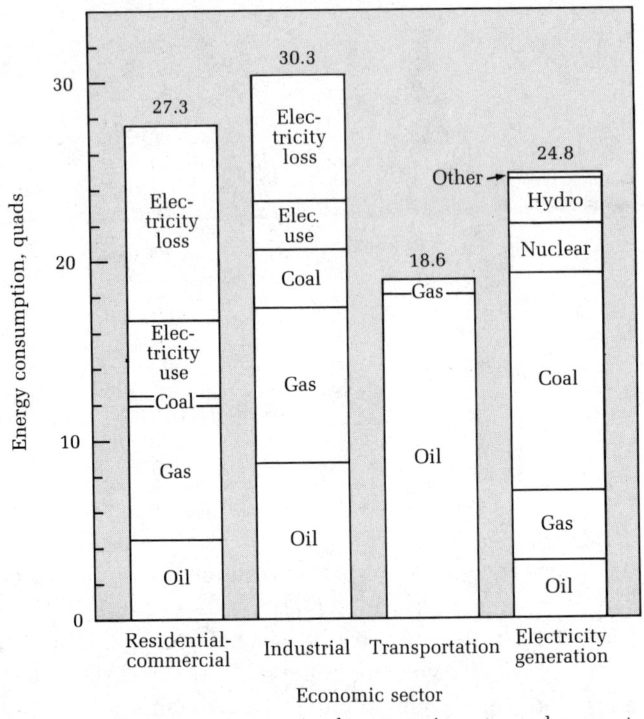

Figure 9.3 U.S. energy consumption by economic sector and energy type for 1980. From U.S. Department of Energy (1981).

TABLE 9.1 U.S. energy consumption by type in 1980.

Source	*Quantity*	*Energy value (quads)*[a]	*Percent of total*
Petroleum	17.0 MMB/D[b]	34.20	44.89
Coal	702.7 MMTons/Y	15.56	20.42
Natural gas	20.1 TCFY[c]	20.50	26.91
Hydroelectric power	276 BKWH/Y[d]	3.12	4.10
Nuclear electric power	251 BKWH/Y	2.70	3.54
Other (includes geothermal, wood, solar, and waste incineration)	5.5 BKWH/Y	0.11	0.14
Total		76.20	100.00

From U.S. Department of Energy (1981).

[a] 1 Quad = 1 quadrillion (10^{15}) BTU or 1.055 billion GJ.

[b] MMB/D = millions of barrels per day (1B = 42 gal).

[c] TCFY = trillions of standard cubic feet per year.

[d] BKWH/Y = billions of kilowatt-hours per year.

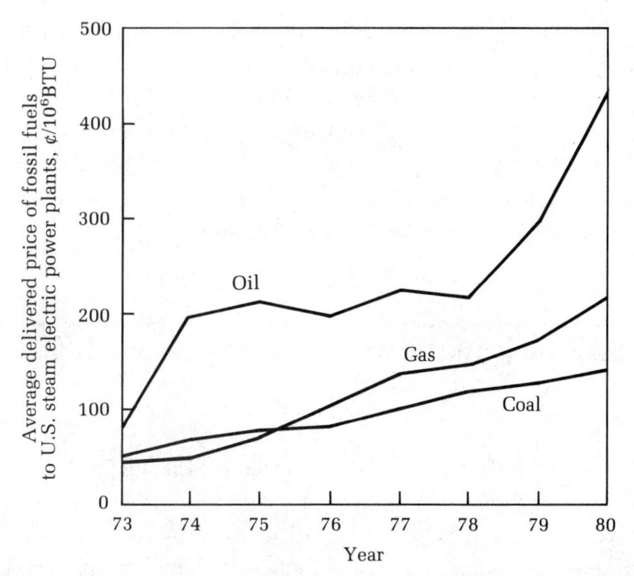

Figure 9.4 Recent history of costs of fossil fuels delivered to U.S. electric plants. From U.S. Department of Energy (1981).

Fossil fuels, which were so crucial to the country's rapid growth following World War II, are still crucial to the normal functioning of our society today.

In the early 1970s the price of oil in particular, but also that of the other fossil fuels, began to escalate very rapidly. The recent price history of the fossil fuels is demonstrated by their costs to U.S. electric plants as shown in Figure 9.4. The major jump in 1973–74 corresponds to the Arab oil embargo in that period. Because of these recent large price increases, there have been substantial cutbacks in recent years in the rate of growth in energy consumption; in fact, petroleum consumption peaked in 1978 and has been declining since.

Preliminary data indicate that in 1982, overall U.S. energy consumption declined 4.3%, the third year in a row that domestic consumption dropped (*Chemical Week*, 1983). Also, in 1982, oil production worldwide averaged 52.8 million barrels per day, down from the peak of 62.5 million in 1979 (*Chemical Week*, 1983). A major reason for these reductions was the worldwide recession of 1981–82, but another important reason was improvements in energy conservation made during the late 1970s.

Despite large reductions in the per capita usage of energy and despite large capital investments in energy conservation devices, as a nation we still use a lot of petroleum. As an example, in one week to drive all of our transport vehicles—cars, trucks, buses, and so forth—we use enough energy to light up the city of New York for 10 years. Also, despite intense efforts to reduce our dependence upon fossil fuels and develop energy resources such as nuclear, geothermal, solar, and hydroelectric, fossil fuels still accounted for 92% of all the energy use in 1980. Because of their importance, therefore, we will briefly describe the fossil fuels.

9-3 Fossil fuels in the world and the United States

Petroleum

Crude oil or crude petroleum is a mixture of hydrocarbons covering a very wide boiling range and chemical type. Crude oil primarily contains carbon and hydrogen, but also usually contains some sulfur, nitrogen, and trace metals. Oil has been found in underground deposits in many areas in the world, but some areas are far richer than others. Saudi Arabia has the largest known reserves of oil in the world, accounting for nearly 25% of the world's total. The U.S. has substantial (but declining) oil reserves; however, our consumption outpaces our production, and in 1980 we imported 5.18 million barrels per day of crude (U.S. Department of Energy, 1981). In 1982, imports were down significantly from this level due to conservation efforts, fuel switching, and a recessionary economy.

After the crude oil is recovered from underground, it must be transported (usually by pipeline or tankers) to refineries where it is rendered into useful products. Once at the refinery (Figure 9.5), the crude oil is distilled into rough-cut fractions that typically correspond to light ends (such as ethane and propane), light naphthas, medium naphthas, heating oils, medium process oils, heavy process oils, and bottoms (tar and asphalt). During the downstream refining processes, the oil is further split into narrower and narrower boiling range streams which can then be blended back together to form specific products, such as gasoline, heating oil, diesel fuel, motor oils and so on.

In addition to simply separating components and recombining them in a proper fashion, another major function of an oil refinery is to chemically change the less useful types of oil that occur naturally into more valuable and useful products. Whereas natural gasoline, diesel fuel, and heating oil may make up 30–45% of a particular

Figure 9.5 A small section of a large modern petroleum refinery. (Photo courtesy of Exxon Co., U.S.A.)

crude oil, demand for these products corresponds to about 75% of all the crude oil that is refined. Therefore, there are units in the refinery that convert the medium and heavy process oils into gasoline and heating oil. There are also units to upgrade the octane of crude naphthas to make them suitable for gasoline. Hundreds of different products are made in the refinery, including some chemical feedstocks, oils for use in food processing, solvents, lubricating oils, and fuels of all different types and grades.

In the last decade, crude oils have become more sour (higher sulfur content) and environmental restrictions have required less sulfur in the products. Thus, many refineries have installed desulfurization units and sulfur recovery units, allowing the sulfur that is removed from the oil to be recovered in elemental form and sold. This has added markedly to the supplies of sulfur available to the world. In fact, in 1977 over 2 million metric tons of sulfur were recovered and sold in the U.S. by oil refineries. This is a good example of changing a process to meet environmental demands which then results in a salable by-product. Example problem 9.1 illustrates the magnitude of sulfur recovery that can occur in a large refinery.

Example Problem 9.1

A refinery processes 400,000 B/D of sour crude that contains 0.7% by weight sulfur. The density of the crude oil is 0.85 kg/L. The refinery produces a variety of desulfurized products such as gasoline, kerosene, jet fuel, heating oils, and low-sulfur fuel oils. However, some sulfur remains in some of the heavier products (residual fuel oil, asphalt, coke) and some is emitted to the environment during the refining process. On the average, about 60% of all the entering sulfur is recovered and sold as elemental sulfur. How much sulfur is produced from this refinery (kg/day), and how much money does the sale of by-product sulfur bring in each year if the sulfur sells for $120/metric ton?

Solution

$$\frac{400{,}000\text{ B}}{\text{day}} \times \frac{42\text{ gal}}{\text{B}} \times \frac{3.785\text{ L}}{\text{gal}} \times \frac{0.85\text{ kg}}{\text{L}} \times 0.007 \times 0.60$$

$$= 2.27 \times 10^5 \text{ kg/day sulfur}$$

$$2.27 \times 10^5 \frac{\text{kg sulfur}}{\text{day}} \times \frac{1\text{ metric ton}}{1000\text{ kg}} \times \frac{\$120}{\text{metric ton}} \times \frac{365\text{ days}}{\text{yr}}$$

$$= \$9.9\text{ million}$$

Natural gas

Natural gas is a mixture of very light hydrocarbons, principally methane, with small amounts of ethane, propane, and butanes. It may also contain large amounts of H_2S and CO_2. Natural gas is found in many, but not all countries in the world, and the United States is fortunate to have large quantities of natural gas (although gas resources are also being depleted).

Natural gas, being a very light hydrocarbon, is usually not stored by users. It is produced in the fields, processed to remove the H_2S, CO_2, and hydrocarbon liquids and then transported via pipeline to the using locations. Natural gas is used in many major industries, as well as in homes and businesses. Its primary use in the United States is as fuel because of its convenience and its clean-burning nature. It is also the primary feedstock for the production of ammonia. In this country gas production appeared to peak in the early 1970s and has been declining slightly since that time. During the period 1978–80, due to substantially higher prices, gas exploration increased markedly, and production increased enough to halt the past trend of declining reserves. Like oil, though, gas reserves are finite and will eventually run out. We will have to develop suitable substitutes before that happens.

Example Problem 9.2

A natural gas processing plant in Mississippi processes 100 million standard cubic feet (SCF) of gas per day. The gas is very sour, containing 25 mole % H_2S. Assume that the processing plant is 95% efficient in capturing the H_2S to produce elemental sulfur. (Note: in the natural gas industry, 1 lb-mole = 379.5 SCF at 60°F). Estimate the production rate of sulfur in metric tons per day from this plant.

Solution

$$100 \times 10^6 \frac{\text{SCF}}{\text{day}} \times \frac{1 \text{ lb-mol}}{379 \text{ SCF}} \times \frac{1 \text{ kg-mol}}{2.2 \text{ lb-mol}} \times 0.25$$

$$= 3.00 \times 10^4 \frac{\text{kg-mol}}{\text{day}} H_2S$$

$$3.00 \times 10^4 \frac{\text{kg-mol } H_2S}{\text{day}} \times 0.95 \times \frac{34 \text{ kg } H_2S}{1 \text{ kg mol } H_2S} \times \frac{32 \text{ kg S}}{34 \text{ kg } H_2S} \times \frac{1 \text{ MT}}{1000 \text{ kg}}$$

$$= 911 \text{ MT/day sulfur}$$

Coal

Coal, a very abundant fossil fuel is a heavier hydrocarbon than oil or gas in that it is solid and has the highest carbon-to-hydrogen ratio. Coal is produced from both surface and underground mines and transported by rail to using locations. Typically, the major use of coal in the U.S. is for generating electricity. Coal was the principal fuel for industry and electricity generation during the early part of this century, but the low price and clean-burning nature of oil and gas displaced it. Recently, decreased availability of oil and gas and the lower cost of coal has caused a resurgence in the use of coal.

Comparisons among the fossil fuels

When we compare the advantages and disadvantages of the fossil fuels, we must consider not only their prices but also their availability, both short-term and long-term. In addition, versatility, nearness of

TABLE 9.2 Comparisons among the fossil fuels.

	Gas	*Oil*	*Coal*
Cost	Med	High	Low
Ease of use	Easy	Med	Hard
Versatility	Med	High	Med
Availability (domestic)	Med	Low	High
Environmental effects	Low	Med	High
Current use (dependence)	High	High	High

supplies, pollution effects, transportation, and other areas of comparison may be important.

Presently coal is the cheapest and most abundant of the three, while oil is the most expensive and appears to be the scarcest. Oil is the most versatile because of its ability to provide not only industrial fuels but also motor fuels, chemicals, plastics, lubricating oils, and solvents. Coal and gas are used principally as burner fuels, with coal limited to large industries. Coal production and use has the largest impact on the environment, producing air, water, and land pollution. Coal and gas are primarily domestic resources, whereas much of our oil is imported. Coal transportation depends on the railroads, although recently several large coal slurry pipelines have been proposed. Table 9.2 summarizes some of these comparisons.

Despite the great advantages of fossil fuels, it must be stated that there are serious problems with fossil fuel combustion, principally acid precipitation and CO_2 build-up. Also, oil and gas reserves are being depleted rapidly. Neither are coal reserves infinite. However, an immediate halt to the use of these fuels and an abrupt switch to nuclear, solar, or other alternate forms of energy, is impossible. The fossil fuels represent a bridge to the future in that their continued use today is necessary while we carefully develop the best alternatives for tomorrow. Recall that our society is very dependent on electricity. Before we discuss alternative sources of energy (which also must be useful for generating electricity), we shall first discuss electrical power generation.

9-4 Electrical power generation

An electrical current can be made to flow in a wire by moving the wire relative to a magnetic field. All electric generators take advantage of this simple principle, usually by rotating a magnet inside a coil of wires. A large generator is turned by a turbine, which is coupled to

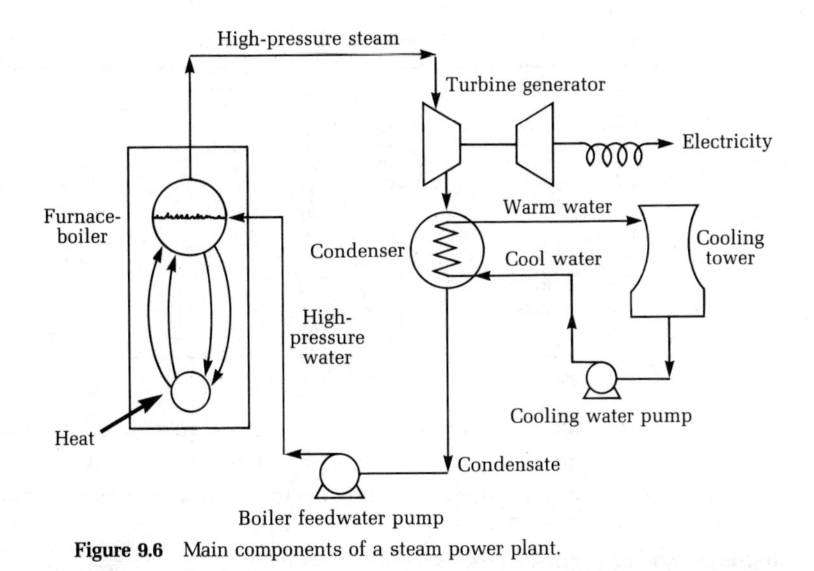

Figure 9.6 Main components of a steam power plant.

it through a shaft. A turbine can be turned by water pressure or by combustion gases, but usually turbines are turned by steam.

In the conventional steam power plant, the fuel is burned in a furnace and the heat from combustion is transferred through steel pipes which contain flowing water. The water, under pressure, is heated and returns to a steam drum where it flashes into steam. The steam is then circulated through another section of the furnace, and picks up additional heat to a desired temperature and pressure (thus it is called superheated steam). The steam then is routed to a turbine where it expands, pushing the turbine blades and turning the turbine which is coupled to the generator. The low-pressure steam that exits the turbine is condensed, and the water goes back to the furnace-boiler for another cycle.

The thermal efficiency of the plant is enhanced by preheating the air used in the combustion of the fuel and by preheating the makeup water that is continuously sent into the steam system. All in all, heat recovery is quite good but is clearly limited by the laws of thermodynamics (see the next section) and properties of materials. The diagram in Figure 9.6 represents the main components of a steam power plant. Notice that any heat source can be used, whether it be gas, oil, coal, wood, sunlight, municipal solid waste, or nuclear fission. In Table 9.3, we have detailed the history of electricity production by primary energy source. There are many items of interest to be found

TABLE 9.3 U.S. net electricity production by primary energy type.

	Electricity production, billions of kWh							
Energy source	*1960*	*1965*	*1970*	*1975*	*1977*	*1978*	*1979*	*1980*
Coal	430	520	660	853	985	976	1075	1162
Oil	60	80	209	289	358	365	304	246
Gas	120	260	400	300	306	305	329	346
Hydro	160	200	250	300	220	280	280	276
Nuclear	—	—	30	173	251	276	255	251
Geothermal and wastes	—	—	1	3.4	4.1	3.3	4.4	5.5
Total	770	1060	1550	1918	2124	2205	2247	2286

From U.S. Department of Energy (1981).

in Table 9.3, but particularly note the recent upsurge in the use of coal. Because coal is so important to the generation of electricity in the United States, we next discuss coal-fired power plants.

Coal-fired power plants

There are many grades or types of coal in use today. These different coals can have quite different heating values, sulfur content, ash content, and differences in other parameters. All of these differences must be accounted for in any detailed design work, and can influence the final choice of a system for combustion. Typically, however, eastern coals have higher heating values and higher sulfur content than western coals.

There are two basic types of coal combustion systems: fixed-bed and air-suspension. In a fixed-bed system, coal is placed on a stationary or moving grate and air flows upward through the bed of coal. Ash falls through the grate or is carried out with the combustion gases. A stoker supplies coal to the grate as needed. Fixed-bed systems are usually used in small applications such as residential or commercial boilers.

In an air-suspension system, the coal is pulverized to a fine powder and blown into the furnace mixed with air. The fine coal particles burn very rapidly allowing for rapid rates of heat release, steam production, and subsequent power production. This high rate of heat release is required for large steam-electric plants, thus making the pulverized coal furnace the standard of the power-generating industry (see Figure 9.7). Because of the small size of the particles, much of the ash remains suspended in the air and must be removed by electrostatic precipitation, fabric filtration, or such similar systems.

In either type of coal combustion system, certain pre- and post-combustion facilities are required. The coal must be received by rail

Figure 9.7 Aerial view of an operating 325-MW coal-fired power plant. (Courtesy of Philadelphia Electric Company. Photo by Charles H. Peatross, Jr.)

car, unloaded, and transferred to a storage pile. The coal is transferred to the mills, where it is pulverized, then blown into the furnace. A substantial portion of the ash falls out of the furnace as slag, which is quenched and removed. A majority of the ash remains suspended in the flue gas, and must be removed and disposed of. The most recent federal regulations for coal-fired power plants require almost complete removal of particulates (usually around 99% or more), and substantial desulfurization of the flue gas (even for low-sulfur coals). The type of flue-gas desulfurization system most widely used today is lime or limestone scrubbing, which results in a sludge that must be disposed of. The following example problem illustrates the enormous materials-handling problems associated with a large coal-fired power plant.

Example Problem 9.3

The major inputs and outputs of a 1000-MW coal-fired power plant are shown schematically in Figure 9.8. Assume that the plant has a thermal efficiency of 39% and operates at an average annual load

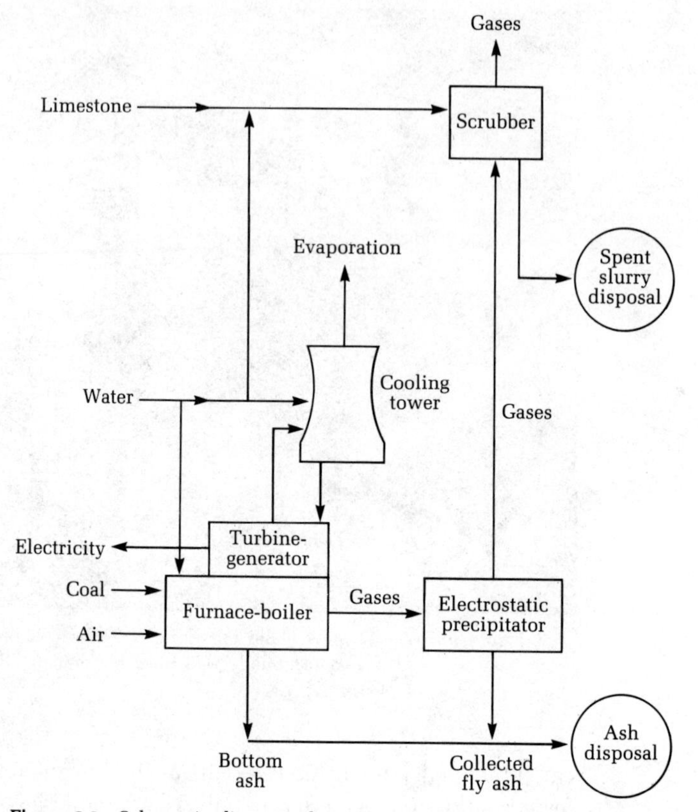

Figure 9.8 Schematic diagram showing major input and output streams for a coal-fired power plant.

factor (proportion of total capacity) of 0.65. The coal has the following properties: heating value is 11,000 BTU/lb, sulfur content is 3.0%, ash content is 10%. The plant meets the new standards for emissions of ash (0.03 lb/million BTU heat input) and SO_2 (0.6 lb/million BTU heat input). Assume that all of the collected SO_2 is produced as a $CaSO_4$ sludge which is 40% solids. The chemistry is as follows:

$$CaCO_3 + SO_2 \longrightarrow CaSO_3 + CO_2$$
$$CaSO_3 + \tfrac{1}{2}O_2 \longrightarrow CaSO_4$$

Estimate the annual rates of (1) coal usage, (2) ash collection, (3) SO_2 release, (4) spent slurry, and (5) limestone requirements (assuming 100% calcium purity and 100% calcium utilization). Give all answers in metric tons per year.

Solution

1. $$\frac{1000\ \text{MW}}{0.39} \times 0.65 \times \frac{1000\ \text{kW}}{1\ \text{MW}} \times \frac{24\ \text{hr}}{\text{day}} \times \frac{365\ \text{day}}{\text{year}} \times \frac{3412\ \text{BTU}}{\text{kWh}}$$

$$= 4.98 \times 10^{13}\ \text{BTU/yr}$$

$$4.98 \times 10^{13}\ \frac{\text{BTU}}{\text{yr}} \times \frac{1\ \text{lb}}{11{,}000\ \text{BTU}} \times \frac{0.454\ \text{kg}}{\text{lb}} \times \frac{1\ \text{MT}}{1000\ \text{kg}}$$

$$= 2.056 \times 10^{6}\ \frac{\text{MT}}{\text{yr}}\ \text{coal}$$

2. $$\text{Ash input} = \frac{0.10\ \text{lb ash}}{\text{lb coal}} \times \frac{1\ \text{lb coal}}{11{,}000\ \text{BTU}} = \frac{9.09\ \text{lb ash}}{\text{million BTU}}$$

$$\text{Ash output} = 0.03\ \text{lb/million BTU}$$

Therefore,

$$\text{Ash collected} = 9.09 - 0.03 = 9.06\ \text{lb/million BTU}$$

$$= 9.06\ \frac{\text{lb}}{10^{6}\ \text{BTU}} \times 4.98 \times 10^{13}\ \text{BTU/yr}$$

$$\times\ 4.54 \times 10^{-4}\ \text{MT/lb}$$

$$= 2.05 \times 10^{5}\ \text{MT ash/yr}$$

3. $$\frac{0.6\ \text{lb}\ SO_2}{\text{million BTU}} \times 4.98 \times 10^{13}\ \text{BTU/yr} \times 4.54 \times 10^{-4}\ \text{MT/lb}$$

$$= 1.36 \times 10^{4}\ \text{MT}\ SO_2\ \text{emitted/yr}$$

4. Using sulfur as the material-balance component:

$$\text{S input} = 0.03(2.056 \times 10^{6}) = 6.17 \times 10^{4}\ \text{MT/yr}$$

$$\text{S output} = 1.36 \times 10^{4} \times 0.5 = 0.68 \times 10^{4}\ \text{MT/yr (as S)}$$

Therefore,

$$\text{S collected} = 5.49 \times 10^{4}\ \text{MT/yr (as S)}$$

$$\text{Calcium sulfate collected} = 5.49 \times 10^{4}\ \text{MT/yr of S} \times \frac{136\ \text{MT}}{32\ \text{MT}}$$

$$= 2.33 \times 10^{5}\ \text{MT/yr as}\ CaSO_4$$

But this slurry is only 40% solids, so

$$\text{Spent slurry} = \frac{2.33 \times 10^5}{0.4} = 5.83 \times 10^5 \text{ MT/yr}$$

5. Using calcium as the component of interest:

$$2.33 \times 10^5 \text{ MT/yr } CaSO_4 \times \frac{40 \text{ MT}}{136 \text{ MT}} = 6.85 \times 10^4 \text{ MT/yr of Ca}$$

Limestone is $CaCO_3$, therefore limestone requirement is:

$$6.85 \times 10^4 \text{ MT/yr Ca} \times \frac{100 \text{ MT } CaCO_3}{40 \text{ MT Ca}}$$

$$= 1.71 \times 10^5 \text{ MT/yr as limestone}$$

In reality, the limestone requirement would be higher because limestone would not be 100% pure, nor would it be 100% utilized.

The cost of power from a coal-fired power plant depends on two factors: the capital investment (and its associated fixed charges) and the operating costs (primarily fuel costs). Although both sets of costs vary considerably depending on location and many other factors, the costs for construction of a 1000-megawatt plant in 1980 dollars is approximately $600/kW or $600 million total (Cooper, 1981). Despite this enormous investment, the cost of power from the coal-fired plant constructed in 1980 would be less than the cost for a comparably sized oil-fired power plant because of the large difference in the cost of the fuel.

As the size of the plant decreases, the capital costs per kilowatt increase rapidly because there are certain necessary facilities, such as coal rail car unloading and ash and sludge treatment/disposal facilities, which are required for each plant regardless of its size. Thus, larger plants can spread these fixed costs over a larger base and reduce the unit costs of construction. Furthermore, the costs of a coal-fired power plant or any large capital project depend on the timing of construction. During the latter part of the 1970s, inflation was an extremely important factor. Many projects that were initially estimated to cost in the range of $100–200 million finally cost 2 or 3 times that amount due to inflation between the time of conception of the project and time of actual construction.

In spite of large capital costs and even larger fuel costs, the cost of electricity could be reduced drastically if the overall thermal efficiency could be increased from its typical value of 35–40%. Why can't we do this? Some reasons are explored in the next section.

9-5 Some thermodynamic considerations

Heat engines

Basic thermodynamic considerations dictate that when heat is converted to useful work, the efficiency of the process is limited by the temperatures involved. The second law of thermodynamics may be stated thus: it is impossible to convert completely into useful work the heat which is absorbed in a cyclical process. This can be demonstrated through the analysis of a Carnot cycle, which was proposed by Sadi Carnot in 1824. Picture a perfect machine, a heat engine, operating between two temperatures such that it is transferring heat from a high-temperature region (T_H) to a low-temperature region (T_C) and in so doing, is doing some useful work. This heat engine cycle is depicted in Figure 9.9.

The ideal cyclic process is approximated by a power plant as follows (most of the following discussion and derivation of equations follows closely the approach of Smith and Van Ness, 1959):

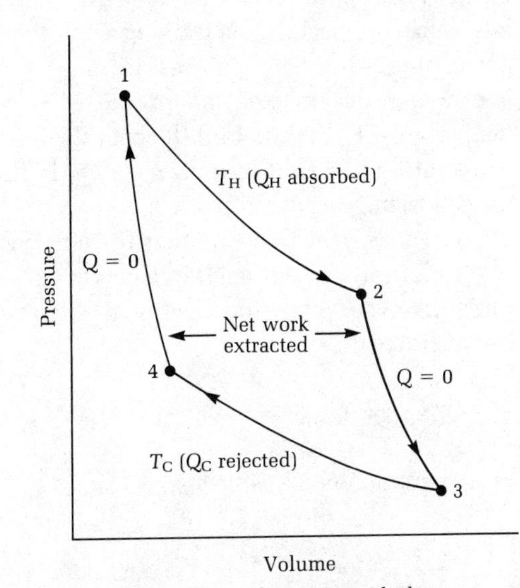

Figure 9.9 Carnot heat engine cycle using an ideal gas.

1. Part of the heat from combusting the fuel or heat from a nuclear reaction is used to boil water into high-pressure, high-temperature steam.
2. Useful work is obtained from this steam by turning a turbine.
3. The exhaust steam from the turbine is condensed at a low temperature and low pressure.
4. The condensate is pumped back to the boiler, thus completing the cycle.

In process 1, much heat is transferred into the steam; call this heat Q_H. In process 2, which is adiabatic (no heat transfer), energy as shaft work is extracted from the steam. In 3, an amount of heat Q_C is transferred to the cooling water. The last step is also adiabatic but requires the addition of a small amount of energy to the water (for pumping).

Since the cyclic process must have a net zero energy change,

$$\Delta E_{cycle} = 0 = Q_H - Q_C - W \tag{9.1}$$

or

$$Q_H - Q_C = W \tag{9.2}$$

where W is the net work taken from the steam during the whole cycle. Of the heat Q_H absorbed at the high temperature, part is converted into work and part, Q_C, is transferred at the low temperature. In a real power plant, the temperatures are not constant for processes 1 and 3; however, assume that all the heat Q_H is absorbed at a constant temperature T_H and that all the heat Q_C is transferred at the constant temperature T_C. The ratio of W to Q_H is the efficiency of the engine for converting heat into work.

Carnot's principle was that the efficiency of converting heat into work in an ideal heat engine depends only on the absolute temperatures involved. Since the fractional efficiency is W/Q_H, we can state this mathematically as

$$W/Q_H = f(T_H, T_C) \tag{9.3}$$

From equation (9.2), we get

$$\frac{Q_H - Q_C}{Q_H} = f(T_H, T_C) \tag{9.4}$$

or

$$Q_C/Q_H = 1 - f(T_H, T_C) = f'(T_H, T_C) \tag{9.5}$$

So, the ratio of heat released at T_C (the low temperature) to that absorbed at T_H (the high temperature) is a function only of the absolute temperatures. Thermodynamic analysis of two reversible heat engines in series leads to the conclusion that $f'(T_H, T_C)$ is nothing more than T_C/T_H. Thus, equation (9.5) becomes

$$Q_C/Q_H = T_C/T_H \tag{9.6}$$

and

$$W/Q_H = 1 - (T_C/T_H) \tag{9.7}$$

Equation (9.7) shows that the only conditions under which the efficiency of a heat engine can approach unity are those for which T_H approaches infinity or T_C approaches zero. (Note that the efficiency we are talking about is the thermal efficiency and not a ratio of actual work obtained to ideal or reversible work.) Since in the real world T_C never approaches absolute zero, and T_H does not approach infinity, the maximum possible thermal efficiency is much less than 1.0. In practice, T_H is limited by the properties of materials (roughly 1000°K for high-quality steel tubes in boilers producing steam at 2500 psi and 800°K), and T_C is limited by readily available cooling in the environment (about 300°K). The key point is that all heat engines must discard a certain amount of heat to the surroundings (usually to cooling water or to the atmosphere), and this discarded heat cannot be converted into work by the engine.

Example Problem 9.4

Calculate the maximum possible efficiency of a reversible heat engine operating between 800°K and 300°K.

Solution

$$\frac{W}{Q_H} = 1 - \frac{300}{800} = 0.625$$

Other sources of efficiency reductions

Another important distinction to make at this time is the difference between the higher heating value (HHV) and the lower heating value (LHV) of fuels. This difference comes about because of the way in which heating value is measured in a calorimeter. When a small amount of fuel is combusted in the calorimeter at the standard temperature, 25°C, the products of combustion are carbon dioxide gas and water as a liquid. In the real world, combustion processes result in carbon dioxide gas and water vapor. For several practical reasons, the combustion products cannot be cooled down enough to condense the water vapor and recover that heat, so the heat of vaporization of water is "lost." The heat that is available to the process is the lower heating value of the fuel. The lower heating value is more realistic for determining the useful heat that is available in a real-world process, but the higher heating value is much more widely known. Basing calculations on the higher heating value results in an inevitable reduction in thermal efficiency of about 10%.

Example Problem 9.5

Propane combustion in a calorimeter occurs according to the reaction

$$C_3H_8(g) + 5O_2(g) \longrightarrow 3CO_2(g) + 4H_2O(l)$$

From tests done in the calorimeter, the higher heating value of propane was determined to be 50.0 kJ/g. Calculate the lower heating value of propane given that the heat of vaporization of water is 2.45 kJ/g of H_2O.

Solution For 1 gram of propane combusted, 1.636 grams of water is produced. Thus, the HHV of propane includes the heat of vaporization of the water, which is $(1.636)(2.45) = 4.01$ kJ. So the LHV $= 50.0 - 4.01 = 46.0$ kJ/g propane.

From the above example problem, we can see that the LHV is 92% of the HHV for propane:

$$\frac{46.0}{50.0} = 0.92$$

Thus, the maximum possible thermal efficiency of any device using propane as fuel will be 92% if the efficiency is based on HHV. In the next example problem, it is shown that the efficiency of a real device must be lower than 92%.

Example Problem 9.6

An industrial furnace burns propane as a fuel. Calculate the overall thermal efficiency of the furnace (based on higher heating value) if the furnace is 95% efficient in transferring heat.

Solution From example 9.5, the LHV of propane is 46.0 kJ/g and the best thermal efficiency of a perfect device is 92%. The real efficiency of this furnace is at most

$$(0.95)(0.92) = 0.874 \text{ or } 87.4\%$$

Another reason that real processes do not operate at their theoretical maximum efficiency is that the combustion gases leaving a fossil-fuel power plant must leave at a temperature above the dew point of the acid gases (primarily SO_3) contained in the exhaust. If the SO_3 condenses with water vapor (which can happen below 125–150°C), then corrosion of heat-exchange equipment will be rapid and extensive. Therefore, these gases are generally released at temperatures above about 125–150°C to prevent excessive maintenance costs. Because of this, a sizable amount of sensible heat is lost and is not recoverable in a practical sense.

There is yet another reduction from the theoretical maximum efficiency. In order to ensure complete combustion of fuel, a small but significant amount of excess air is admitted to the furnace. The excess amount of air usually ranges from 5 to 15% in a well-controlled furnace but may be as high as 50–100% in a loosely controlled furnace. This excess air requires heat in order to reach the temperature of the exit gas stream. This is another thermodynamic loss. Overall, the actual thermal efficiency of a modern coal-fired power plant is approximately 40%, based on the higher heating value of the fuel, and the overall thermal efficiency of a nuclear power plant is roughly 32%. The following example problem illustrates the large amounts of heat involved with these thermal efficiencies.

Example Problem 9.7

The electricity demand of a certain region requires that a 2000-megawatt plant be built. Calculate the rate of heat released to the environment from a 40% efficient coal-fired plant and from a 32% efficient nuclear plant.

Solution The energy balance for this problem reduces to input rate = output rate for either plant. The output rate is the useful power plus the rate of heat released to the environment, and the input is determined by the thermal efficiency.

Mathematically:

$$E_{in} = E_{elec} + E_{heat}$$

and

$$E_{elec} = (\text{Efficiency}) \times E_{in}$$

For the coal-fired plant:

$$\begin{aligned} E_{heat} &= E_{elec}\left(\frac{1}{0.40} - 1\right) \\ &= 3000 \text{ MW} \\ &= 3.0 \times 10^9 \text{ J/s} = 2.46 \times 10^{11} \text{ BTU/day} \end{aligned}$$

For the nuclear plant:

$$\begin{aligned} E_{heat} &= E_{elec}\left(\frac{1}{0.32} - 1\right) \\ &= 4250 \text{ MW} \\ &= 4.25 \times 10^9 \text{ J/s} = 3.48 \times 10^{11} \text{ BTU/day} \end{aligned}$$

From the previous example problem, we see that the nuclear plant releases 41% more heat to the environment than the coal-fired plant. This difference is further exaggerated if neither plant has a cooling tower, but both use "once-through" cooling. All of the waste heat from the nuclear plant goes into the water, but only a portion of the waste heat from the coal-fired plant does. About 15% of the waste heat from a fossil-fuel plant goes directly to the atmosphere with the stack gases.

9-6 Alternative sources of energy

In outlining any plans for alternative energy sources, we must consider the present infrastructure, that is, how our society is presently established to produce, transport, and use various forms of energy. At present, our society is still very dependent on fossil fuels in general and on liquid fuels in particular. Ultimately, however, these fossil fuels will run out. Before that time, we need to have developed reliable alternate forms of energy in sufficient quantities to avoid hardships and dislocations. Therefore, in this section, we briefly describe some of the potential energy sources for the future and point out some of their advantages and disadvantages. In particular, we wish to address the present and future capabilities of the United States to utilize these different forms of energy.

Nuclear

Nuclear energy is derived from two major processes: fission and fusion. Fusion energy is still very much in the research stage, so we will not discuss fusion to any great extent. It will suffice to say that fusion involves the fusing of two light atomic nuclei, with the consequent release of enormous amounts of energy per unit mass. Fusion requires temperatures in the range of those in the sun, and thus the main technological challenge is to sustain a controlled fusion reaction.

Nuclear fission, as we saw in Table 9.3, is a major source of heat for conventional steam electric power plants. Fission is the splitting apart of heavy nuclei into fragments, usually radioactive, with the consequent release of large amounts of energy. As much energy is released by the complete fissioning of 1 gram of uranium-235 as by the complete combustion of 3.3 tons of coal or 700 gallons of gasoline. The energy released by the fission process is collected as heat by water that circulates through the core of the nuclear reactor. Eventually, that heat produces steam to turn a turbine.

Two types of reactors are used in this country, the boiling-water reactor (BWR) and the pressurized-water reactor (PWR). Basically, BWRs and PWRs operate on the same principles. The fuel is an oxide of uranium which has been fired in a ceramic fashion and formed into a pellet. Uranium-235 (U-235) is the fissionable isotope and is the only one that participates in the reactions. U-235 makes up approximately 0.7% of naturally occurring uranium. However, the uranium in the fuel pellets has been enriched by concentration of U-235 to 3–4%.

The main difference between the PWR and the BWR is in the way heat is transferred from the reactor to the steam system. In a BWR,

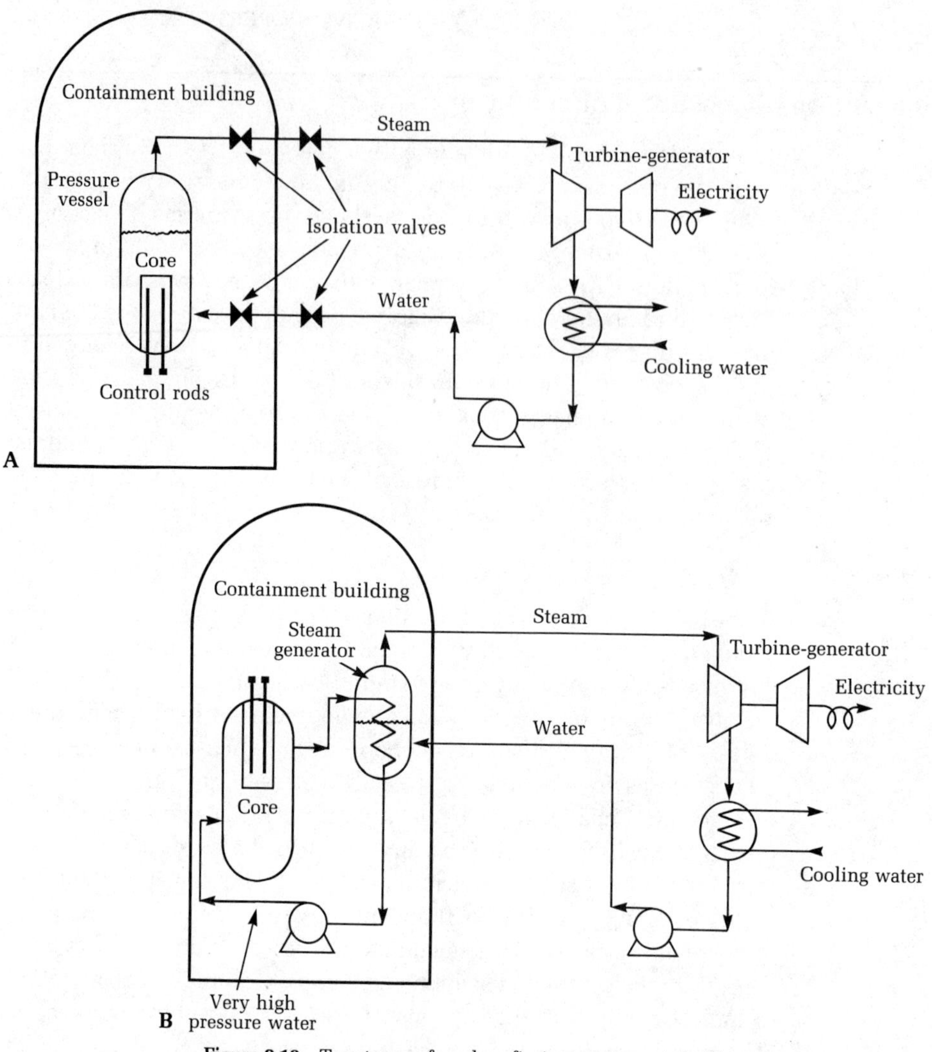

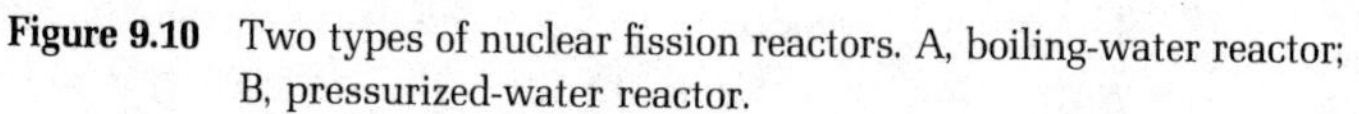

Figure 9.10 Two types of nuclear fission reactors. A, boiling-water reactor; B, pressurized-water reactor.

the water that circulates through the core boils directly. The steam then goes to the turbine generator system, is condensed by cooling water, and is recycled back into the core. In a PWR, the reactor water is pressurized so that when it goes through the hot core it will not boil. From the core the hot pressurized water passes into a heat exchanger where it causes a different stream of water to boil. The steam produced goes through a turbine, is condensed, and is recycled to this separate steam generator. These two types of reactors are

Figure 9.11 An aerial view of nuclear power plant under construction. The large cones are cooling towers. (Courtesy of Philadelphia Electric Company. Photo by Charles H. Peatross, Jr.)

pictured in Figure 9.10. An aerial photograph of a nuclear power plant under construction is shown in Figure 9.11.

Nuclear plants cost more and take longer to construct than coal plants, partly because of the additional regulatory measures that are involved. However, uranium as a fuel produces much more energy per unit cost than coal and so the lower cost of fuel tends to offset the higher capital cost of the plant.

Fission probably should not be listed as an alternative source of energy since it is a proven technology. As was seen previously, nuclear fission today provides a significant part of U.S. electrical energy. In 1976, nuclear power provided about 10% of the total U.S. electricity demand, and it was projected that it could supply 50% by 1990. However, after a mechanical breakdown at the Three Mile Island power plant in Harrisburg, Pennsylvania in March 1979, there apparently was a drastic change in thinking by the general public about the safety of nuclear energy. In addition, the government has been indecisive for years on the issues of radioactive waste handling and disposal. Now it remains to be seen if nuclear energy will resume its former growth potential. Despite the problems of nuclear power plants, they may well be less harmful in the long run than fossil fuel plants.

Hydroelectric power

Hydroelectric power is an extremely attractive form of energy because it produces no pollutants. Basically, hydroelectric power is generated by allowing water to fall through a spillway, turning a turbine which is coupled to an electric generator. Of course, in order to take advantage of hydropower, one must dam a river, which certainly has more than a minor impact on the local environment. However, hydropower does not entail the long-term continuous release of pollutants from burning a fossil fuel, nor the generation of radioactive wastes from a nuclear plant. The problem is simply that there are no more major dammable sites in the United States, so the potential for increased contributions of hydroelectric power to our total electric supply is small. Hydroelectric power contributed 12% of our 1980 electricity production.

Solar energy

Solar energy can be "captured" in two ways. First, solar energy is used to heat water or air in individual homes. This is useful from the point of view that a relatively dispersed, low-level source of energy (sunlight) can be collected, concentrated, and used on a local scale. When solar energy is used in this fashion, individuals cut back their use of higher level energy (electricity, gas, oil) in performing simple heating tasks. Solar energy has been used by industry in this fashion to generate steam for process use (*Chemical Week*, 1982). Solar energy is not always reliable in this form because several days in a row of rain or cloudy weather interrupts the supply. This is particularly true in northern latitudes, where heating demands are greater and insolation (the incident solar radiation) less. Thus, solar energy must usually be backed up by installed conventional heating, which is an extra cost.

The other major use of solar energy is as electricity. Here, there are two approaches. In the first, sunlight strikes a solar cell (which is formed from the junction of a P-type and N-type semiconductor) and causes a current to flow. The cell is connected to a device which can make use of direct current. Of course, in a commercial solar collector, many thousands of these individual cells would be hooked up in a series-parallel arrangement to provide the design voltage and current requirements. Photovoltaics are not yet commercialized for large-scale application; however, they are gaining wide use in specialized applications such as lighting or communication installations in remote places.

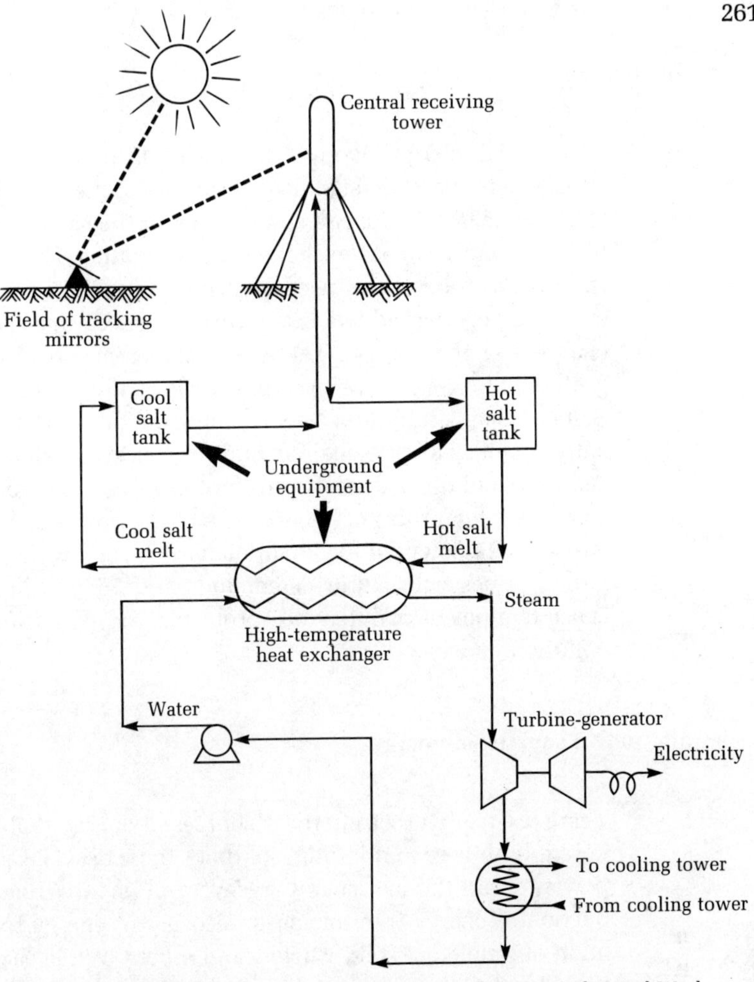

Figure 9.12 A solar energy system to generate electricity (*Chemical Week*, 1981).

The second approach to converting solar energy into electricity is to use a field of tracking mirrors to reflect the sun's rays to a central receiving tower. The concentrated solar radiation heats a working fluid such as molten salt to temperatures as high as 565°C (*Chemical Week*, 1981). The hot fluid is passed through a heat exchanger to boil water and produce steam, which then generates electricity via a conventional steam turbine cycle. This process is depicted in Figure 9.12.

It is difficult to predict future growth in the use of solar energy. In the near future it will probably not make a large-scale contribution, but for 20 years from now forecasts vary greatly. Certainly, as an essentially limitless, very clean source of energy, it holds great promise.

Geothermal energy

Geothermal energy involves drilling wells to an area of the earth's crust where there are significant deposits of steam or hot water. Molten rock (magma) deposits can be used in theory but would require a high-temperature heat exchanger underground. The fluid is piped to the surface where it is used to drive turbines in a standard power cycle. The condensed water is returned underground or it can be discharged on the surface. There are some commercial deposits where geothermal energy is being tapped, especially in the western United States, Iceland, Italy, and New Zealand, but on an overall basis, it presently provides a very small amount of energy, as shown in Table 9.3. As these and other deposits are further developed we may gain more significant quantities of energy. One of the problems with geothermal energy is its potential for pollution. Geothermal waters commonly are strong brines with sulfur compounds and mineral impurities, which cause equipment maintenance problems and may cause air or water pollution as well.

Other alternative sources of energy

There are many other alternate sources of energy that have been proposed, both here and in other parts of the world. These include wind power, ocean thermal energy, recovery of energy from the tides, production of energy from biomass, recovery of energy from the incineration of municipal solid wastes, and others. While each of these may provide a local supplement to conventional energy sources, at the present time none appears to be able to supply significant quantities for the United States. The one exception may be municipal solid waste (MSW) incineration. In large urban areas, MSW incineration is often attractive economically, due to higher costs of conventional land disposal. However, recent research has shown that usually MSW incineration is economically unattractive for electricity generation unless low-pressure process steam can also be sold to nearby businesses or industries. MSW incineration will be discussed more fully in the next chapter.

Lately, one hears of synthetic fuels as an alternate energy source. Synthetic fuels really should be classified as a conversion of one fossil fuel to another. Typically, "synthetic fuels" refers to liquid or gaseous fuels, produced from coal.

9-7 Synthetic fuels

The sharply rising world prices for oil and natural gas in recent years has sparked renewed interest in developing substitutes derived from coal, oil shale or other non-conventional fossil fuels. Over 75 synfuels projects were proposed to the U.S. Synfuels Corporation for assistance in funding in 1980–81; however, many of these were cancelled when funding was denied. Several of the projects are sufficiently well planned and designed at this stage that it seems possible that three to six of these coal-based synfuel projects will be operational by 1990.

The major factors to be considered in any of these projects are the desired products, the technology with which to produce them, the capital and operating costs involved, and environmental effects. Some of the coal-based projects are designed primarily to produce medium and high-BTU-content gas for use as a clean-burning fuel by industry or by a steam electric power plant. Other projects involve liquefying the coal to produce such products as gasoline, heating oil, other fuel oils, and petrochemical feedstocks. Most of the technology is still in the developmental stage; however, coal gasification today appears more feasible than liquefaction.

Coal gasification

There are several coal gasification projects currently proposed or under construction in the U.S. today. Some will produce a medium-BTU gas for industrial burning; however, others are planned that will produce liquid fuels and/or chemicals. Unit processes have been developed that can be added onto a coal-gasification project to produce methanol and even a high-quality unleaded gasoline. An example of a process block diagram showing these operations is in Figure 9.13.

The major reactions in gasification involve heating the coal in the presence of a limited amount of oxygen to form carbon monoxide. Some of the carbon monoxide is reacted with water to produce carbon dioxide plus hydrogen. The hydrogen and carbon monoxide are concentrated and can be reacted to further products or can be burned directly as fuel. The advantages in gasification lie in producing clean, easy-to-use fuels, allowing a downstream user to burn them without pollution-control equipment. All of the ash and sulfur in the coal are removed in the gasification process, and the sulfur can be sold as a by-product.

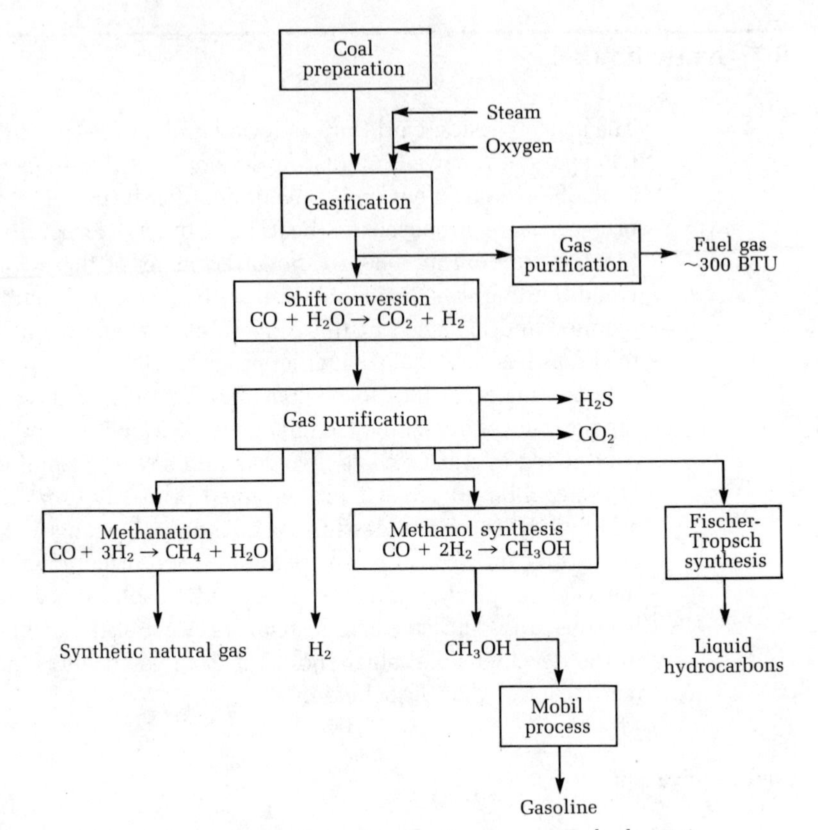

Figure 9.13 Options in the coal gasification process (Fryback, 1981).

Coal liquefaction

Several competing processes are currently under development for coal liquefaction. In one of these processes, the H-coal process (Eccles and De Vaux, 1981), the coal is crushed and slurried with an oil, pumped to high pressure (about 200 atm) and reacted with hydrogen in the presence of a catalyst. The products of liquefaction are typically a gasoline-type stream, a fuel-oil-type stream, a heavier oil, and in some processes, a solid coke product. In the liquefaction process, the products are not as clean as with gasification because some sulfur and ash remain in them. Example projects include that sponsored by W. R. Grace in Kentucky to produce gasoline, by Tennessee Eastman in Kingsport, Tennessee to produce a chemical feedstock, and by the Koppers Company in Oakridge, Tennessee to produce gasoline. A typical process flow sheet is presented in Figure 9.14.

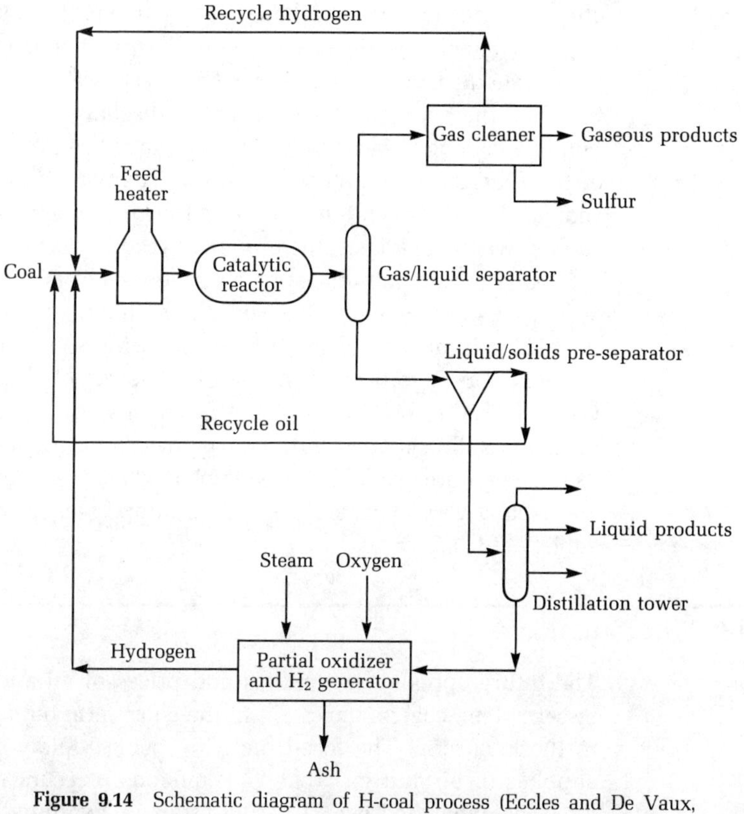

Figure 9.14 Schematic diagram of H-coal process (Eccles and De Vaux, 1981).

9-8 Pollution problems associated with energy processing

Water-pollution problems associated with power plants, refineries, synfuels plants, and so forth include thermal pollution, suspended particulates, dissolved chemicals, and toxic materials. The control measures for suspended particulates and specific dissolved chemicals are fairly standard. Such things as settling ponds and filters are easily applied technology that can be used in a specific situation as needed. Ion exchange and chemical neutralization and precipitation can be used to remove specific contaminants. Some of these processes are discussed elsewhere in this text.

One of the problems with large energy-conversion plants is thermal pollution, the discharge of excessive amounts of heat. If the plant is located on a river, the river may be locally heated and aquatic life affected in the region immediately downstream. If the plant is located on a bay or inlet on the coast, the local ecosystem may be altered by the heated discharge water. As we have seen, waste heat is inevitable; however, there are choices of where to discharge the heat. One solution is a cooling tower. Cooling towers have been pictured in several of the schematic flow diagrams in this chapter. A cooling tower exchanges heat between the water and the atmosphere and recycles the cooling water. The heat, then, is more readily dissipated throughout a large region in the atmosphere and does not affect any local ecosystems, as it does with direct discharge into a body of water.

Air pollution problems from energy-processing facilities are potentially very significant because of the sizes of the operations (as we have seen in previous examples). The major air pollutants of concern are sulfur oxides, nitrogen oxides, and particulates. Control technologies are available but are not 100% efficient. In addition, they are usually very expensive to build and operate. Air pollution is the subject of Chapter 10.

9-9 Future outlook

The future outlook for supplies and prices of oil and gas and, to a lesser extent, coal, is uncertain in the short term but apparently clear in the long term. The fossil fuels are being rapidly depleted. Finite supplies of oil and gas that took millions of years to produce will likely be exhausted over a period of about two hundred years (mid-1800s to perhaps the middle of the next century). Coal may last somewhat longer.

Obviously, in our economic system, as the supply of a commodity declines, its price increases. In the 1970s the world experienced two periods when demand briefly exceeded supply of oil, and prices rose dramatically both times. These brief instances are but a foreshadowing of things to come. However, the key question is when exactly will the severe shortages begin? The answer is not clear.

In the short view, supplies and prices are subject to severe swings with local new discoveries and particularly with the world economy. Currently, there is an excess supply of oil, created by a combination of factors in the 1979–83 period: excessive price increases, a sudden drop in supplies caused by the Iranian revolution, a worldwide recession, a marked effort by consumers to conserve, and development of other fuels. The result has been a dramatic cutback in oil production from the foreign oil cartel and price cuts. In 1979, forecasts were for constantly increasing oil prices, but now most experts are predicting stable supplies and prices through the year 2000 (Cook, 1981).

However, the apparent good news for oil-consuming nations may not be the blessing that it seems. Already we are witnessing the demise of many of the synthetic fuel projects which seemed so urgent in the late 1970s. Certainly we should use this respite to plan alternate fuel projects carefully and to choose them on a cost-effective basis. However, we should be careful not to completely abandon synfuels and other alternate energy sources such as solar and nuclear fusion, just because oil and gas are cheaper now. If we are not careful, our reprieve may run out on us.

9-10 Problems

9.1 The U.S. energy consumption in 1980 was 83 quads, of which oil supplied about 45%. The United States imported about 38% of all its oil needs in 1980. If the average price of imported oil in 1980 was $34 per barrel, calculate the total cash outflow in 1980 to foreign countries due to imported oil.

9.2 In 1977, the United States produced (and used) 7.2 MMB/D of gasoline. If the specific gravity of gasoline is 0.7, calculate the mass of sulfur sold in the gasoline in 1977. Assume that the sulfur content of gasoline was 0.05%. If all the sulfur produced SO_2 when the gasoline was burned in autos, calculate the emissions of SO_2 from this source (in kg).

9.3 In 1977, the United States produced 1.75 MMB/D of residual (No. 6) fuel oil and imported another 1.36 MMB/D. If the average weight percent of sulfur in the imported oil was 1.2%, and the sulfur content of domestic No. 6 was 1.6%, and the average densities of both domestic and imported were 330 lb/B, calculate the mass of sulfur sold in residual in 1977. If all the sulfur produced SO_2, calculate the emissions of SO_2 from this source (in kg).

9.4 What is the net power output for the following series of operations? Give your answer in kW.

Natural gas is burned in a furnace at the rate of 1000 standard m^3 per hour. The gas has an energy content of 1.066×10^4 kWh/m^3. The furnace is 87% efficient at transferring heat to steam. The steam is transported via a pipe with 3% heat losses to a turbine. The turbine is 56% efficient in converting the flow of heat from the steam into mechanical power of turning a generator. The generator is 98% efficient in converting the mechanical power into electricity.

9.5 A coal-fired power plant produces 400 MW of electrical power. Twelve percent of the input energy goes out the stack with the exit gases, 5%

is consumed internally, and 43% goes out with the cooling water; the rest of the input energy comes out as electricity.

1. Draw a diagram representing the energy flows for this plant. Include numerical values for each energy stream (units of MW).
2. The coal used in this plant has a heating value of 25,000 kJ/kg. Calculate the input rate of coal in kg/hr.
3. Calculate the amount of heat discharged to the water in one day (24 hours). Give your answer in kJ.

9.6 What percentage of U.S. electricity demand was satisfied by hydropower in 1980? By nuclear?

9.7 Calculate the percentage increase in electrical energy generated from coal from 1978–80 and the percent decrease in electricity generated from oil.

9.8 Calculate the maximum possible efficiency of a geothermal system to generate electricity which operates between temperatures of 700°F and 100°F.

9.9 Calculate the maximum possible efficiency of an ocean thermal power generation system which operates between 80°F and 40°F.

9.10 Calculate the maximum possible efficiency of a solar energy power generation system (see Figure 9.12) which achieves a temperature of 1000°F and exhausts heat at 100°F.

9.11 Show by calculation why a nuclear power plant is inherently less efficient than a fossil-fuel power plant. The maximum steam temperature in a nuclear plant is limited by the materials in the core to about 590°K while the maximum steam temperature in a fossil fuel plant is limited by the boiler tubes to about 810°K. The cold temperature (limited by environmental coolants) is about 300°K in both cases.

9.12 Consider the gigantic 2000-MW plant of example problem 9.7. Assume that 20% of the heat losses of the coal-fired plant go directly to the atmosphere. Calculate the cooling water circulation rate required for the coal-fired plant and the nuclear plant. Assume that cooling water is available at 85°F and can be returned at 115°F. If the losses of water from cooling tower evaporation and blowdown equal just 2% of the daily circulation rate, calculate the daily loss of water that must be resupplied to the cooling system for the coal and the nuclear plant.

9.13 A utility has proposed a 400-MW coal-fired power plant at an inland location. Coal would be brought in by unit trains. An average unit train is 100 cars long and each car carries 80 metric tons of coal. If the plant is 40% efficient and the coal has a heat content of 22,000 kJ/kg, calculate how many unit trains are needed per year to supply this plant if it runs at an average of 90% of its rated capacity each year.

9.14 The electricity consumption in the average American home is about

11,000 kWh per year. Assuming an average thermal efficiency of 37% for the power plant, calculate the annual fuel required to generate this amount of electricity. Give answers for Illinois bituminous coal (pounds), No. 6 fuel oil (gallons) (assume HHV similar to crude oil), and natural gas (standard cubic feet). Refer to Table 3.1 for heating value.

References

Chemical Week. 1981. Molten-salt solar units are getting a closer look. November 25, 1981.

Chemical Week. 1982. Tapping the sun for process steam. September 22, 1982.

Chemical Week. 1983. Washington newsletter. March 23, 1983.

Cook, W. J. 1981. Why oil prices won't go up. *Newsweek*, September 14, 1981: 74.

Cooper, C. D. 1981. Evaluation of the feasibility of a coal-fired electricity generating plant/municipal wastes incinerator for the City of Kissimmee. Orlando: University of Central Florida.

Eccles, R. M. and De Vaux, G. R. 1981. Current status of H-coal commercialization. *Chemical Engineering Progress* 77 (5): 80.

Exxon Co. 1977. *Energy Outlook: 1978–1990.* Houston: Exxon.

Fryback, M. G. 1981. Synthetic fuels: Promises and problems. *Chemical Engineering Progress* 77 (5): 39.

Smith, J. M. and Van Ness, H. C. 1959. *Introduction to Chemical Engineering Thermodynamics*, 2nd ed. New York: McGraw-Hill.

U.S. Department of Energy. 1981. *Monthly Energy Review.* U.S. Department of Energy, DOE/EIA-0035 (81/05).

10

Air Pollution

When our ancestors first discovered fire they also "discovered" air pollution, particularly if that first fire was built inside a cave without a chimney! Air pollution is tied directly to the use of energy and the concentration in cities of increasing numbers of people. Although air pollution is a historic problem—the use of coal in London was banned temporarily in 1272 to help clean up the city's air (Wark and Warner, 1981)—air pollution did not become a global problem until the advent of the industrial revolution.

Since the early 1800s, human population has increased from about 0.5 billion to over 4 billion, or about one order of magnitude. World energy consumption has increased perhaps by several orders of magnitude, with much greater than average increases occurring in the industrialized centers. With such great increases in the *rate of emissions*, and with no change in the atmosphere's natural *rates of dispersion and dilution*, more frequent *accumulations* of pollution have occurred, resulting in occasionally harmful levels of pollutants in the air. (Note the appropriateness of the basic concepts of materials balance.) In this chapter we briefly address some major air pollution problems facing the industrialized world and review the basics of some engineering approaches to the control of these problems.

10-1 The major air pollutants

There are five major primary air pollutants and one secondary one (so called because it is derived, by photochemical reaction in the atmosphere, from two of the primary pollutants). The five major primary pollutants are particulates, sulfur oxides (SO_x), nitrogen oxides (NO_x), volatile organic compounds (VOC), and carbon monoxide (CO). The major secondary pollutant is termed photochemical oxidants (measured as ozone).

Particulates are very small diameter solids or liquids which remain suspended in exhaust gases and can be discharged into the atmosphere. Particulates are emitted primarily from fossil-fuel combustion sources and mineral processing. Particulate effects include reduction in visibility due to smog and haze, and soiling of buildings and other materials. Also, particulates have been connected with human health effects because they can be breathed in and retained in the lungs.

Sulfur oxides are produced whenever a fuel that contains sulfur is burned. The main source is fossil-fuel combustion, although nonferrous metal smelting is also a major source. The primary sulfur oxide is SO_2; however, some SO_3 is formed in the furnace. Also, it is thought that SO_2 is oxidized to SO_3 in the atmosphere. The SO_2 and SO_3 can form acids when they hydrolyze with water, and the acid can then have detrimental effects on the environment. In addition, SO_2 has been associated with human health problems directly and can cause damage to plants and animals.

Nitrogen oxides are formed whenever any fuel is burned in air at a high enough temperature. The nitrogen and oxygen in the air combine to form NO and NO_2. Also, nitrogen atoms present in some fuels can contribute to nitrogen oxide (NO_x) emissions. Total U.S. emissions of NO_x are about equally distributed between stationary sources and mobile sources (vehicles). Nitrogen oxides contribute to smog and haze, can be injurious to plants and animals, and can have an effect on human health. Nitrogen oxides have also been implicated in acid precipitation. Furthermore, NO_x reacts with VOC in the presence of sunlight to form photochemical oxidants.

Volatile organic compounds (VOC) include any organics with an appreciable vapor pressure such that they vaporize when exposed to air. The major sources of VOC are automobiles and other mobile sources, from which small amounts of unburned fuel are exhausted to the air. Petroleum production, refining, and marketing account for substantial VOC emissions, as does evaporation of solvents (such as those in oil-based paints). Some VOCs are carcinogenic, but the major problem with VOCs is that they react in the atmosphere to form oxidants.

Carbon monoxide is a colorless, odorless, tasteless, poisonous gas that results from the incomplete combustion of any carbonaceous fuel. Usually, power plants and other large furnaces are designed and operated carefully enough to ensure fairly complete combustion and do not emit much CO. Thus, the major sources are transportation: automobiles, trucks, buses, airplanes, and other vehicles exhausted over 85 million metric tons of CO in 1977 (Wark and Warner, 1981). CO reacts with the hemoglobin in blood to prevent oxygen transfer. Depending on the concentration of CO and length of exposure, effects on humans may range from slight impairment of some psychomotor functions to death.

Photochemical oxidants are not emitted from a source per se, but are formed by complex reactions in the atmosphere involving VOC and NO_x and sunlight. Typically, photochemical smog can be a problem in urban areas in warm climates. The ozone and other oxidants that form by the photochemical reactions attack plants and materials, and cause eye, nose, and throat irritation. The discussion of the previous paragraphs is summarized in Table 10.1. In Table 10.2, consolidated estimates of U.S. emissions of air pollutants are presented.

One gaseous "pollutant" not mentioned above that is produced in great quantity by fuel combustion is carbon dioxide (CO_2). Because CO_2 has no short-term toxic or irritating effects, and because it is abundant in the atmosphere and is necessary to plant life, it is normally not considered to be a pollutant. However, from material balance considerations, we know that excess emissions of CO_2 into the atmosphere will result in an increase in its overall concentration. It has been projected that an increase in the mean atmospheric CO_2 concentration from its present value of about 340 ppm to about 600 ppm will occur by the early part of the next century. The potential consequences of the global warming which will likely result from such an increase include dramatic effects on the earth's climate and weather conditions (Hansen and others, 1981).

10-2 Atmospheric dispersion

The release of pollutants into the atmosphere is a time-honored technique for "disposing" of them. One of the reasons that our local air did not become completely unusable long ago is the atmosphere's ability to quickly disperse high concentrations of pollutants to lower,

TABLE 10.1 Summary of major air pollutants—causes, sources, and effects.

Pollutant	Cause	Source	Effects on: Human health	Plants, animals	Materials	Other
Particulates	Burning coal, crushing, grinding	Power plant boilers, construction industrial processes	Bronchitis, emphysema, cancer, etc.	Damage to leaf structure, toxic effects	Soiling, corrosion	Haze, smog
SO_x	Burning any fuel with sulfur; processes using liquid SO_2 or H_2SO_4	Boilers and furnaces, industrial processes, smelting	Acts with particulates (synergism)	Necrosis, chlorosis	Corrosion	Forms H_2SO_4 in atmosphere, haze, smog
VOC	Incomplete burning of fuels; evaporation of solvents	Motor vehicles, industrial processes	Small	Small	None	Reacts with NO_x to form photochemical smog
NO_x	Reaction of N_2 with O_2 at high temperature	Boilers, furnaces, vehicles	Respiratory irritant	Small	Corrosion	Reacts with VOC to form photochemical smog and forms HNO_3 in atmosphere
CO	Incomplete combustion of carbon fuels	Vehicles and industrial processes	Poisonous; reacts with blood hemoglobin	None on plants, deadly to animals	None	Eventually oxidizes to CO_2
Photochemical oxidants; ozone	Chemical reaction of VOC, NO_x, and sunlight		Irritating and damaging to lungs, eyes, nose, and throat	Severe damage to plants	Corrosion, oxidation, bleaching	Smog formation

TABLE 10.2 National U.S. emission estimates, 1975 (10^6 tons/yr).

Source category	*Particulates*	*SO_x*	*NO_x*	*VOC*	*CO*
Transportation (total)	1.3	0.8	10.7	11.7	77.4
Highway	0.9	0.4	8.2	10.0	67.8
Non-highway	0.4	0.4	2.5	1.7	9.6
Stationary fuel combustion (total)	6.6	26.3	12.4	1.4	1.2
Electric utilities	3.5	21.0	6.8	0.1	0.3
Other	3.1	5.3	5.6	1.3	0.9
Industrial processes (total)	8.7	5.7	0.7	3.5	9.4
Chemicals	0.2	1.0	0.3	1.6	3.3
Petroleum refining	0.1	0.9	0.3	0.9	2.2
Metals	1.3	3.2	0	0.2	2.8
Mineral products	4.5	0.6	0.1	0	0
Other	2.6	<0.1	<0.1	0.8	1.1
Solid waste (total)	0.6	<0.1	0.2	0.9	3.3
Miscellaneous (total)	0.8	0.1	0.2	13.4	4.9
Forest wildfires	0.4	0	0.1	0.6	3.3
Forest managed burning	0.1	0	<0.1	0.2	0.5
Agricultural burning	0.1	0	<0.1	0.1	0.6
Coal refuse burning	0.1	0.1	0.1	0.1	0.3
Structural fires	0.1	0	<0.1	<0.1	0.1
Organic solvents	0	0	0	8.3	0
Oil and gas production and marketing	0	0	0	4.2	0
Total	18.0	32.9	24.2	30.9	96.2

From Environmental Protection Agency (1976).

more or less harmless concentrations. Of course, the atmosphere's ability to disperse pollutants is not infinite and varies from quite good to quite poor depending on meteorological and geographical conditions. Thus, with the onset of the industrial revolution and the subsequent exponential growth of energy consumption, we have had to begin installing pollution-control equipment to prevent indiscriminate abuse of the atmosphere. However, we still rely heavily on atmospheric dispersion for final disposal of many pollutants.

Harmful effects occur when pollutants build up to high concentrations in a local area. The accumulation of pollutants in any localized region is a function of emission (input) rates, dispersion (output) rates, and generation or destruction rates (by chemical reaction). The dispersion of pollutants is almost entirely due to natural conditions like geography and local meteorological conditions such as wind, rainfall, and atmospheric stability (the tendency of air to not mix in the vertical direction). If there is a strong wind or good vertical mixing in the atmosphere, pollutants will be dispersed quickly into a large volume of air, resulting in low concentrations. If the wind is weak and there is an inversion (a layer of warm air above a layer of cooler air, which prevents the vertical mixing of air in the region), then pollutant concentrations can increase.

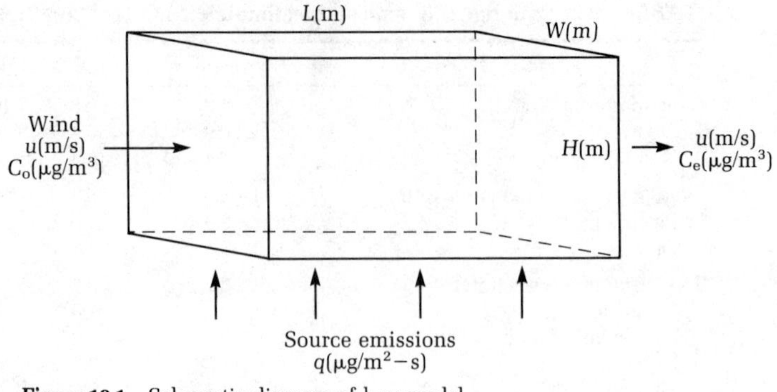

Figure 10.1 Schematic diagram of box model.

The box model

The simplest model that can be developed to predict pollutant concentrations under an inversion is called a box model (Wanta, 1968). In this model we picture an area being covered by a rectangular box (see Figure 10.1). The height of the box is determined by the inversion layer and the box is aligned with the wind direction. Pollutants can be carried into the box from upwind or be emitted from ground-based sources. Pollutants are carried out of the box by the wind. If the air in the box is assumed to be well mixed (like a CSTR), then the analysis is straightforward. At steady-state conditions, a material balance on a non-reactive pollutant yields

$$0 = uWHC_o + qLW - uWHC_e \tag{10.1}$$

where terms are as depicted in Figure 10.1. Equation (10.1) can be solved for the outlet concentration as follows:

$$C_e = C_o + \frac{qL}{uH} \tag{10.2}$$

Example Problem 10.1

A town is covered by an inversion layer 500 m above the ground. The town's area is approximately rectangular with dimensions 9 km by 12 km. The incoming air is clean and the wind is blowing at 3 m/s parallel with the 12-km side of the town. The emissions of SO_2 in this town come from a number of small sources and average

0.008 mg/m^2-s. Calculate the steady-state concentration of SO_2 in the air surrounding the town.

Solution
$$C_e = 0 + \frac{3 \times 12{,}000}{3 \times 500} = 64\ \mu g/m^3$$

Another use of materials balances to calculate air pollutant concentrations is given in the following example. Note, however, that because of the physical situation we cannot assume that the air is well mixed throughout the length of the tunnel. We shall have to make our balance on a differential element as shown below.

Example Problem 10.2

A 2000-foot one-way two-lane rectangular tunnel filled with cars is blocked by an accident at the exit. The longitudinal ventilation system involves one large supply fan at the entrance. Derive an equation to predict the steady-state concentration of CO as a function of position in the tunnel. Calculate the distance where the CO concentration surpasses 40 mg/m^3. Assume that the tunnel fills with cars, all with engines idling. One car occupies 20 linear feet in the tunnel and emits 2.8 grams per minute of CO with the engine idling. The tunnel is 30 feet wide and 12 feet tall. The fan blows 100,000 ft^3/min of fresh air into the tunnel entrance. The fresh air has 2.0 mg/m^3 of CO in it.

Solution The problem is diagrammed schematically below.

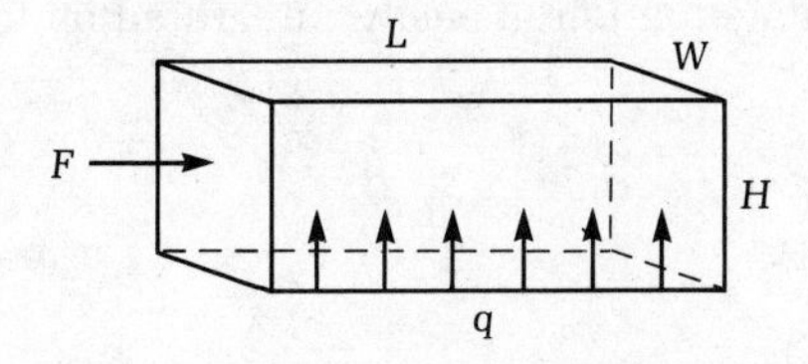

Assume that emissions q can be represented as a continuous distributed source along the length of the tunnel.

$$q = \frac{1 \text{ car}}{\text{lane-20 ft}} \times 2 \text{ lanes} \times \frac{2.8 \text{ g CO}}{\text{minute-car}} = \frac{0.28 \text{ g CO}}{\text{ft-minute}}$$

Choose an arbitrary volume element in the tunnel located x feet from the entrance.

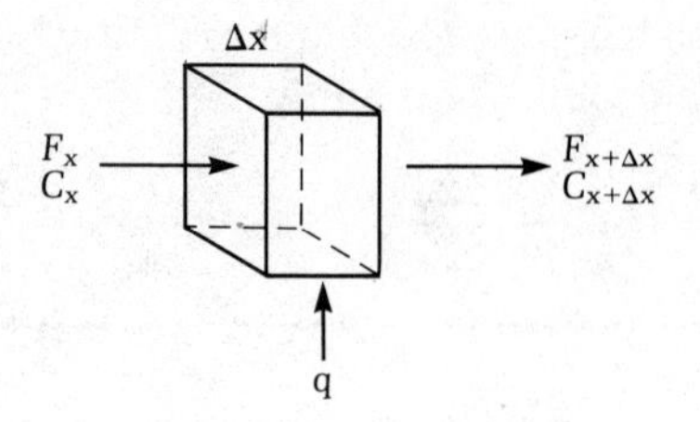

A steady-state material balance for CO on this volume element (assuming it is well-mixed vertically and horizontally) is

$$0 = F_x C_x + q\Delta x - F_{x+\Delta x} C_{x+\Delta x}$$

Since F is essentially constant, we get

$$-F(C_x - C_{x+\Delta x}) = q\Delta x$$

Dividing by Δx we get

$$F\left(\frac{C_{x+\Delta x} - C_x}{\Delta x}\right) = q$$

Taking the limit as $\Delta x \to 0$ yields the definition of a derivative:

$$F\left(\frac{dC}{dx}\right) = q$$

This equation is easily integrated as follows:

$$F\int_{C_0}^{C_L} dC = q\int_0^L dx$$

or

$$F(C_L - C_0) = qL \quad \text{or} \quad C_L = C_0 + \frac{q}{F}L$$

Solving for L, we get

$$L = F/q(C_L - C_0)$$

Substituting:

$$L = 100{,}000\,\frac{\text{ft}^3}{\text{min}}\,(40 - 2)\ \text{mg/m}^3 \times \frac{1\ \text{m}^3}{35.3\ \text{ft}^3} \Bigg/ \left(\frac{0.28\ \text{g}}{\text{ft-min}} \times \frac{1000\ \text{mg}}{\text{g}}\right)$$

$$= 384\ \text{feet}$$

The Gaussian dispersion model

Releases of large amounts of pollution (as from a large power plant) are directed through tall stacks to allow some "time and space" for the pollutants to disperse before reaching ground level (the ground-level case is important to us because that is where we live). Although the equations to describe the dispersion of pollutants in the open atmosphere can be developed from materials-balance considerations, it is somewhat beyond the scope of this text to do so. Rather, we will qualitatively describe this process and at the end present the equation that is widely used to calculate concentrations at points downwind from the source.

Pollutants released into the atmosphere travel with the mean wind. They also spread out in both the horizontal and vertical directions from the center line of the plume. The rate of spread in each direction is a complex function of micrometeorological conditions, the characteristics of the pollutants, and local geographical and cultural features. Figure 10.2 schematically portrays the behavior of a plume as it it emitted into the air and is bent over by a steady wind.

It has been shown that the spread of pollutants can be approximated by a Gaussian or normal distribution (Turner, 1970). The equation that models the normal dispersion of a gaseous pollutant from an elevated source is given below in a form that predicts the steady-state concentration at a point (x, y, z):

$$C = \frac{Q}{2\pi u \sigma_y \sigma_z} \exp\left(-\frac{1}{2}\frac{y^2}{\sigma_y{}^2}\right)\left(\exp\left(-\frac{1}{2}\frac{(z-H)^2}{\sigma_z{}^2}\right) + \exp\left(-\frac{1}{2}\frac{(z+H)^2}{\sigma_z{}^2}\right)\right) \qquad (10.3)$$

where

C = steady-state concentration at a point (x, y, z), $\mu\text{g/m}^3$

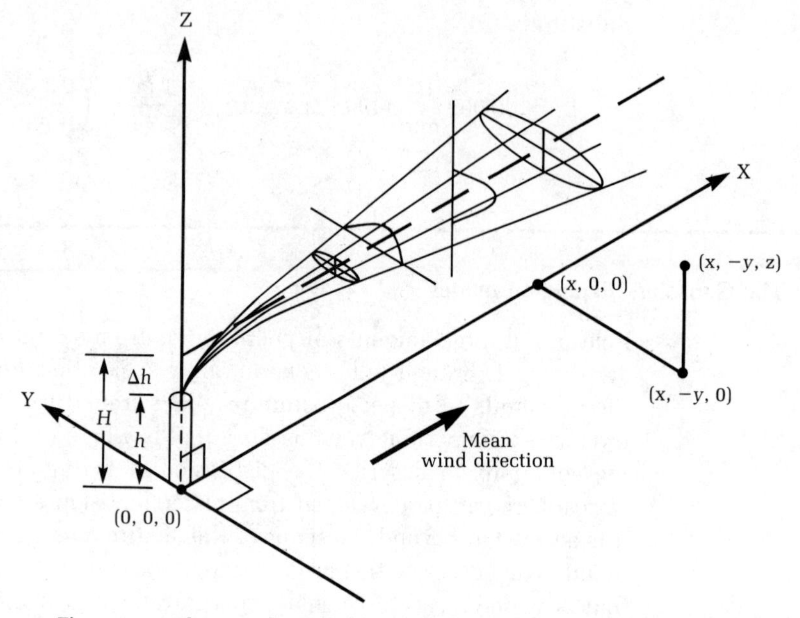

Figure 10.2 The spreading of a bent-over plume, showing coordinate axes (from Turner, 1970).

Q = emissions rate, μg/s
σ_y, σ_z = horizontal and vertical spread parameters, m (these are functions of distance x and atmospheric stability)
u = average wind speed at stack height, m/s
y = horizontal distance from plume center line, m
z = vertical distance from ground level, m
H = effective stack height ($H = h + \Delta h$, where h = physical stack height and Δh = plume rise), m

Refer to any current text on air pollution (Wark and Warner, 1981) for a review of the development of equation (10.3) and the assumptions implicit in its use. It is important to keep in mind some general relationships indicated by equation (10.3):

1. The downwind concentration at any location is directly proportional to the source strength Q.
2. The downwind ground-level concentration is generally inversely proportional to wind speed. (H also depends on wind speed in a complicated fashion which prevents a strict inverse proportionality.)

3. Because σ_y and σ_z increase as the downwind distance x increases, the plume center-line concentration continuously declines with increasing x. However, *ground level* concentrations increase, go through a maximum, and then decrease as one moves away from the stack.
4. The dispersion parameters σ_y and σ_z increase with increasing atmospheric turbulence (instability). Thus, unstable conditions decrease downwind concentrations (on the average).
5. The maximum ground-level concentration calculated from equation (10.3) decreases as effective stack height increases. The distance from the stack at which the maximum concentration occurs also increases.

The Gaussian dispersion equation is extremely important in air-pollution work. It is the basis for almost all of the computer programs developed by the U.S. Environmental Protection Agency for atmospheric dispersion modeling. For this reason, a brief discussion of atmospheric stability and an illustration of the use of the Gaussian equation are presented below. However, there is not enough space here to discuss micrometeorology or the many applications of equation (10.3) in any detail. The interested reader is referred to the references for further information.

Atmospheric stability

Air is termed unstable when there is good vertical mixing. This occurs when there is strong insolation and consequent heating of the layers of air near the ground (Williamson, 1973). The warm air rises and is replaced by cooler air, which in turn is heated and rises. This process is good for diluting pollutants and carrying them away from the ground.

Stable air results when the surface of the earth is cooler than the air above it (such as on a clear, cool night). Then the layers of air next to the earth are cooled and no vertical mixing can occur. In stable air pollutants tend to stay near the ground.

For convenience, atmospheric stability has been broken into six categories, arbitrarily labeled A through F, with A being the most unstable. Previous researchers have correlated values of the dispersion coefficients with atmospheric stability and, of course, distance away

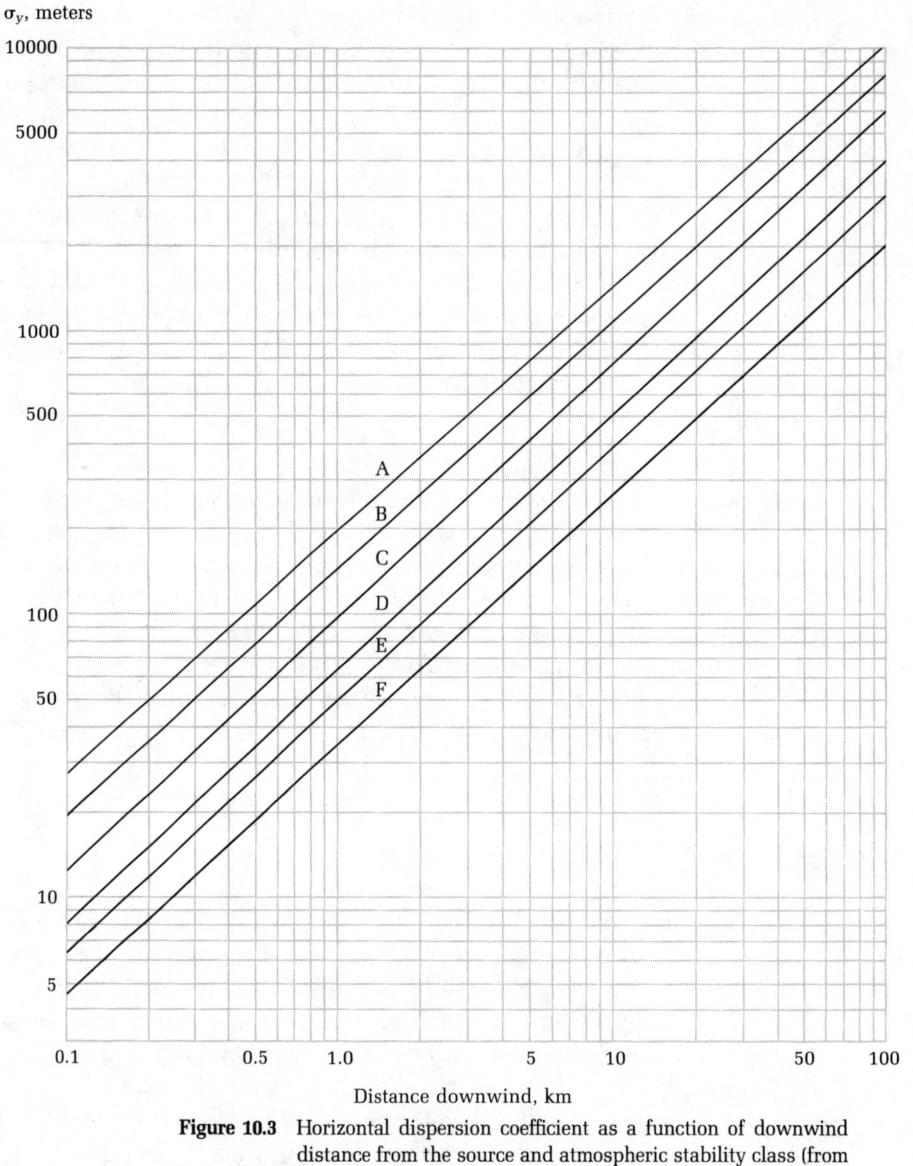

Figure 10.3 Horizontal dispersion coefficient as a function of downwind distance from the source and atmospheric stability class (from Turner, 1970).

from the source. These correlations are presented in Figures 10.3 and 10.4.

There are ways to estimate the stability class of the atmosphere based on the angle of the sun and extent of cloud cover. Once the stability classification is determined, then Figures 10.3 and 10.4 can be

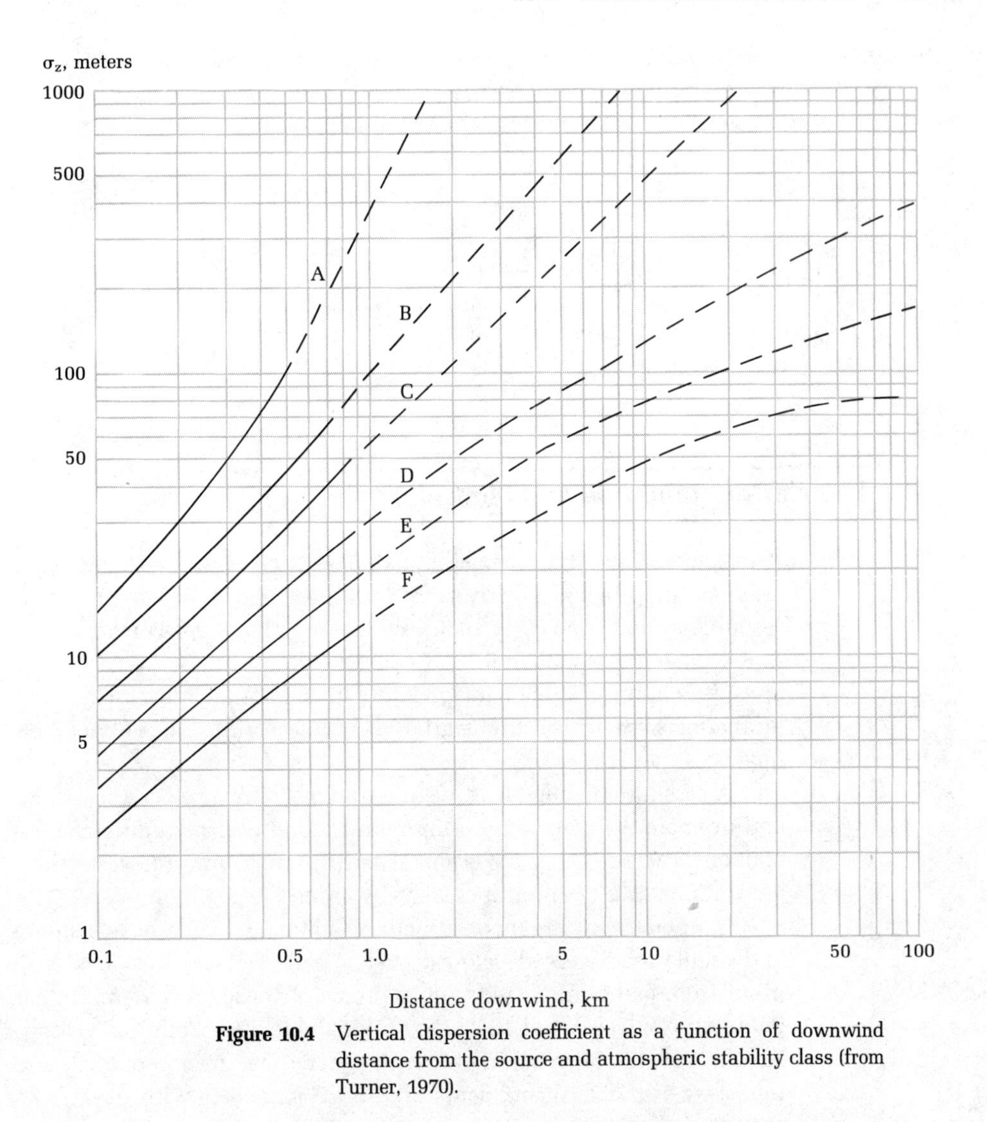

Figure 10.4 Vertical dispersion coefficient as a function of downwind distance from the source and atmospheric stability class (from Turner, 1970).

used to obtain values for the dispersion parameters, and equation (10.3) can be used to predict the concentration anywhere downwind from the source.

Example Problem 10.3

Calculate the ground-level downwind center-line concentration of SO_2 10 km from the source. The effective stack height is 100 m, the wind speed is 6 m/s, and the emissions rate is 0.25 kg/s. The stability class is D.

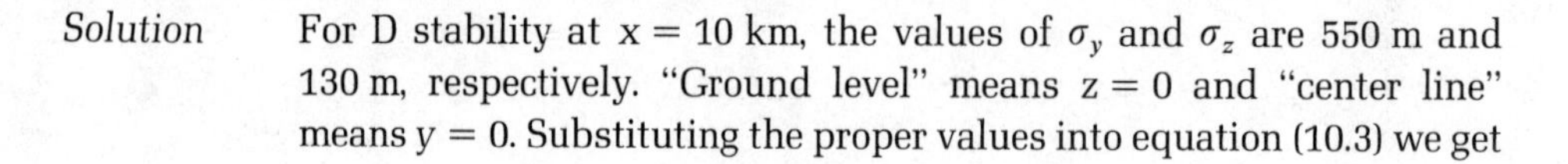

Solution For D stability at x = 10 km, the values of σ_y and σ_z are 550 m and 130 m, respectively. "Ground level" means z = 0 and "center line" means y = 0. Substituting the proper values into equation (10.3) we get

$$C = \frac{0.25(10)^9\ \mu g/s}{2\pi \frac{6m}{s} 550\ m\ 130\ m} (1)\left(\exp\left(-\frac{1}{2}\frac{100^2}{130^2}\right) + \exp\left(-\frac{1}{2}\frac{100^2}{130^2}\right)\right)$$

$$= 92.7(1.49)$$

$$= 138\ \mu g/m^3$$

10-3 Legislative standards and controls

It has been said that "something that is everybody's responsibility often ends up being nobody's responsibility." So it was with our air resource up until the early 1950s, when the federal government began to recognize its responsibility. Because air knows no political boundaries, it is appropriately managed only at the national level. However, local efforts are critical to the attainment and maintenance of good air quality.

Air-pollution standards were established to protect the health and promote the well-being of individuals and of communities. These standards were set by government with input from professional organizations as a result of increased awareness of pollutants and their effects upon living organisms, including humans. Federal legislation and regulations were developed over a period of two decades with input from many interested groups. Some of these laws were the Air Pollution Control Act of 1955, the Motor Vehicle Air Pollution Control Act of 1965, the far-reaching Clean Air Act Amendments of 1970 and the Clean Air Act Amendments of 1977. Compliance with these laws require not only proper environmental engineering design and operation of pollution-abatement equipment but careful analysis and accurate measurements of specified pollutants and environmental quality parameters.

There are two types of standards. Ambient air quality standards (AAQS) deal with concentrations of pollutants in the outdoor atmosphere. Source performance standards (SPS) apply to emissions of pollutants from specific sources. AAQSs are always written in terms of concentration ($\mu g/m^3$ or ppm), while SPSs are written in terms of mass emissions per unit of time or unit of production (g/min or kg of pollutant per ton of product produced).

AAQSs were set by the Environmental Protection Agency for the major pollutants based on two criteria. The primary standards were established to protect the public health, whereas secondary standards were set to protect the public well-being. These standards are presented in Table 10.3. It should be noted that some states have set their own standards, which are stricter than those listed.

TABLE 10.3 National ambient air quality standards.*

Pollutant	*Averaging time*	*Primary standard*	*Secondary standard*
Suspended particulate matter	Annual geometric mean	75 $\mu g/m^3$	60 $\mu g/m^3$
	24 hr	260 $\mu g/m^3$	150 $\mu g/m^3$
Sulfur dioxide	Annual average	80 $\mu g/m^3$	
	24 hr	365 $\mu g/m^3$	
	3 hr		1300 $\mu g/m^3$
Hydrocarbons (VOC) (Corrected for methane)	3 hr (6–9 A.M.)	160 $\mu g/m^3$	Same
Nitrogen dioxide	Annual average	100 $\mu g/m^3$	Same
Carbon monoxide	8 hr	10 mg/m^3	Same
	1 hr	40 mg/m^3	Same
Photochemical oxidants as ozone	1 hr	235 $\mu g/m^3$	Same

From Code of Federal Regulations (1982).

*Standards, other than those based on annual average or annual geometric average, are not to be exceeded more than once a year.

Example Problem 10.4

The National Ambient Air Quality Standards are presented in Table 10.3 for CO and NO_2, in terms of mg/m^3 and $\mu g/m^3$. Calculate the concentrations in ppm at 25°C and atmospheric presssure.

Solution From the ideal gas law,

$$\frac{P_1V_1}{T_1} = \frac{P_2V_2}{T_2}$$

$$\frac{1.0 \text{ atm.} \times 22.41 \text{ L/mol}}{(273^\circ + 0^\circ\text{C})} = \frac{1.0 \text{ atm. } V_2}{(273^\circ + 25^\circ\text{C})}$$

$$V_2 = \frac{(298)(22.41)}{273} = 24.46 \text{ L/mol}$$

The density ρ of CO at atmosphere pressure and 25°C is

$$\rho = \frac{\text{Mol. wt. of CO in g/mol}}{24.46 \text{ L/mol}}$$

$$= \frac{12 + 16}{24.46} = 1.145 \text{ g/L or } 1.145 \text{ mg/mL}$$

$$\text{Concentration of CO} = \frac{10 \text{ mg/m}^3}{1.145 \text{ mg/mL}} = 8.7 \frac{\text{mL}}{\text{m}^3} = 8.7 \text{ ppm}$$

Note that with gases, ppm is always on a volume (or molar) basis. Note also that we could have used equation (1.3) from Chapter 1 to do this calculation. Similarly, the concentration of NO_2 can be calculated as 0.05 ppm.

Source performance standards (or emissions standards) are numerous because of the variety of sources. Some examples are given in Table 10.4. The new source performance standards listed in Table 10.4 were derived either from materials balance considerations or from actual field tests at a number of industrial plants. One use of these standards is as emissions factors to estimate emissions from plants yet to be constructed. Another use is to estimate emissions from sources that are impossible to measure. The next three example problems illustrate these uses.

Example Problem 10.5

Calculate the daily SO_2 emissions from a 200 ton per day sulfuric acid plant which emits at the maximum allowable rate.

Solution The standard is a maximum of 2 kg SO_2 per metric ton of sulfuric acid.

$$\frac{2 \text{ kg SO}_2}{\text{MT acid}} \times \frac{1 \text{ MT}}{1.102 \text{ T}} \times 200 \text{ T/day} = 363 \text{ kg/day of SO}_2$$

Example Problem 10.6

Calculate the daily emissions of particulates and SO_2 from a 1000-MW coal-fired power plant which meets the performance standards listed in Table 10.4. Assume the coal has a heating value of 11,000 BTU/lb and meets the SO_2 standard of 1.2 lb/million BTU heat input. The plant is 39% efficient,

TABLE 10.4 Selected examples of new source performance standards (NSPS).

1. Steam electric power plants (1980 NSPS)
   a. Particulates: 0.03 lb/million Btu of heat input (13 g/million kJ)
   b. NO_x: 0.02 lb/million Btu (86 g/million kJ) for gaseous fuel
      0.30 lb/million Btu (130 g/million kJ) for liquid fuel
      0.60 lb/million Btu (260 g/million kJ) for anthracite or bituminous coal
   c. SO_2: 0.20 lb/million Btu (86 g/million kJ) for gas or liquid fuel. For coal-fired plants, the SO_2 standard requires a scrubber that is at least 70% efficient and may be more than 90% efficient depending on the percent sulfur in the coal. The maximum permissible emissions rate is 1.2 lb SO_2 per million BTU of heat input; the permissible emissions rate may be less depending on the coal sulfur content and the scrubber efficiency required.
2. Solid waste incinerators. Established for new incinerators with a charging rate in excess of 50 tons/day. Particulate emission standard is a maximum 2-hr average concentration of 0.08 gr/dscf (0.18 g/m^3) corrected to 12 percent carbon dioxide.
3. Nitric acid plants. Standard is a maximum 2-hr average nitrogen oxide emission of 3 lb/ton of acid produced (1.5 kg/metric ton), expressed as nitrogen dioxide. Applicable to any unit producing 30 to 70 percent nitric acid by either pressure or atmospheric pressure processes
4. Sulfuric acid plants. Applies to plants employing the contact process. Standard is a maximum 2-hr average emission of SO_2 of 4 lb/ton of acid produced (2 kg/metric ton). An acid mist standard, a maximum 2-hr emission of 0.15 lb/ton of acid produced (0.075 kg/metric ton), also is established
5. Petroleum refineries. Particulates from fluid catalytic cracking unit catalyst regenerator may not exceed 50 mg/m^3 (0.022 gr/dscf). Opacity restricted to 20 percent or less except for 3 min in any hour. If auxiliary fuels are burned in an incinerator waste-heat boiler, particulates in excess of above shall not exceed 0.043 kg/million kJ (0.10 lb/million BTU). In addition, no discharge of CO in excess of 0.050 percent by volume, and no release of H_2S in excess of 230 mg/m^3 (dry normal) (0.10 gr/dscf)
6. Iron and steel plants. Particulate discharges may not exceed 50 mg/m^3 (0.022 gr/dscf) from basic oxygen furnaces, and the opacity must be 10 percent or less except for 2 min in any hour
7. 1982 passenger cars:
   a. CO: 3.4 g/mile
   b. NO_x: 1.0 g/mile
   c. VOC: 0.41 g/mile

From Wark and Warner (1981).

Solution First calculate the heat input rate for a 39% efficient plant:

$$E_{in} = \frac{1000\text{MW}}{0.39} \times \frac{1000 \text{ kW}}{\text{MW}} \times \frac{24 \text{ hr}}{\text{day}} \times \frac{3412 \text{ BTU}}{\text{kWh}}$$

$$= 2.10 \times 10^{11} \text{ BTU/day}$$

$$\text{Particulates emitted} = \frac{0.03 \text{ lb}}{10^6 \text{ BTU}} \times \frac{2.10 \times 10^{11} \text{ BTU}}{\text{day}} \times \frac{1 \text{ ton}}{2000 \text{ lb}}$$

$$= 3.2 \text{ tons/day}$$

$$SO_2 \text{ emitted} = \frac{1.2 \text{ lb } SO_2}{10^6 \text{ BTU}} \times 2.10 \times 10^{11} \frac{\text{BTU}}{\text{day}} \times \frac{1 \text{ ton}}{2000 \text{ lb}}$$

$$= 126 \text{ tons/day}$$

Example Problem 10.7

Estimate the average daily emissions of NO_x and VOC in a city in which daily traffic can be approximated by 100,000 cars traveling about 40 miles each. Assume the cars emit at an average rate equal to 1.5 times the new car standards.

Solution

The new car standards for NO_x and VOC are 1.0 and 0.41 g/mile, respectively. Thus emission rates are 1.5 g/mile and 0.62 g/mile, respectively.

$$\frac{1.5 \text{ g NO}_x}{\text{car-mile}} \times \frac{40 \text{ miles}}{\text{day}} \times 100{,}000 \text{ cars} \times \frac{1 \text{ kg}}{1000 \text{ g}}$$

$$= 6000 \text{ kg/day NO}_x$$

$$\frac{0.62 \text{ g VOC}}{\text{car-mile}} \times \frac{40 \text{ miles}}{\text{day}} \times 100{,}000 \text{ cars} \times \frac{1 \text{ kg}}{1000 \text{ g}}$$

$$= 2480 \text{ kg/day VOC}$$

10-4 Pollution control technology

Mobile sources contribute to large air pollution problems in many cities in the world today. However, emissions from recent model cars are now controlled to much lower levels than from older models. So average highway emissions are declining each year as older cars are replaced by newer ones. Some control techniques are catalytic converters, exhaust-gas recirculation, two-stage combustion, fuel injection, and careful control of the engine air-to-fuel ratio. Collectively, mobile sources emit large amounts of pollutants, but individually each is so small that control devices are limited in size and scope. On the other hand, engineers have developed several large, interesting, and important pollution control devices for industrial-scale sources. These devices fall into two broad categories—those that remove particles and those that remove gases.

Particulate control devices

There are several control devices for removing particulates from exhaust gases before they are emitted into the atmosphere. These include cyclones, electrostatic precipitators, baghouses, and scrubbers. In the following few pages we give a brief description of each device, then compare their efficiencies for removing particulates.

A cyclone is designed to remove particulates by causing the entire gas stream to spin in a vortex. The centrifugal force acting on the larger particles flings them toward the wall of the cyclone where they impinge and then fall to the bottom of the cyclone. The gas then flows out through the top of the cyclone and the particles can be removed from the bottom. Figure 10.5 presents a schematic diagram of a standard cyclone.

An electrostatic precipitator removes particulates from a gas stream by creating a high voltage drop between electrodes. As the gas with particles passes between the electrodes, gas molecules are ionized, ions stick to the particles, and the particles acquire a charge. The charged particles are attracted to and collected on the oppositely charged plates while the cleaned gas flows through the device. During

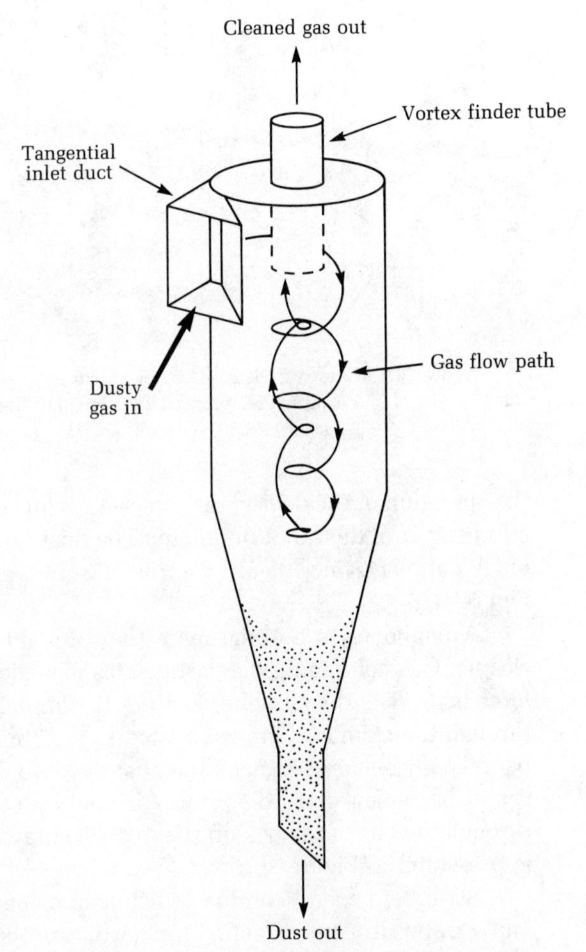

Figure 10.5 Schematic flow diagram of a standard cyclone.

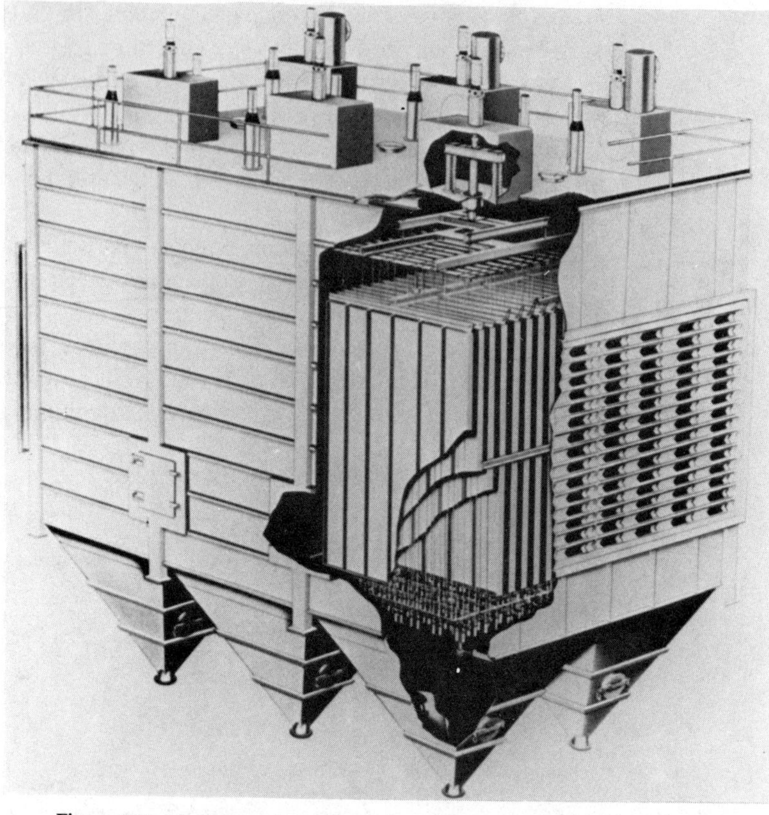

Figure 10.6 Cutaway view of an electrostatic precipitator (courtesy of Western Precipitation Division, Joy Manufacturing Co., Los Angeles, CA).

the operation of the device, the plates are rapped periodically to shake off the layer of dust that builds up. The dust is collected and disposed of. A cutaway view of an electrostatic precipitator is presented in Figure 10.6.

A baghouse is nothing more than a giant multiple-bag vacuum cleaner. Gas containing the particulates is made to flow through cloth filter bags. The dust is filtered from the stream, and the gases pass through the cloth and are exhausted to the atmosphere. The bags are periodically shaken to knock the dust down to the bottom from where it can be removed for disposal. Air could also be blown backward through the bags to knock off the dust. A cutaway view of a baghouse is presented in Figure 10.7.

Scrubbers operate on the principle of collision between a particle and a water droplet, collecting the particle in the larger, heavier water droplet. The water droplets fall through the upward-flowing gases,

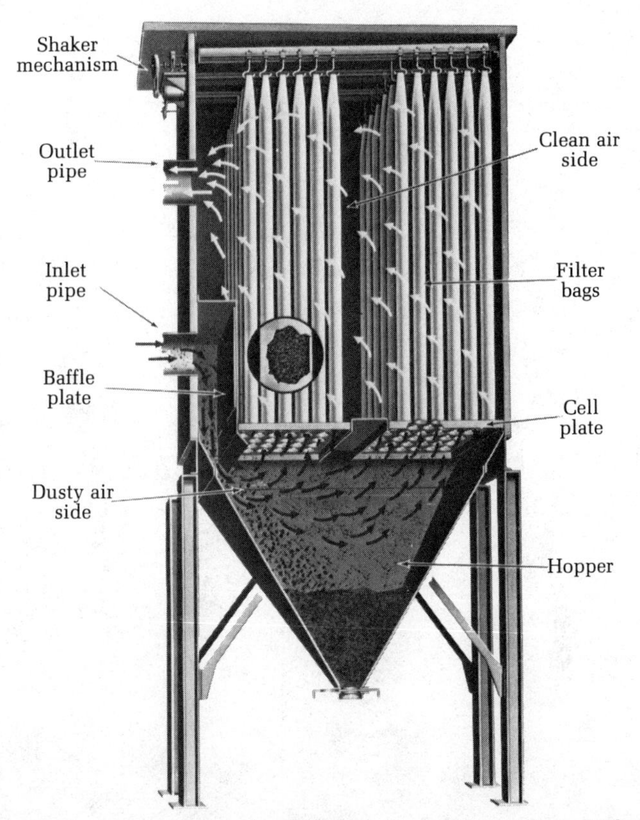

Figure 10.7 Cutaway view of a shaker-type baghourse (courtesy of Wheelabrator Frye, Inc., Mishawaka, IN).

colliding with and removing particulates, and accumulate in the bottom of the scrubber. Later the stream of water with particulates can be treated to remove the particulates. Figure 10.8 presents a cutaway view of one of the many types of spray tower scrubbers.

In comparing particulate control devices we should consider collection efficiency and cost (both capital and operating) simultaneously. As engineers, we always want to try to accomplish our objectives in the most cost-effective manner. In general, the cyclone does only a moderate job of removing particles but is the cheapest device. The other devices are all quite good at removing particulates but are expensive. However, each application is slightly different from all others. Proper engineering analysis and design are required to ensure the right choice for the job. Figure 10.9 compares the collection efficiencies of these devices with each other over a range of particle sizes. Collection efficiency is defined as

$$\eta = \frac{\text{Mass rate of particles collected}}{\text{Mass input rate of particles}} \times 100\% \tag{10.4}$$

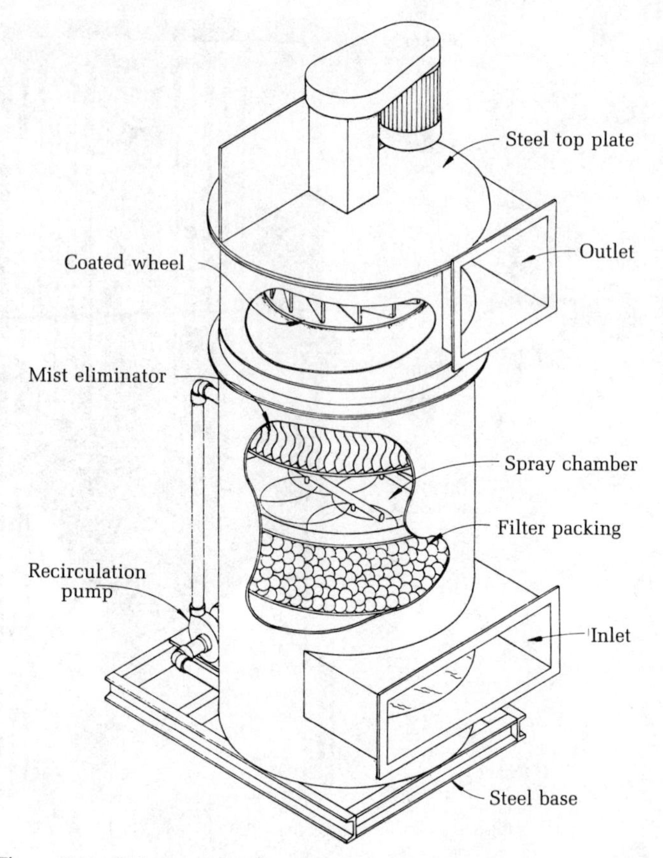

Figure 10.8 Cutaway view of a spray tower scrubber (courtesy of Midwest Air Products Co., Owosso, MI).

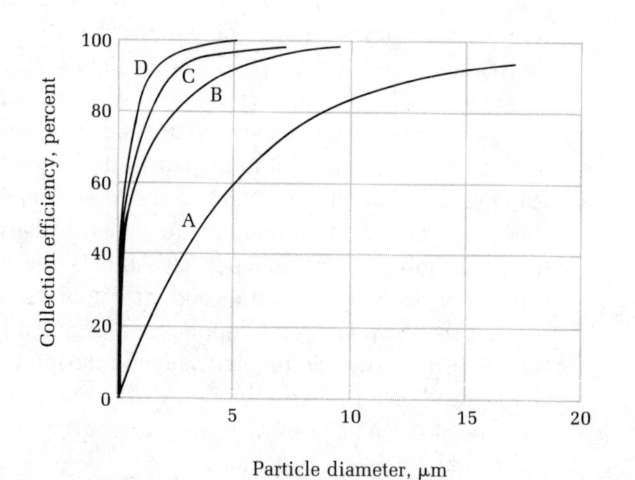

Figure 10.9 Typical collection efficiencies for various types of particle collectors: A, ordinary cyclone; B, spray tower; C, electrostatic precipitator; D, baghouse.

Two important observations can be made from Figure 10.9. First notice that in general, as particle size decreases, efficiency decreases. Second, notice that in general the baghouse is the most efficient device over the widest range of particle sizes; the electrostatic precipitator is the second, and the scrubber is the third. The efficiencies of these three devices are all high, but cyclones typically have lower efficiencies.

Example Problem 10.8

Calculate the overall efficiency of a particulate control system composed of a cyclone (75% efficient) followed by an electrostatic precipitator (95% efficient).

Solution The overall system looks like this:

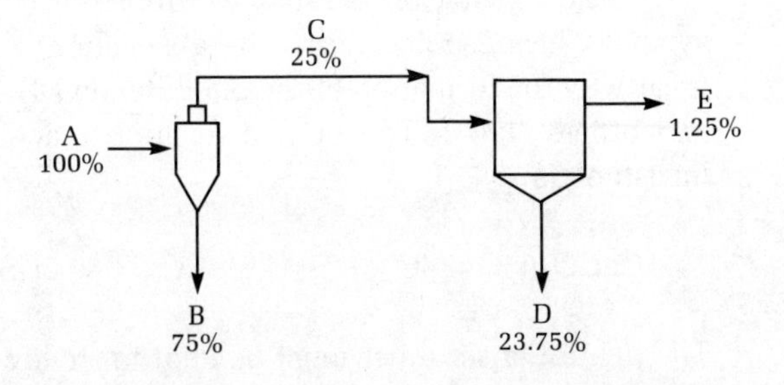

where streams A, B, C, D, and E represent particulate flow rates into and out of the equipment. The overall collection efficiency is the sum of B and D, or 98.75%.

In example 10.8, we had to add the collection efficiencies because of the definition of this term. Let us define penetration as:

$$Pt = 1 - \frac{\eta}{100} \tag{10.5}$$

Then, it should be obvious that penetration is the "efficiency" of particle pass-through and that overall penetration for two devices in series is

$$Pt_{\text{overall}} = Pt_1 \times Pt_2 \tag{10.6}$$

Thus, the overall efficiency of collection for two devices in series is

$$\eta_{\text{overall}} = 1 - Pt_{\text{overall}} \tag{10.7}$$

Equation (10.7) allows us to solve example problem 10.8 directly.

Gaseous emissions control devices

Another major use of scrubbers besides removing particulates is to absorb a pollutant gas from a mixture of gases. The rate and extent of absorption is commonly assisted by chemical reaction in the absorbing medium. A widespread example is the scrubbing of SO_2 from power-plant combustion gases by an alkaline solution. The most widely used system in this country uses a lime or limestone slurry to scrub the SO_2, producing a $CaSO_3$ plus $CaSO_4$ sludge which must be discarded.

The chemistry of limestone scrubbing is complex and only a brief synopsis is presented here. The absorption of SO_2 gas is enhanced greatly by the formation of the bisulfite ion HSO_3^- in solution (Henzel and others, 1981). The SO_2 absorption reaction in the presence of limestone is

$$CaCO_3(s) + 2SO_2 + H_2O \longrightarrow Ca^{+2} + 2HSO_3^- + CO_2(g) \tag{10.8}$$

The pH in the scrubber must be kept fairly low (less than about 6) to prevent precipitation in the scrubber itself. The possible plugging of the scrubber is the reason for a separate vessel, the effluent hold tank (see Figure 10.10) in which the precipitation reactions occur. The main precipitation reaction is

$$CaCO_3 + 2HSO_3^- + Ca^{+2} \longrightarrow 2CaSO_3 + CO_2 + H_2O \tag{10.9}$$

With excess oxygen in the scrubber there is some oxidation of sulfite, so some of the sludge produced is $CaSO_4$. In fact, because $CaSO_4$ can be dewatered much more easily than $CaSO_3$, and because it is a more stable sludge in a landfill, a popular process modification, called forced oxidation, has been developed to purposely oxidize the sludge before disposal. The net reaction here is

$$CaSO_3 + \tfrac{1}{2}O_2 \longrightarrow CaSO_4 \tag{10.10}$$

Lime-limestone scrubbing is a "throwaway" process because of

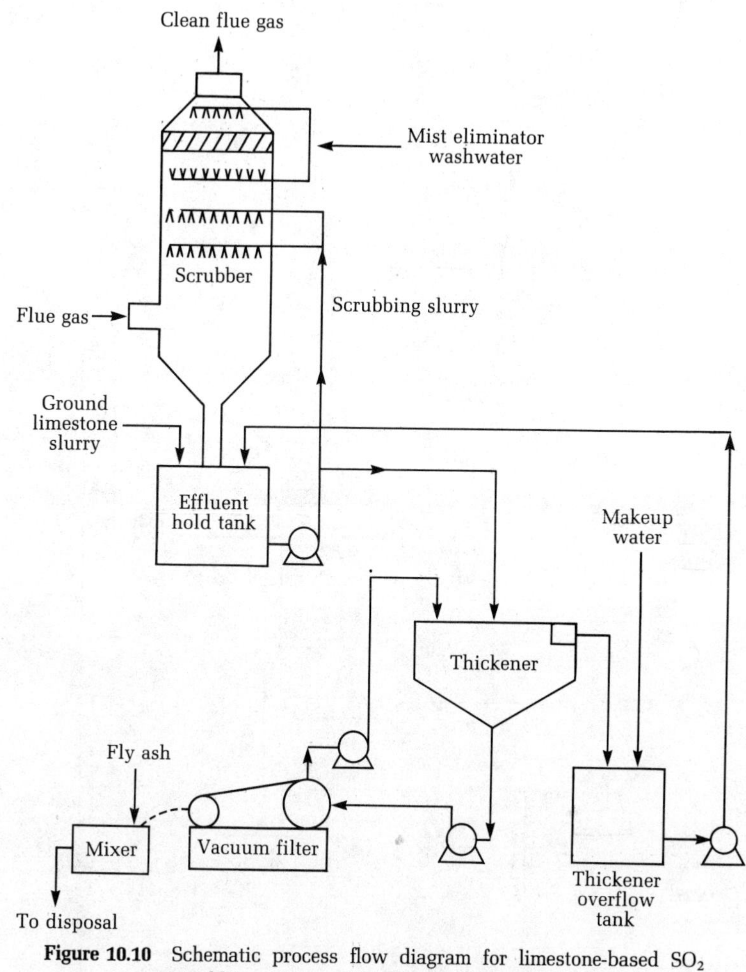

Figure 10.10 Schematic process flow diagram for limestone-based SO_2 scrubbing system (Henzel and others, 1981).

the discarded sludge. This process accounts for about 92% of all scrubbing systems in the United States because it is the cheapest and easiest to use (Henzel and others, 1981). However, there are many other processes, some of which recover the SO_2 as either H_2SO_4 or elemental sulfur, both of which are valuable products.

Other countries use recovery-type processes. In Japan, for example, SO_2 scrubbing with recovery is mandated by the government, and results in significant volumes of sulfuric acid, which is marketed within that country and exported. As sulfur supplies decline and become more expensive, these regenerative processes should become more widely used in the future.

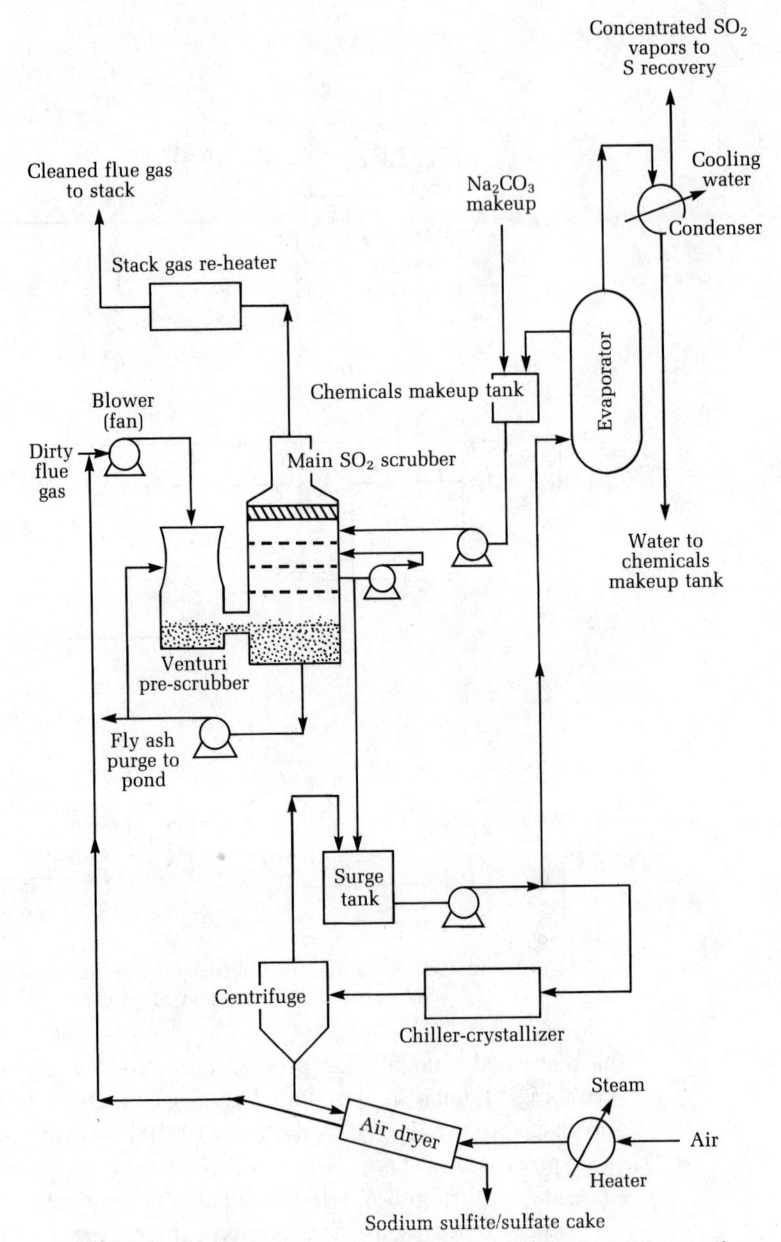

Figure 10.11 Schematic process flow diagram of Wellman-Lord SO_2 scrubbing and recovery system (Environmental Protection Agency, 1977).

Figure 10.12 A view of a Wellman-Lord-type flue gas desulfurization unit on a coal-fired power plant. (Environmental Protection Agency, 1977).

One of the several processes for SO_2 scrubbing with recovery of sulfur values is the Wellman-Lord systems, as depicted in Figures 10.11 and 10.12. In this process, the SO_2 is scrubbed with a solution of sodium sulfite to form sodium bisulfite (Environmental Protection Agency, 1977):

$$SO_2 + Na_2SO_3 + H_2O \longrightarrow 2NaHSO_3 \tag{10.11}$$

Some oxidation of sulfite occurs, forming sodium sulfate, which must be purged from the system:

$$NaSO_3 + \frac{1}{2}O_2 \longrightarrow Na_2SO_4 \tag{10.12}$$

Makeup sodium is provided by sodium carbonate. The sodium bisulfite solution is heated in an evaporator to drive off SO_2 and water vapor:

$$2NaHSO_3 \longrightarrow Na_2SO_3 + SO_2 + H_2O \tag{10.13}$$

The water and concentrated SO_2 gas are separated by condensing the water. The concentrated SO_2 gas can be further reacted to useful products. It can be reduced to elemental sulfur (the desired product if the sales market is quite distant), or it can be oxidized to sulfuric acid (if markets are close by).

No matter what type of process is used, SO_2 scrubbing is a complex, large-scale, expensive undertaking. This type of chemical processing is unfamiliar to most power-plant operators, and most utilities have been slow to accept SO_2 scrubbing. However, federal law requires it on all new coal-fired power plants, and it certainly does reduce acid gas discharges to the atmosphere.

Gases such as SO_2 can be scrubbed out of the exhaust gases with alkaline aqueous solutions, and this is currently being done at a number of power plants throughout the world. But a gas like NO_2 or NO, which is relatively insoluble, is very difficult to remove once it has been formed. The principal control mechanism for nitrogen oxides thus is to minimize their formation by the proper design and operation of burners and furnaces.

Volatile organic compounds usually are controlled by one of two methods: carbon adsorption or vapor incineration. Both are effective, and while carbon adsorption is a bit more complicated, it usually allows for recovery and reuse of the organics. Incineration may be preferred for highly toxic compounds.

In the carbon adsorption process, the contaminated air stream is passed through a bed of granular activated charcoal. The organic molecules are adsorbed onto the surface of the highly porous carbon pellets, while the cleaned air flows through. When all the surfaces of the carbon have been covered, the air stream is switched to an identical bed. While the air is being treated on the new bed, the old bed

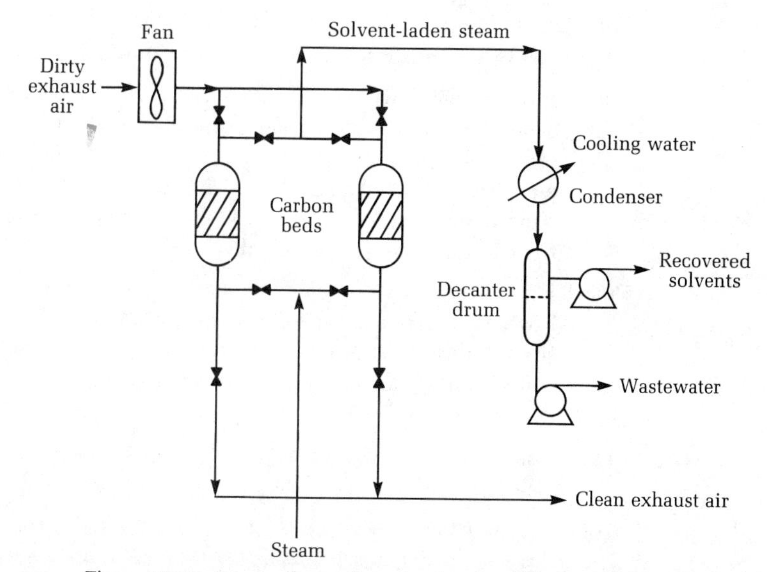

Figure 10.13 Process flow schematic of carbon-adsorption system.

is regenerated by passing steam through it. The hot steam desorbs the organics and carries them out of the bed, thus renewing the carbon for another cycle of adsorption. The steam and organic vapors are separated by condensation and decanting the two immiscible liquids. Figure 10.13 presents a schematic diagram of a carbon adsorption system.

Vapor incineration is a very simple process. The contaminated air stream is heated by burning a fuel gas in air. In the hot air stream, the VOCs are oxidized by reaction with the oxygen already present in the air stream. The end products of the reactions are CO_2 and water vapor (and others if other elements are present). It is only necessary to provide enough temperature, time, and turbulence (the "three Ts" of incinerator design) to ensure good performance of the incinerator. Because the operating temperature is in the range of 650°C to 870°C, good designs always provide for heat recovery.

10-5 Problems

10.1 Assume that gasoline is combusted with 99% efficiency in a car engine with 1% remaining in the exhaust gases as VOC. If the engine exhausts 16 kg of gases (mol. wt. = 30) for each kilogram of gasoline (mol. wt. = 100), calculate the fraction VOC in the exhaust. Give your answer in ppm.

10.2 In 1975 what fraction of U.S. NO_x emissions were contributed by highway transportation? by electric utilities?

10.3 Wind blows down a trapezoidal valley at 8 m/s. The valley depth is 800 m, the floor width is 1000 m, and the width at the top is 2000 m. A smelter emits SO_2 at a steady rate of 100 kg/day. The valley is capped by an inversion. Calculate the steady-state concentration a long way downwind from the smelter, where the pollutant is uniformly spread across the width and height of the valley.

10.4 In problem 3 above, consider the same smelter to be on level, open ground. The stack is 200 m tall and the plume rise is 100 m. The wind is 4 m/s and the stability class is C. Estimate the ground-level SO_2 concentration 6000 m directly downwind. What is the concentration 300 m off the center line at this same x value?

10.5 The secondary AAQS for SO_2 is 1300 $\mu g/m^3$ (for a 3-hour averaging time). Calculate the equivalent concentration in ppm.

10.6 A power plant uses 10,000 kg/hr of 3.00% sulfur coal. Approximately 90% of the SO_2 formed must be removed prior to release of the stack gases. A limestone system is to be used to remove the SO_2. Estimate the minimum limestone requirements for this plant (kg/hr).

10.7 A particulate removal system consists of a cyclone followed by an electrostatic precipitator. The cyclone is 65% efficient and the ESP is 95% efficient. Calculate the overall efficiency of the system.

10.8 A particulate removal system must achieve 99.4% overall efficiency. Calculate the required efficiency of a baghouse if it is preceded by an 80% efficient cyclone.

10.9 Assuming compliance with federal NSPS, predict the rate of emissions of NO_x, and SO_x, and particulates from a coal-fired power plant producing 800 MW of electrical power at an overall thermal efficiency of 40%.

10.10 Which produces more SO_x per unit amount of electricity produced, a coal-fired plant with a 90% efficient SO_2 scrubber or an oil-fired plant with no scrubber? Assume the coal is Illinois bituminous (see Table 3.1) with 3.5% sulfur. Assume the oil has a heat content similar to that of crude oil and contains 0.9% sulfur, and has a specific gravity of 0.92.

10.11 Calculate the annual emissions of CO, NO_x, and VOC from an older car which travels 10,000 miles in a year. Assume it emits at 10 times the federal standards for new cars.

10.12 Estimate the daily emissions of particulates from a solid waste incinerator emitting at 0.18 g/m^3. The incinerator burns 50 tons/day and exhausts gases in a ratio of 20 kg gases per kilogram of feed. The gases exit at 200°C and 1 atm. Assume that the gases have an average molecular weight of 30 and contain 12% CO_2 as emitted.

10.13 Estimate the daily emissions of CO and particulates from a refinery fluid catalytic cracking unit emitting at 80% of the maximum standards. The total gas flow is 20,000 m^3/min (after correction to dry normal conditions).

10.14 The diagram given below illustrates a coal-fired power plant with an SO_2 scrubber in which an overall SO_2 removal efficiency of 85% is required. Rather than treat the entire stream to 85% removal, the company proposes to treat part of the flue gas to 95% removal, and to bypass the remainder. The reblended steam must still satisfy the overall removal requirement. Calculate the fraction of the flue gas stream that can be bypassed around the scrubber.

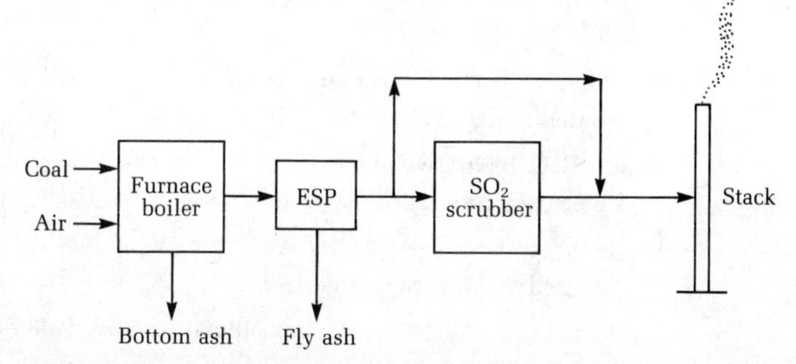

10.15 Referring to the diagram in problem 14, the gas temperature required in the stack for good buoyancy is at least 80°C (which corresponds to an enthalpy of 380 kJ/kg). The gases bypassing the scrubber have an enthalpy of 420 kJ/kg, and the gases coming out of the scrubber have an enthalpy of 360 kJ/kg. For a bypass percentage of 10%, calculate the enthalpy of the mixed stack gases. Ignoring the water which evaporates in the scrubber, will the blended stack gases have the enthalpy required for good buoyancy?

10.16 A carbon adsorption system has been proposed to treat a stream of contaminated air flowing at 17,090 ft^3/min (T = 104°F, P = 1 atm) that contains 1000 ppm hexane vapor. Assume that we can recover 96% of the inlet hexane as liquid hexane, which can be reused. (Some hexane vapor is lost into the exhaust air when we switch beds, and some hexane liquid is lost into the wastewater stream.) If hexane costs $1.50 per liquid gallon (which has a density of 2.50 kg/gal), calculate how much money we can potentially save by this operation. Give your answer in $/year assuming that the plant runs 24 hours per day, 240 days per year.

References

Code of Federal Regulations. 1982. Title 40, Part 50: National Primary and Secondary Ambient Air Quality Standards. Washington, D.C: Office of Federal Register.

Hansen, J., Johnson, D., Lacis, A., Lebedeff, S., Lee, P., Rind, D., and Russell, G. 1981. Climate impact of increasing atmospheric carbon dioxide. *Science* 213:4511.

Henzel, D. S., Laseke, B. A., Smith, E. O., and Swenson, D. O. 1981. *Limestone scrubbers: User's handbook*, U.S. Environmental Protection Agency EPA-600/8-81-017.

Turner, D. B. 1970. *Workbook of atmospheric dispersion estimates.* Washington, D.C: Environmental Protection Agency Publication AP-26.

U.S. Environmental Protection agency. 1977. First progress report: Wellman-Lord SO_2 recovery process—Flue gas de-sulfurization plant. EPA-625/2-77-011.

U.S. Environmental Protection Agency. 1981. National air quality and emissions trends report, 1975, EPA-450/1-76-002.

Wanta, R. C. 1968. Meteorology and air pollution. In *Air Pollution*, 2nd ed. A. C. Stern, ed. New York: Academic Press.

Wark, K. and Warner, C. F. 1981. *Air pollution–Its origin and control*, 2nd ed. New York: Harper & Row.

Williamson, S. J. 1973, *Fundamentals of Air Pollution.* Reading, MA: Addison-Wesley.

11

Solid Wastes and Residues

11-1 Types, quantities, and characteristics

The term solid wastes has been used to include residential, commercial, and industrial wastes and many different states of matter. Items that have been identified as solid wastes include household garbage, grass clippings, tree branches, worn-out stoves, sludges from sewage treatment plants, "empty" drums that contained hazardous chemicals, and calcium sulfate/calcium sulfite sludges from SO_2 scrubbers. Most authors separate all solid wastes into three categories: municipal solid wastes, industrial solid wastes, and agricultural and mining wastes. Municipal solid wastes (MSW) include materials from household and commercial enterprises, and encompass such items as food wastes, paper wastes, glass, metal, plastic, rubber, leather, wood, and small amounts of miscellaneous materials. Table 11.1 presents a typical breakdown of municipal solid waste by category.

The amount of MSW generated in the United States was estimated to be approximately 130 million metric tons in 1976, or enough to fill the New Orleans Superdome from floor to ceiling twice a day, weekends and holidays included (Council on Environmental Quality, 1978). Industrial wastes were estimated by the Environmental Protection Agency to total about 340 million metric tons in 1977. Both categories of solid waste are projected to continue to grow in the future; however, the industrial wastes are probably growing slightly

TABLE 11.1 Components of municipal solid wastes generated in the United States, 1971.

	Total disposed	
Components	*Tons, millions*	*Percent*
Paper	47.3	37.8
Glass	12.5	10.0
Metal (all kinds)	12.6	10.1
Ferrous	—	9.1
Aluminium	—	0.7
Other nonferrous	—	0.3
Plastic	4.7	3.8
Rubber and leather	3.4	2.7
Textiles	2.0	1.6
Wood	4.6	3.7
Food	17.7	14.2
Subtotal	104.8	83.9
Yard wastes	18.2	14.6
Miscellaneous inorganics	2.0	1.5
Total	125.0	100.0

From Tchobanoglous and others (1977).

faster than municipal solid wastes because sludge and other residues from pollution control devices are increasing as the regulations on discharges of air and water pollutants become stricter.

Industrial wastes include such things as sludges, hazardous chemicals, processing waste streams, which although not solid wastes can be ponded for storage and settling, and a variety of other types of wastes that are generated by industry. Approximately 10–15% of all industrial solid wastes may be classified as hazardous, according to the Environmental Protection Agency; that is, they pose a particular threat to public health or environmental well-being unless they are properly handled, transported, treated, stored, and disposed of (Environmental Protection Agency, 1975). Hazardous wastes will be dealt with in more detail in Chapter 12.

Agricultural and mining wastes can be very voluminous depending on the area of the country being studied. Much of these wastes (which include spoiled fruits and vegetables, cattle manure and other animal manure, corn stalks and other left-over portions of crops, and mine tailings) are disposed of on-site and quantities are difficult to estimate. As an example, however, in 1968 agricultural wastes from California (a major food-producing state) amounted to about 35 million tons or nearly 10 lb/person/day for every man, woman, and child in the state (Tchobanoglous and others, 1977).

11-2 Municipal solid wastes—Generation and collection

The United States today is a throwaway society. Most of our consumer products or food products are packaged in plastic wrap or paper containers; we generate a tremendous quantity of materials that are simply discarded. The problems associated with managing solid wastes today are complicated because of the great diversity of types of materials that are discarded from the average U.S. household. In addition, the widely dispersed and large number of sources of these wastes requires a complex system for collection. In fact, the collection portion of a municipal solid waste management system accounts for the greatest percentage of total cost in disposal. In Table 11.2 we present some historical and projected growth rates for solid waste generation.

The activities of a solid waste management system have been classified into several different phases. The first is waste generation, which we have just discussed. The second (which is really included in the waste generation from a functional point of view) is on-site storage. That is, as a household generates wastes it accumulates them in garbage cans until the regular garbage pickup. The next step, collection, will be discussed separately and in more detail later in this section because of its importance in the overall cost of any solid waste management system. After collection the MSW is transferred and transported to a disposal site; the final step, then, is disposal.

Recently there have been several MSW processing systems developed that recover materials and energy from MSW. However, the processing operation is not 100% efficient, and there are materials from this step that need to be disposed of. Ultimately these residues are disposed of in a sanitary landfill or into the ocean in some manner. Because of the widespread use of sanitary landfills, this particular phase will be discussed in some detail later. Because of the economic

TABLE 11.2 U.S. municipal solid waste generation rates

Year	*lb/person/day*
1968	3.3
1973	3.7 (3.5)*
1980 (estimated)	4.3 (3.8)
1985 (projected)	4.8 (3.8)
1990 (projected)	5.0 (3.7)

From various sources.

* Numbers in parentheses are estimates of net disposal after recycling and recovery.

and environmental impacts of MSW processing and recovery, a section is devoted to recovery later in this chapter.

Now let us return to the collection of MSW. The collection phase, which accounts for approximately 80% of the total cost, is the most crucial part of the overall solid waste management system. The collection phase is so expensive because it is very labor intensive. That is, one, two, or three people are involved in a crew that services at most several hundred residences before the truck is filled and must be taken to a transfer point or a final dump site. The optimization of municipal solid waste collection is not a trivial problem and involves several considerations including route optimization, type-of-service optimization, and transfer station locations. The reason for transfer stations is that the types of trucks that are well suited for start-stop collection service are not well suited for long-distance hauling, and vice versa. A photograph of an MSW collection truck discharging wastes at a transfer station is shown in Figure 11.1.

There are several different types of collection systems, which vary with the type of truck and type of workers needed with each truck. Systems also vary as to whether or not they collect commercial and industrial solid waste or strictly residential solid waste. It is beyond the scope of this text to go into detail about different types of collection vehicles or optimization of collection systems.

Figure 11.1 A municipal solid wastes collection truck unloading at a transfer station.

11-3 Sanitary landfill

The term sanitary landfill refers to an area of land in which municipal solid waste is deposited, compacted, and covered by a layer of earth in daily operations (Brunner and Kelly, 1972). The disposal site is large enough to handle the projected volume of waste over a certain period of time (its design life). Some time after closure the land may become available for some other use. The design of the sanitary landfill depends upon its intended ultimate use, thus this factor must be taken into account at the very beginning. In many areas (such as parts of New York City), parks and other facilities have been built on old sanitary landfills.

Several factors must be considered in evaluating disposal by sanitary landfill. Some are availability of land, distance from transfer stations to the site, soil conditions, surface water hydrology, underlying geologic conditions, local rainfall, meteorological conditions, and potential uses (Tchobanoglous and others, 1977). All of these factors are overlaid by the consideration of economics, and the ultimate decision is a combination of economic, technological, and political considerations. Figure 11.2 depicts two approaches to sanitary landfilling.

Example Problem 11.1

Estimate the required landfill area for a small city of population 25,000 assuming that the following conditions apply: solid waste generation rate is 4.0 lb/person/day, compacted density of the solid waste in the landfill is 800 pounds per cubic yard, and average depth of compacted solid waste is 8 feet.

Solution

$$\text{Generation rate} = \frac{4\text{ lb}}{\text{person-day}} \times 25{,}000\text{ people} \times \frac{\text{ton}}{2000\text{ lb}}$$

$$\times \frac{365\text{ day}}{\text{yr}} = 18{,}250\,\frac{\text{tons}}{\text{yr}}$$

$$\text{Required volume} = 18{,}250\,\frac{\text{tons}}{\text{yr}} \times \frac{\text{yd}^3}{800\text{ lb}} \times \frac{2000\text{ lb}}{\text{ton}} = 45{,}625\,\frac{\text{yd}^3}{\text{yr}}$$

$$\text{Required area} = \frac{45{,}625\text{ yd}^3}{\text{yr}} \times \frac{27\text{ ft}^3}{1\text{ yd}^3} \times \frac{1}{8\text{ ft}} \times \frac{1\text{ acre}}{43{,}560\text{ ft}^2}$$

$$= \frac{3.5\text{ acres}}{\text{yr}}$$

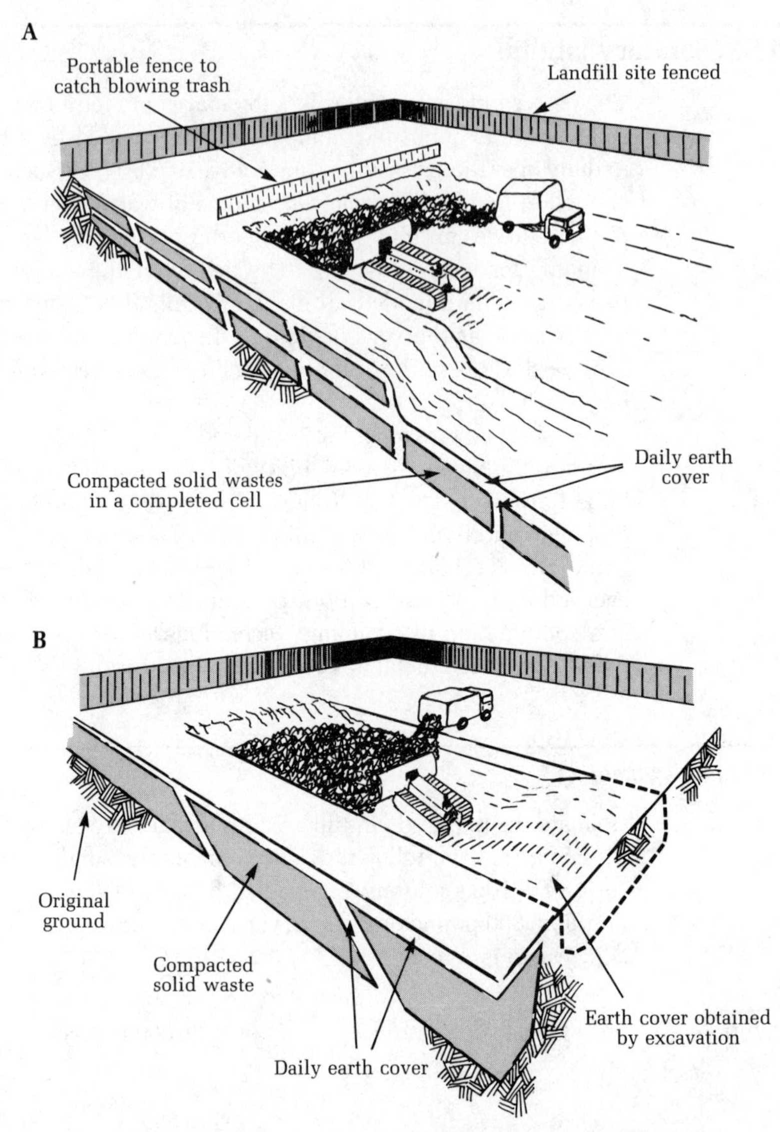

Figure 11.2 Two methods of operation for a sanitary landfill: A, area method; B, trench method (from Brunner and Kelly, 1972).

If a design life of 20 years is desired, then a minimum of 71 acres is needed for landfill (assuming no growth in population or generation rate). The actual acreage required will be greater than that calculated above because land is also required for access roads, buildings, and utilities. Often this allowance ranges from 20% to 40% of the total land area.

The purposes of covering the wastes with a layer of earth include preventing the waste from being blown by the wind, preventing access to the solid wastes, and allowing soil bacteria to quickly begin to decompose the biodegradable portions of waste. Initially organic components undergo bacterial decomposition by aerobes; soon, however, the oxygen is depleted and anaerobic processes become important. The overall rate at which organic materials decompose depends on the moisture content, the types of wastes that are present, and other conditions such as initial soil bacterial activity, temperature, and humidity (Tchobanoglous and others, 1977). As the solid wastes decompose, gases are formed such as ammonia, carbon dioxide, carbon monoxide, hydrogen, hydrogen sulfide, and methane. Furthermore, as decomposition occurs, the wastes settle and the land begins to sink, which must be accounted for when planning its final use.

Perhaps the most serious problem from ordinary sanitary landfills is that of leachate. Leachate is contaminated water that begins with rainfall seeping through a landfill and leaching out potentially hazardous substances, carrying them downstream into a groundwater or a surface-water supply. Leachate may contain BOD, TOC, COD, ammonia, phosphorus, nitrates, alkalinity, and various heavy metals. The quantity of leachate is directly proportional to the amount of external water entering the landfill. Table 11.3 gives some data on leachate composition.

TABLE 11.3 Composition of leachate from landfills.

	Value, mg/L	
Constituent	*Range*	*Typical*
BOD_5 (5-day biochemical oxygen demand)	2,000–30,000	10,000
TOC (total organic carbon)	1,500–20,000	6,000
COD (chemical oxygen demand)	3,000–45,000	18,000
Total suspended solids	200–1,000	500
Organic nitrogen	10–600	200
Ammonia nitrogen	10–800	200
Nitrate	5–40	25
Total phosphorus	1–70	30
Ortho phosphorus	1–50	20
Alkalinity as $CaCO_3$	1,000–10,000	3,000
pH (in pH units)	5–8	6
Total hardness as $CaCO_3$	300–10,000	3,500
Calcium	200–3,000	1,000
Magnesium	50–1,500	250
Potassium	200–2,000	300
Sodium	200–2,000	500
Chloride	100–3,000	500
Sulfate	100–1,500	300
Total iron	50–600	60

From Tchobanoglous and others (1977).

If the landfill can be designed to exclude external water or to provide a slope to the landfill such that most of the rainfall runs off on the surface rather than seeping down through the wastes, then the problems of leachate can be minimized. If the landfill is located in an area that is environmentally sensitive to leachate, then the leachate should be collected and treated. This can be accomplished by underlaying the landfill with a porous drain system (gravel) with an impervious layer (clay) beneath it. Thus any leachate can be collected and treated before being discharged from the landfill site. There are also methods to control gases that form in the landfill; they may be vented at a safe location or collected and burned in a waste-gas burner.

11-4 Alternatives to landfilling

Several alternatives have been proposed to the landfilling of MSW. These include incineration, resource recovery, pyrolysis, and several biochemical processes. Although the latter two categories hold potential for the future, and in fact have been thoroughly discussed (Vesilind and Rimer, 1981), their use is not widespread today. Simple incineration and resource recovery (which usually includes incineration of a processed portion of the MSW) are used in several locations in the United States today.

Simple incineration

A simple incineration system includes a receiving building, one or more incinerators, waste heat boilers (optional), and an ash disposal system (Cooper, 1981). In the operation of a modular incinerator installation, raw MSW is dumped onto the floor of a receiving building. Large items that are easily separated (such as stoves) are moved aside. Everything else is scooped up by a front-end loader and pushed into one of the several hoppers for feeding into a furnace. In a 100-ton-per-day plant, there may be four to eight separate furnaces to allow for flexibility in maintenance and operations.

Once a charge of MSW is emplaced in the morning, it is ignited by natural gas burners. The waste burns relatively slowly, giving off incompletely oxidized gases and particles. These unburned materials pass into an afterburner chamber where additional fuel is combusted with additional air to raise the temperature several hundred degrees above the main chamber temperature and complete the burn. In most package installations, this is thought to be sufficient for pollution control. One pollutant not controlled by this strategy is hydrogen chloride, which is produced when many types of plastics are burned, and large incinerators should provide scrubbers to remove HCl before the gases are discharged to the atmosphere. Hydrogen chloride can also cause severe corrosion problems in boilers.

In an intermittent operation, at the end of the day the ash is allowed to cool. The next morning, water is sprayed on the ash, and then the ash is removed to a landfill. Packaged systems are also available with a semicontinuous automatic ash removal (see Figure 11.3). This latter approach allows for continuous steam (and power) production and may be preferable to daily start-ups and shutdowns.

Auxiliary fuel must be used in a MSW incinerator both for initial start-up and to maintain good afterburning conditions. Auxiliary fuel requirements range from about 500 to 2000 BTU per pound of MSW depending on the net heating value of the waste and the excess air in the furnace (Parker and Pech, 1978). Heat recovery can vary widely with the design and operation of the system; in any case a package boiler is usually used to heat water or to generate steam. Although most applications utilize the steam directly for process heat, higher pressure steam could be generated, then routed through a turbine generator to produce electricity. Because of the various inefficiencies, steam production ranges from 1.0 to 3.5 lb steam/lb MSW (Parker and Pech, 1978). A well-operated packaged incinerator with process heat recovery may operate at roughly 40% thermal efficiency. If a small turbine generator is included, but no process use made of the steam,

Figure 11.3 Cutaway view of a package system for continuous solid wastes incineration with heat recovery (courtesy of Consumat Systems, Inc., Richmond, VA).

the overall efficiency would be fairly low (15–20%). If a nearby industry could use low-pressure steam for process heat, the overall efficiency (including electricity generation) might be as high as 50–60%.

Example Problem 11.2

Estimate the potential annual energy recovery from MSW from the city of example problem 11.1. Assume that the waste as delivered has a heat content of 4500 BTU/lb and that an overall net thermal efficiency of 40% can be achieved (including the use of auxiliary fuel gas). Give an estimate for the present and for 1990, when the gross heating value of the waste will be 5500 BTU/lb, the generation rate will be 5.0 lb/person/day, and the population will be 30,000.

Solution At present:

$$18{,}250\,\frac{\text{tons}}{\text{yr}} \times \frac{2000\ \text{lb}}{1\ \text{ton}} \times \frac{4500\ \text{BTU}}{\text{lb}} \times 0.40 = 6.57 \times 10^{10}\ \text{BTU/yr}$$

In 1990:

$$30{,}000\ \text{people} \times \frac{5.0\ \text{lb}}{\text{person-day}} \times \frac{365\ \text{day}}{\text{yr}} \times \frac{5500\ \text{BTU}}{\text{lb}} \times 0.40$$

$$= 1.20 \times 10^{11}\ \text{BTU/yr}$$

The major inputs and outputs of a modular MSW system are shown schematically in Figure 11.4. Calculations were made to estimate the mass and energy flow rates of the various streams for a typical case, which represents a small- to medium-sized operation. The results of these calculations are presented in Table 11.4.

Resource recovery

The primary form of resource recovery from MSW involves a series of operations in which the wastes are first received, shredded, and air-classified to separate combustibles from noncombustibles (Tennessee Valley Authority, 1976). The lighter (combustible) materials are carried by a stream of air to a secondary shredder where this portion of the wastes is further reduced in size to promote good combustion in a

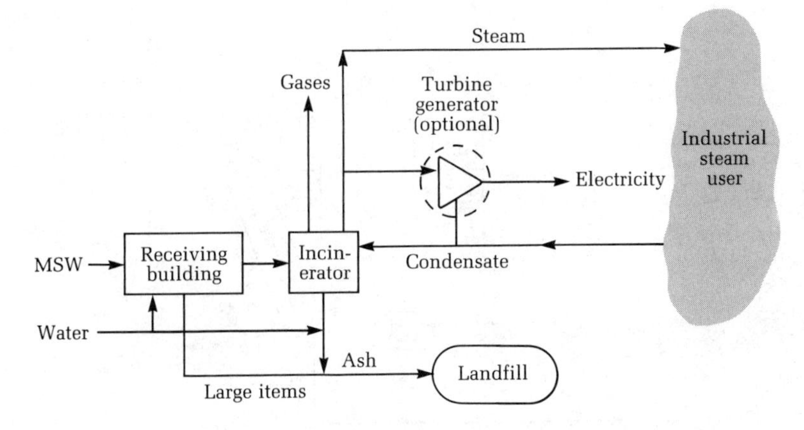

Figure 11.4 Schematic diagram of MSW incineration process (from Cooper, 1981).

power boiler. The noncombustibles go through a series of recovery processes including a magnetic separation for ferrous metal recovery, a rotary screen separator for glass recovery and an aluminum separator, such as an eddy current (electromagnetic) device (Vesilind and Rimer, 1981). The diagram in Figure 11.5 represents the basic process involved.

A very large scale resource recovery project has been proposed by Hooker Chemical Corporation in New York (Schweiger, 1978). It is a technically sophisticated, expensive operation. Wastes are collected from the surrounding community and brought to the plant. Wastes are dumped there for a tipping fee, which is comparable to present-day landfill fees. The wastes are processed with the primary objective of burning them to recover the heating values to produce high-pressure steam and electricity for the industrial complex.

Resource recovery projects are very capital intensive. Thus, under economic conditions which include high interest rates, these projects

TABLE 11.4 Input and output balances for municipal solid waste incineration.

Inputs	
Municipal solid waste, thousand tons/year	16.8
Heat content of MSW, BTU/lb	5100
Auxiliary fuel gas, billion BTU/yr	26.9
Total heat input (MSW + fuel gas), billion BTU/yr	199
Water, million gal/yr	1.7
Outputs	
Ash + large items, thousand tons/year	2.5
Recoverable heat as steam, billions BTU/yr	79
Electricity (optional), millions kWh/yr	5.9

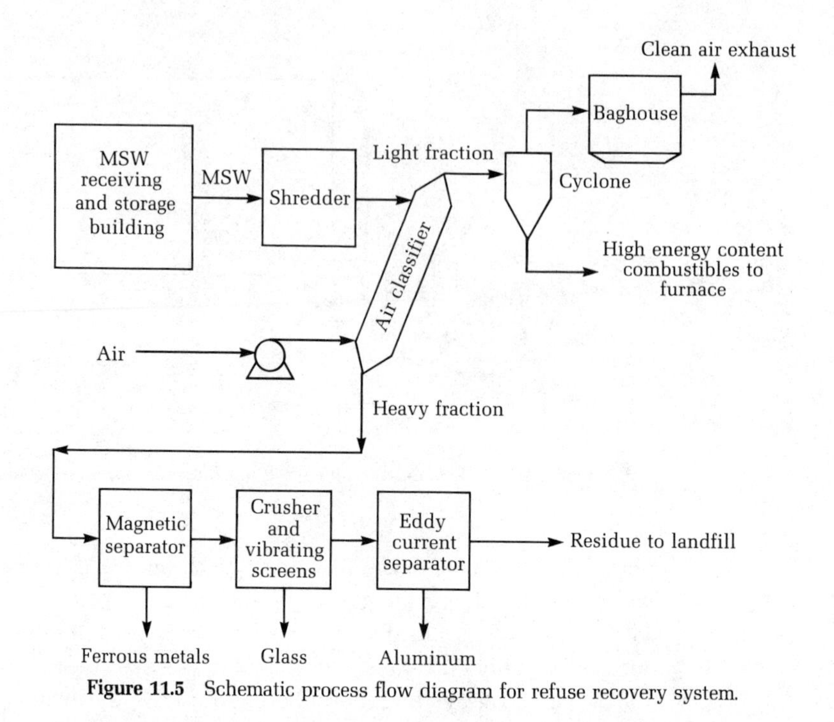

Figure 11.5 Schematic process flow diagram for refuse recovery system.

can be prohibitively expensive. Even simple incineration projects are expensive, and usually require cost recovery (in the form of steam credits) in order to be economically justified.

11-5 Economic analysis of MSW incineration and resource recovery

In analyzing the economics of municipal solid waste incineration with energy recovery, several variables are important. Capital investment (with associated debt service) and auxiliary fuel charges are the two most important costs. Tipping fees (for MSW disposal) and steam sales are the two income generators, with steam sales being the more important.

During the past five years, there have been many MSW incineration projects in the United States. Some are small, simple, relatively inexpensive operations in which MSW is simply loaded into a furnace and burned. Others are large, sophisticated, expensive projects in which the MSW was processed, sorted, and shredded, to produce various recyclables and a refuse-derived fuel (RDF) for burning in a full suspension-type power boiler. Capital costs are the most important single item in the economic analysis for many of these projects;

of course, variable operating costs are important also. Auxiliary fuel, labor, and maintenance are the largest components of these variable costs.

Refuse-derived fuel systems can be extremely expensive; large investments are required both for facilities to process the MSW into RDF and for the facilities to burn the RDF. Burning RDF is not as simple as burning coal because of corrosion problems (Krause and others, 1976). Reilly and Powers (1980) collected cost data on many MSW projects in this country, including RDF projects and simple projects. Tables 11.5 and 11.6 present some of their data for these complex systems (we have added the last column, "Unit costs"). Notice that to get an average overall capital cost for the complete process (MSW to electricity) we must add the costs from Tables 11.5 and 11.6, resulting in a figure of $57,100 per ton-per-day (TPD) of capacity. Also, notice the wide variability in the reported data.

Table 11.7 presents data from Reilly and Powers (1980) for simpler systems that burn unprocessed solid wastes (note that the smaller

TABLE 11.5 Capital costs for production of RDF from MSW.

Location	*Capacity (TDP)*	*Product*	*Capital cost ($ million)*	*Unit cost ($ thous/TPD)*
Baltimore, MD	600–1,500	RDF	8.4	9.8
Bridgeport, CT	1,800	Eco-Fuel II	53	46.1
East Bridgewater, MA	550	Eco-Fuel II	10–12	20.0
Lane County, OR	500	RDF	2.1	4.2
Madison, WI	400	RDF	2.5	6.2
Monroe County, NY	2,000	RDF	60.2	30.1
Newark, NJ	3,000	RDF	70	23.3
New Orleans, LA	700	RDF	9.1	13.0
Tacoma, WA	500	RDF	2.5	5.0
				Average = 17.5

From Reilly and Powers (1980).

TABLE 11.6 Capital costs for burning RDF to produce steam or electricity.

Location	*Capacity (TPD)*	Product	*Capital cost ($ million)*	*Unit cost ($ thous/TPD)*
Ames, IA	200	Electricity	6.19	31.0
Chicago, IL (SW)	1,000	Electricity	19	19.0
Columbus, OH	2,000	Electricity	127	63.5
Detroit, MI	3,000	Steam and/or electricity	125	41.7
Duluth, MN	400	Electricity	19	47.5
Lakeland, FL	300	Electricity	5	16.7
Milwaukee, WI	1,600	Electricity	22	13.9
Portsmouth, VA	2,000	Electricity	144.9	72.4
Akron, OH	1,000	Steam	51	51.0
				Average = 39.6

From Reilly and Powers (1980).

TABLE 11.7 Capital costs for burning unprocessed MSW.

Location	*Capacity (TPD)*	Product	*Capital cost ($ million)*	*Unit cost ($ thous/TPD)*
Waterwall incinerators				
Braintree, MA	250	Steam	2.8	11.2
Chicago, IL (NW)	1,600	Steam	23	14.4
Harrisburg, PA	720	Steam	8.3	11.5
Norfolk, VA	360	Steam	2.2	6.1
Oceanside, NY	750	Steam	9	12.0
Portsmouth, VA	160	Steam	4.5	28.1
Saugus, MA	1,200	Steam	50	41.7
Albany, NY	750	Steam	22	29.3
Modular incinerators				
Auburn, ME	100	Steam	3.2	21.3
Crossville, TN	60	Steam	1.11	18.5
Oversburg, TN	100	Steam	2	20.0
Genessee Township, MI	100	Steam	2.0	20.0
North Little Rock, AR	100	Steam	1.45	14.5
Osceola, AR	50	Steam	1.1	22.0
Pittsfield, MA	240	Steam	6.2	25.8
Salem, VA	100	Steam	1.9	19.0
Siloam Springs, AR	16	Steam	0.4	25.0
			Average =	20.0

From Reilly and Powers (1980).

systems are all modular incinerators). From these data, an average capital cost of $20,000/TPD of capacity was obtained. Even though much lower in cost than an RDF facility, a simple MSW incineration system is still expensive too. Because of the high capital cost, revenues are required both from steam sales and disposal fees in order to make these systems economically attractive.

Example Problem 11.3

Estimate the capital cost for a simple incineration system to handle the wastes of example problem 11.2. Size the system for 1990 data. Assuming the plant would be constructed in 1985, escalate the average cost (Table 11.7) from 1980 dollars to 1985 dollars using a 9% per annum inflation factor.

Solution For 1990 data, assuming the plant operates (5 × 52) = 260 days per year,
the capacity of the plant is:

$$30{,}000 \text{ people} \times \frac{5.0 \text{ lb}}{\text{person-day}} \times \frac{365 \text{ day}}{\text{yr}} \times \frac{1 \text{ yr}}{260 \text{ days}} \times \frac{1 \text{ ton}}{2000 \text{ lb}}$$

$$= 105 \text{ TPD}$$

In 1980 dollars, the capital cost is

$$105 \text{ TPD} \times \frac{\$20{,}000}{\text{TPD}} = \$2.1 \text{ million}$$

In 1985 dollars, the cost is projected to be

$$\$2.1 \text{ million} \times (1.09)^5 = \$3.2 \text{ million}$$

The previous example problem demonstrated how to estimate the investment required for a relatively small, simple incinerator system. The complete annual operating cost of a simple incineration facility such as the one of example problem 11.3 includes costs for labor, supervision, overhead, auxiliary fuel, utilities, boiler water treatment, maintenance, taxes, insurance, ash disposal, depreciation, and interest on invested capital. The next example problem illustrates the fact that these operating costs require that charges be assessed for dumping the MSW at the incinerator and that steam be sold in order to generate revenues equal to these operating costs.

Example Problem 11.4

Consider the data of example problems 11.1–11.3. If the variable annual operating cost of an incinerator is \$11 per ton of waste (in 1980 dollars), and the 1980 tipping fee is \$6.00 per ton, estimate the price that must be charged for steam produced to break even on the incinerator. Steam generation requires roughly 10^6 BTU/1000 lb steam. For simplicity, do all work in constant 1980 dollars. Assume straight-line depreciation over 15 years with zero salvage value, and assume an interest rate of 11%.

Solution

Invested capital (\$ thou)	2100
Capital charges (\$ thou/year)	
Interest	231
Depreciation	140
Variable costs (labor, etc.)	
$\frac{\$11}{\text{ton}} \times 18{,}250$ tons/yr	201
Total costs (\$ thou/year)	572
(\$/annual ton)	31
Revenues (\$ thou/year)	
Tipping fees	110
Steam revenues required	462
Total revenues	572

Steam price required to break even is calculated as follows:

$$\$462{,}000 \times \frac{1 \text{ yr}}{6.57 \times 10^{10} \text{ BTU}} \times \frac{10^6 \text{ BTU}}{1000 \text{ lb steam}}$$

$$= \$7.03/1000 \text{ lb steam}$$

As can be seen from the previous example, a price of roughly \$7.00/1000 lb steam is required. This is higher than the cost of fuel to generate steam in a fuel-fired boiler in many parts of the country, although it may be competitive with the total cost of steam (which includes investment in a new boiler and all associated fixed and variable costs).

11-6 Conclusion

As with our liquid and gaseous wastes, treatment and disposal of our solid wastes is an expensive but necessary proposition. People generate solid wastes, and people in large cities (particularly in affluent countries such as the United States) generate enormous quantities of solid wastes. Currently, the most widely used form of disposal of MSW is to place it into sanitary landfills. As land availability decreases, and the prices of materials and energy increase, we may see a gradual switch to resource recovery operations.

11-7 Problems

11.1 A city is considering MSW incineration as an alternative to landfilling. The city has a 1982 population of 70,000 people; it is growing at 2.0%/year compounded. The wastes have an average heat content of 4600 BTU/lb. A new simple incineration plant with heat recovery could come on line in 1986. Calculate the design feed rate to the plant in 1986 and the steam generation rate (assume 50% thermal efficiency and 1000 BTU/lb steam).

11.2 List some of the major cost considerations for the above plant. Estimate the capital and annual operating costs for this plant from information presented in this chapter. (Estimate the AOC in 1986 and 1990.)

11.3 If the above plant were to be designed for complete resource recovery, estimate the annual recovery rates of iron, glass, and aluminum. Estimate the capital cost from data presented in this chapter.

11.4 Define: sanitary landfill; resource recovery.

11.5 Do some detective work in your own community. Find out the population, then obtain estimates of the annual amount of MSW collected and disposed. Does this amount include commercial refuse as well as MSW? Find out where the landfill is located. Does your community have an incinerator?

11.6 Estimate the landfill area required for MSW disposal for a town with a stable population of 40,000. The new landfill must last for 20 years. Include a 25% land allowance for buildings and access roads. State any assumptions you must make.

11.7 A thousand kilograms of the soluble poison potassium cyanide is deposited in a sanitary landfill. Given the following assumptions, calculate the average concentration of cyanide ion in the leachate. Assumptions: landfill area = 40,000 m^2; rainfall for one month = 15 cm; 60% of the rainfall runs off the landfill and does not percolate through; all of the KCN has been leached away by the end of the month. Give your answer in mg/L.

11.8 Leachate from an MSW landfill has a BOD_5 of 10,000 mg/L. It flows directly into a clean stream flowing at 20 m^3/min. Immediately downstream from the mixing point, the stream BOD_5 is 300 mg/L. Calculate the flow rate of the leachate.

11.9 The city wishes to treat the leachate of problem 8 before discharging it to the stream. Calculate the percent reduction in BOD_5 required to achieve 4 mg/L in the stream after mixing.

References

Brunner, D. R. and Kelly, D. J. 1972. Sanitary landfill design and operation. Washington, DC: Environmental Protection Agency Office of Solid Waste Management, SW-65ts.

Cooper, C. D. 1981. Evaluation of the feasibility of a coal-fired electricity generating plant/municipal wastes incinerator for the City of Kissimmee. Orlando: University of Central Florida.

Council on Environmental Quality. 1978. *Environmental quality—The Ninth Annual Report of the Council on Environmental Quality.* Washington, DC.

Environmental Protection Agency. 1975. *Hazardous wastes.* Washington, DC: Environmental Protection Agency Publication SW-138.

Krause, H. H., Vaughan, D. A. and Boyd, W. K. 1976. Corrosion and deposits from combustion of solid waste—Part IV: Combined firing of refuse and coal. *Engineering for Power* 98, 3.

Parker, F. G. and Pech, J. W. 1978. Tennessee Valley Authority solid waste management technology assessment.

Reilly, T. C. and Powers, D. L. 1980. Resource recovery systems, Part II: Environmental, energy, and economic factors. *Solid Waste Management/Refuse Removal Journal* 23, June 1980.

Schweiger, B. 1978. Refuse-derived fuel wins out over oil and coal at new industrial power plant. *Power* 122 (5):33–40.

Tchobanoglous, G., Thiesen, H. and Eliassen, R. 1977. *Solid wastes engineering principles and management issues.* New York: McGraw-Hill.

Tennessee Valley Authority. 1976. Resource recovery from municipal solid waste—Feasibility study. Tennessee Valley Authority Project No. 3023.

Vesilind, P. A. and Rimer, A. E. 1981. *Unit operations in resource recovery engineering.* Englewood Cliffs, NJ: Prentice-Hall.

12

Hazardous Wastes

12-1 Introduction

Common sense tells us that hazardous wastes are those wastes which are dangerous to us or to some other part of our biosphere. However, this general statement neglects several important questions. Exactly what are those things in the wastes that are hazardous? Why are they hazardous? In what quantities or concentrations are they hazardous? Where do they come from and where do they go? How can we properly treat and dispose of them? These are certainly key questions for today's society. In this chapter, we shall try to answer some of them.

In 1980 several billions of metric tons of solid wastes were generated and disposed of in the United States. Of this total the Environmental Protection Agency (1980) estimates that 57 millions tons could be classified as hazardous. Over the past decades, thousands of industries throughout the country discarded or buried waste chemicals that were toxic or had toxic constituents. Most did so in the naive belief that once these wastes were safely tucked away underground, they would no longer pose a threat to society. Unfortunately, recent experiences, such as that at Love Canal, have proved that belief false.

Wastes that contain heavy metals or toxic organics such as pesticide residues persist and remain toxic in the environment for long times. When one small geographic area accumulates a large quantity of wastes, such as in any of a number of abandoned chemical dump sites around the country, the wastes can leach into and contaminate public drinking water supplies. Several examples documented by the

Figure 12.1 A hazardous waste dump site (Environmental Protection Agency, 1975).

EPA are cited in the following paragraphs to illustrate some of the very real hazards presented by past practices of unregulated disposal of these wastes. Figure 12.1 shows one of these improper hazardous waste dump sites.

"The water supplies of Toone and Teague, Tennessee, were contaminated in 1978 with organic compounds when water leached from a nearby landfill. When the landfill closed, about 6 years earlier, the site held some 350,000 drums, many of them leaking pesticide wastes. Because the towns no longer have access to uncontaminated ground water, they must pump water in from other locations.

"Ground water in a 30-square-mile area near Denver was contaminated from disposal of pesticide waste in unlined disposal ponds. The waste, from manufacturing activities of the U.S. Army and a chemical company, dates back to the 1943-to-1957 period. Decontamination, if possible, could take several years and cost as much as $80 million.

"About 17,000 drums littered a 7-acre site in Kentucky—which became known as "Valley of the Drums"—about 25 miles south of Louisville, Kentucky. Some 6000 drums were full, many of them oozing their toxic contents onto the ground. In addition, an undetermined quantity of hazardous waste was buried in drums and surface pits. In

1979, EPA analyses of soil and surface water in the drainage area identified about 200 organic chemicals and 30 metals.

"A truck driver was killed in 1978 as he discharged waste from his truck into one of the four open pits at a disposal site in Iberville Parish, Louisiana. He was asphyxiated by hydrogen sulfide produced when liquid wastes mixed [and reacted] in the open pit. The area was surrounded by water and had a history of flooding.

"A fire broke out in 1978 at a disposal site in Chester, Pennsylvania, where 30,000 to 50,000 drums of industrial waste had been received over a 3-year period. The smoke forced closing of the Commodore Barry Bridge and 45 firemen required medical treatment, mostly as a result of lung and skin irritation from chemical fumes. A number of homes are located within three blocks of the site; drummed waste was kept only 20 feet from a natural gas storage tank and liquefied natural gas tanks were about 100 yards away. Waste was emptied directly on the soil of the 3-acre site; some probably drained to the tidal section of the adjacent Delaware River. Waste may even have been dumped into the river.

"Over a 4-month period in 1976, an Indiana family consumed milk contaminated with twice the maximum concentration of polycholorinated biphenyls (PCBs) considered safe by the Food and Drug Administration. The milk came from the family's cow, which had been grazing in a pasture fertilized with the City of Bloomington's sewage sludge. The sludge contained high levels of PCBs from a local manufacturing plant. A Federal law passed in 1976 banned production of PCBs after January 1, 1979.

"The health of some residents of Love Canal, near Niagara Falls, was seriously damaged by chemical waste buried a quarter of a century ago. As drums holding the waste corroded, their contents percolated through the soil into yards and basements, forcing evacuation of over 200 families in 1978 and 1979. About 80 chemicals, a number of them suspected carcinogens, were identified." (Environmental Protection Agency, 1980)

The above examples demonstrate the consequences that can and do result from mismanagement or lack of control of hazardous wastes. In 1976, Congress enacted legislation to help prevent the recurrence of such incidents in the future. Empowered by the legislature, the EPA has promulgated a vast set of regulations aimed at controlling this national problem. Engineers have developed and applied many processes to treating hazardous wastes; some will be presented later in this chapter.

TABLE 12.1 Limits on contaminants in leachate for EP toxicity test.

EPA hazardous waste number	*Contaminant*	*Maximum concentration (mg/L)*
D004	Arsenic	5.0
D005	Barium	100.0
D006	Cadmium	1.0
D007	Chromium	5.0
D008	Lead	5.0
D009	Mercury	0.2
D010	Selenium	1.0
D011	Silver	5.0
D012	Endrin (1,2,3,4,10,10-hexachloro-1, 7-epoxy-1,4,4a,5,6,7,8,8a-octahydro-1, 4-endo, endo-5, 8-dimethano naphthalene)	0.02
D013	Lindane (1,2,3,4,5, 6-hexachlorocyclohexane, gamma isomer)	0.4
D014	Methoxychlor (1,1,1-Trichloro-2, 2-bis(p-methoxyphenyl) ethane)	10.0
D015	Toxaphene ($C_{10}H_{10}Cl_8$) technical (chlorinated camphene, 67–69 percent chlorine)	0.5
D016	2,4-D, (2,4-dichlorophenoxyacetic acid)	10.0
D017	2,4,5-TP Silvex (2,4,5-trichlorophenoxypropionic acid)	1.0

From *Federal Register* (1980).

12-2 Kinds of hazardous waste

The federal government defines hazardous waste as any solid, liquid, or contained gaseous waste which may "cause or significantly contribute to an increase in mortality or . . . illness, or pose a substantial present or potential hazard to human health or the environment when . . . improperly . . . managed" (*Federal Register*, May 19, 1980). Thus, the hazard could be due to toxic materials, acids, explosives, or flammables, for example. Under the provisions of two laws passed by Congress in 1976, the Toxic Substances Control Act (TSCA) and the Resource Conservation and Recovery Act (RCRA), the EPA was given authority and responsibility to identify and control the generation, transportation, treatment, and disposal of hazardous wastes. Since that time literally millions of hours of engineering and planning have been spent on defining and developing solutions to the multifaceted problem of hazardous wastes.

Under EPA regulations (*Federal Register*, May 19, 1980) there are four characteristics or classifications of hazard to be considered: ignitability, corrosivity, reactivity, and EP toxicity. Ignitable materials are those liquids with a flash point below 60°C (140°F), or those non-liquids that are "easily ignited and burn vigorously and persistently." Corrosivity, as pertains to a hazardous waste, is an aqueous solution with a pH outside the range of 2.0 to 12.5, or any liquid which exhibits a steel corrosivity of more than 6.35 mm (0.25 inch) per year. Reactive wastes are unstable, form toxic vapors or fumes, or can explode.

As for toxicity, a waste is considered a toxic hazard if it is specifically listed by the EPA. These listings include generic wastes such as spent halogenated solvents, specific source wastes such as spent pickle liquor from steel-finishing operations, and certain commercial products or intermediates which are toxic such as cyanides, tetraethyl lead, benzene, or phenol. Also, waste is considered to have a toxicity hazard (even if it is not included in any of the previous lists) if so indicated by the EP toxicity test. This test involves leaching the waste material with acetic acid for 24 hours and testing the extract for concentrations of eight toxic elements and six pesticides (see Table 12.1). The waste is considered toxic if any of the concentrations in the extract are greater than 100 times the drinking water standards.

12-3 Sites and sources

Various types of dangers are represented in the examples cited earlier: pollution of groundwater, contamination of lakes, rivers, wetlands, or open land, air pollution, fire or explosion hazard, food-chain poisoning, and direct contamination of people with possible immediate or long-term effects. Although all these dangers are real, perhaps the most serious is contamination of drinking water supplies. The EPA has identified 418 sites of high potential for groundwater contamination by leachate from dump sites and has begun cleanup operations at a few of these. A map (Figure 12.2) showing the locations of some of those sites demonstrates the widespread nature of the problem. Twelve states were listed with ten or more hazardous dumps; New Jersey, a heavily industrialized state, had 65 sites on the list. The agency estimates that about 14,000 toxic chemical dumps exist in the United States (Groer, 1982).

Hazardous wastes can come from almost anywhere. We are an industrial society with a strong dependence on chemicals, and we have an excellent system for the manufacture of almost any product and its distribution to any part of the country. Hazardous wastes can be generated in several ways: sludges and residues generated in the normal operations of process industry, accidental release from a

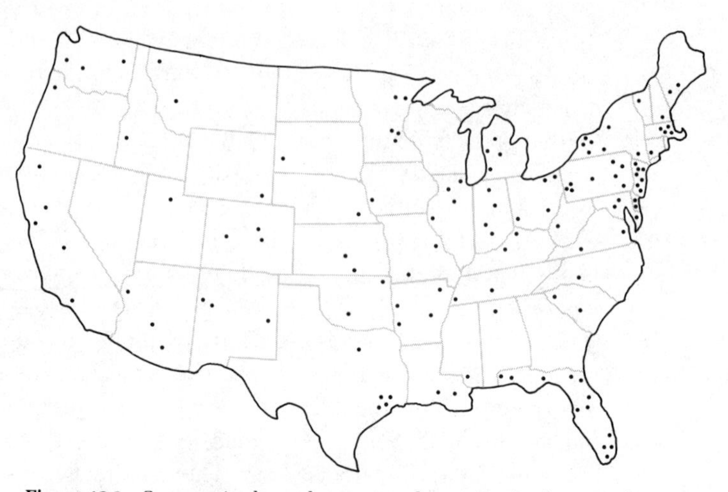

Figure 12.2 Some major hazardous waste dump sites in the United States (from Groer, 1982).

manufacturing plant, spills and leaks during transportation, normal low-concentration air and water discharges from a plant, releases during product usage, and improper disposal of partially used products by consumers. With such an array of potential sources, it is not surprising that hazardous wastes have been found in so many different locations.

12-4 Treatment and disposal of hazardous wastes

There are many different approaches to the treatment and disposal of hazardous wastes, just as there are many different types of wastes. It is worth noting that cleaning up an area after it has been contaminated generally costs much more than properly treating wastes before discharging them into the area.

The EPA hazardous waste management program is designed to track all significant quantities of hazardous wastes from cradle to grave. Standards have been written for generators, transporters, and commercial disposers of hazardous wastes (*Federal Register*, May 19, 1980). A manifest system has been implemented which requires complete reporting of all movements of wastes. Under the RCRA and other legislation, the EPA has been able to provide funds and technical assistance to states for cleanup of disposal sites and for developing state and local hazardous waste management programs. The problem of abandoned dump sites, where owners cannot be found or where they deny any legal liability, is being addressed by the "superfund." The superfund is a fund of money provided by taxes on industries that generate hazardous wastes. The fund was authorized under the

Comprehensive Environmental Response Compensation and Liability Act of 1980. This money can be deployed rapidly to clean up uncontrolled and abandoned sites as well as spills of hazardous substances. The EPA has already selected contractors to begin studies on several specific sites to characterize the wastes and design remedial actions to clean them up (*CDM News*, 1982).

Treatment and disposal of hazardous wastes has become a complex and expensive proposition. Already there are numerous waste-processing plants throughout the United States, and given the current government regulatory outlook, the growth prospects for this relatively new industry seem quite good. In discussing treatment and disposal of hazardous wastes, there appear to be three semi-independent variables which must be considered: the characteristics of the waste, the source of the waste, and the type of process used on the waste (Cooper, 1978).

Waste characteristics include its physical type (gas, liquid, or solid), its chemical composition (heavy metals, chlorinated organics), its hazards (toxic, flammable, corrosive), and its rate of generation (average, peak, frequency). The source of the waste refers to the primary means of escape into the environment. Sources fall into one of four categories: manufacturer, transporter, consumer, or waste processor. A hazardous waste may be generated as a by-product of manufacturing and released in trace amounts continuously, or it may be generated in bulk and stored for later treatment. An accident during transit may transform a chemical feedstock into a hazardous waste. The consumer in discarding half-empty containers of paint or pesticides may be a source of hazardous waste. A different approach must be taken in each of these cases. Finally, the waste processor may also be a source of hazardous wastes if the processes used are inadequate or if the plant experiences operating problems.

Treatment processes used on hazardous wastes fall into three categories. For some substances, the only step that can be taken to reduce danger to the environment is to put them in impervious containers or otherwise isolate them. This is "storage" since the problem remains to be treated whenever the technology to do so is developed. "Treatment" involves chemical, physical, or biological processes to make wastes less toxic, more concentrated, or easier to handle. "Ultimate disposal" means placement of wastes on or in the land, in the ocean, or in some other "last repository."

Because the vast majority of hazardous wastes receive treatment of some kind prior to disposal, perhaps the best framework for discussion is by process. Ottinger and others (1973) categorized all treatment and disposal processes into four major classifications, and we have added a fifth. The classifications are: (1) separation and concentration (physical processes), (2) biological degradation, (3) chemical treatment, (4) ultimate disposal, and (5) energy and material recovery. Each of these classifications is briefly described in the following sections. It is beyond the scope of this text to discuss in detail the many processes in each category that are currently available for treatment of hazardous wastes. We will, however, review the last two categories more thoroughly than the others.

Separation and concentration

Separation and concentration processes are mostly used as pretreatment for gas or liquid effluents. The hazardous component, a small fraction of the total volume of the effluent, is removed from the stream. Alternatively, a large portion of the steam is purified and released, leaving the hazardous component concentrated into a small fraction of the original stream. Some of the more common processes or unit operations used are as follows:

Gas stream	*Liquid stream*
Absorption	Absorption
Adsorption	Adsorption
Electrostatic precipitation	Distillation
Filtration	Electrodialysis
	Evaporation
	Filtration
	Ion exchange
	Reverse osmosis
	Sedimentation

In a majority of these physical treatment processes, there is some sludge, residue, or concentrated waste stream to be disposed of or further treated. Indiscriminate release of the sludge may cause severe environmental damage. Thus, the usual advantage of separation and concentration is only that it reduces the volume of a waste stream so that it may be further processed economically. Separation or concentration unit operations are often followed by another operation involving biological or chemical treatment, but also can be used simply to reduce the volume of material for ultimate disposal.

Biological degradation

Biological treatment processes are primarily used to reduce organic contamination of aqueous streams. Municipal waste treatment plants throughout the United States depend on the natural ability of certain microorganisms to biodegrade organic waste material. In addition to household wastes, many industrial organic wastes (particularly petrochemical related) are treated with biological processes. Some of the biological processes used in treating industrial wastes are as follows:

Activated sludge	Land spreading
Aerobic/anaerobic lagoons	Trickling filter
Anaerobic digestion	Rotating biological contactors

As noted in an earlier chapter, the microorganisms at the heart of any biological process can be very sensitive to any of a number of organic or inorganic compounds that may be present in an industrial wastewater. When treating waters contaminated with toxic organics the problems can be extreme. However, with care, certain bacteria can be acclimated to these toxic organics. In fact, genetically engineered bacteria specifically developed for this type of job may prove to be of significant use in treating certain toxic organics (*Chemical Week*, 1982). Typically, processes to treat continuous industrial wastewater streams require special attention to such environmental factors as pH, temperature, and flow and load variations.

Often, biological processes for industrial wastewaters are operated with much longer mean cell residence times than for municipal wastes. Furthermore, physiochemical processes can be used in combination with biological processes to treat certain toxic wastewaters. An example is an aqueous-phase "dissolved air" oxidation process applied to acrylonitrile plants that has been proposed for treatment of insecticide-manufacturing wastes (Wilhelm and Ely, 1976).

Chemical treatment

Chemical treatment processes effect a chemical change in the hazardous component to render it less hazardous. Chemical changes include the following:

Neutralization
Chemical oxidation/reduction
Precipitation

Neutralization usually refers to the process of adjusting the pH of highly acid or alkaline wastes to near neutrality. In a waste that is only hazardous because of acidity or alkalinity this is a sufficient treatment, but usually neutralization is a preliminary step to other treatments or ultimate disposal.

The term neutralization sometimes also refers to the detoxification of pesticides. Several common insecticides (both carbamates and organophosphates) are readily hydrolyzed in alkaline media to nontoxic products (Atkins, 1972). Oxidation, reduction, and chlorinolysis have been applied to pesticide detoxification, but sometimes result in incomplete destruction of the pesticide or toxic by-products (Munnecke and others, 1976).

Chemical oxidation refers to oxidation of a hazardous material, usually in an aqueous solution, by oxidizing agents such as hypochlorites, peroxides, permanganates, and ozone. Oxidation by gas-phase combustion in air is addressed in a later section on incineration. Chemical oxidation is widely practiced in the plating industry to destroy cyanides and in the pesticide industry for those compounds that are not readily detoxified by alkalai treatment.

Chemical reduction is applied to certain of the heavy metals that are hazardous in the oxidized state but not in the reduced state. Similarly, precipitation reactions not involving oxidation or reduction are often used to remove heavy metals from solution. The sludges resulting from both reduction and precipitation must be disposed of properly to prevent the slow re-oxidation or leaching of the toxic element back into the environment. The next two example problems illustrate chemical oxidation and reduction techniques currently in use by the metal plating industry.

Example Problem 12.1

Cyanides often occur in metal plating wastewaters. A commonly used method of treatment is to oxidize them in two steps using chlorine at high pH. (Why must the pH of a solution with CN^- ions be kept high?) The chemistry is as follows:

$$CN^-(aq) + Cl_2 \longrightarrow CN^+ + 2Cl^-$$

$$CN^+(aq) + 2OH^- \longrightarrow CNO^-(aq) + H_2O$$

$$2CNO^-(aq) + 3Cl_2 + 4(OH)^-(aq) \longrightarrow 2CO_2 + N_2 + 6Cl^- + 2H_2O$$

Calculate the daily addition rate of chlorine gas to completely destroy cyanides in a wastewater stream flowing at 25 L/min which contains 80 mg/L of cyanides.

Solution

$$25\,\frac{\text{L}}{\text{min}} \times \left(80\,\frac{\text{mg}}{\text{L}} - 0\,\frac{\text{mg}}{\text{L}}\right) \times 1440\,\frac{\text{min}}{\text{day}} \times \frac{1\ \text{kg}}{10^6\ \text{mg}}$$

$$= 2.88\,\frac{\text{kg}}{\text{day}}\ \text{of CN}^-\ \text{destroyed}$$

$$2.88\,\frac{\text{kg CN}^-}{\text{day}} \times \frac{5\ \text{kmol Cl}_2}{2\ \text{kmol CN}^-} \times \frac{1\ \text{kmol CN}^-}{26\ \text{kg CN}^-} \times \frac{71\ \text{kg Cl}_2}{1\ \text{kmol Cl}_2}$$

$$= 19.7\,\frac{\text{kg Cl}_2}{\text{day}}$$

In practice we would probably inject 10–20% more Cl_2 than calculated in this example. This excess amount would ensure that water in all parts of the tank had sufficient Cl_2 to do the job in a reasonable time.

Example Problem 12.2

Hexavalent chromium is a very toxic ion often found in the wastewater from electroplating shops as chromic acid, H_2CrO_4. A commonly used treatment method is to first reduce the chromium from the +6 oxidation state to the +3 state (a less toxic form) using SO_2 gas, and then to precipitate $Cr(OH)_3$ using lime or sodium hydroxide. The chemistry is as follows:

$$2H_2CrO_4 + 3SO_2 \longrightarrow 2Cr^{+3} + 3SO_4{}^{-2} + 2H_2O$$

using NaOH:

$$2Cr^{+3} + 6NaOH \longrightarrow 2Cr(OH)_3 + 6Na^+$$

using lime:

$$2Cr^{+3} + 3SO_4{}^{-2} + 3Ca(OH)_2 \longrightarrow 2Cr(OH)_3 + 3CaSO_4$$

Note the increase in sludge production using lime due to the co-precipitation of calcium sulfate.

Calculate the daily addition rate of SO_2 gas (cubic feet per day at 25°C) required to reduce the Cr^{+6} concentration from 300 mg/L (as chromium) to 0.1 mg/L in a wastewater stream of 60 gallons per minute. Also calculate the total mass of sludge produced (in kg/day) if all the Cr^{+3} is precipitated using sodium hydroxide and the final sludge (after filtering) is 30% solids by weight.

Solution

$$(300 - 0.1)\frac{\text{mg}}{\text{L}} \times 60\frac{\text{gal}}{\text{min}} \times \frac{3.785\ \text{L}}{\text{gal}} \times \frac{1\ \text{g}}{1000\ \text{mg}} \times \frac{\text{mol}}{52\ \text{g}} = 1.310\frac{\text{mol}}{\text{min}}$$

The daily addition rate of SO_2 is

$$1.310\frac{\text{mol Cr}}{\text{min}} \times \frac{3\ \text{mol SO}_2}{2\ \text{mol Cr}} \times \frac{24.5\ \text{L SO}_2}{\text{mol SO}_2} \times \frac{1\ \text{ft}^3}{28.3\ \text{L}} \times \frac{1440\ \text{min}}{\text{day}}$$

$$= 2450\frac{\text{ft}^3}{\text{day}}$$

The sludge production (solids only) is

$$1.310\frac{\text{mol Cr}}{\text{min}} \times \frac{1\ \text{mol Cr(OH)}_3}{1\ \text{mol Cr}} \times \frac{103\ \text{g Cr(OH)}_3}{1\ \text{mol Cr(OH)}_3} \times \frac{1\ \text{kg}}{1000\ \text{g}}$$

$$\times \frac{1440\ \text{min}}{\text{day}} = \frac{194\ \text{kg Cr(OH)}_3}{\text{day}}$$

The total mass of sludge production (solids plus retained water) is

$$194\frac{\text{kg Cr(OH)}_3}{\text{day}} \times \frac{1}{0.30} = 647\frac{\text{kg sludge}}{\text{day}}$$

Ultimate disposal

The words "ultimate disposal," when used to describe a technique for handling hazardous materials, can invoke strong emotional arguments from many diverse groups. The reasons for this emotive reaction probably stem from past experiences with so-called ultimate disposal of toxic materials in which serious unanticipated damage was done to the environment. Today, we are more aware of the possible consequences of the disposal of wastes on or in the land, the water, or the atmosphere. Ultimate disposal should be practiced with deliberation.

Many forms of ultimate disposal have lost favor with most regulatory agencies today, but each of the techniques has certain favorable characteristics. As long as the characteristics of the wastes are matched properly with the technique, ultimate disposal is worthy of consideration. For certain wastes, some form of ultimate disposal appears to be mandatory. The techniques of ultimate disposal include the following:

Deep well injection
Dilution or dispersion
Chemical landfill
Ocean dumping

Deep well injection. In theory, deep (3000 to 12,000 feet) wells can provide safe disposal of toxic wastes if certain requirements are met. At the site picked for the deep well, the water table must be underlain by an impermeable stratum (clay, shale) which has beneath it a permeable stratum (sandstone, dolomite, limestone) to use as the receptor of the wastes. The two strata must not be connected in any way. Figure 12.3 presents a schematic of the system.

A well is drilled to the permeable stratum. Two or more casings are used to prevent leakage into the freshwater aquifer, and cement is used to seal the annular space. Injection tubing is inserted, with packers to hold it in place, and the well is perforated at the appropriate level. The space between the tubing and the casing can be filled with an inert fluid and a pressure gauge can be installed. This allows the operator to tell, by a change in pressure, whether leakage from the pipe is occurring (Forrestal, 1975).

The advantages of this method of disposal are that the cost is low, that less pre-treatment is necessary for deep well injection than for surface stream disposal, and that less land is taken up than with a landfill. Several disadvantages to deep-well injection have been found; potential contamination of underground water supplies appears to be the most serious of these disadvantages.

The groundwater can be contaminated in many ways (Forrestal, 1975). Wastes might be injected directly into a freshwater aquifer. The well casing could be fractured by too high an injection rate. If the casing is improperly installed, upward movement of wastes can occur. Leakage can occur through inadequate confining beds, through abandoned deep wells (for instance, petroleum exploration wells) or through hydraulic fractures. Saline water that is displaced by injection may move into a potable aquifer. According to at least one author, a correlation has been drawn between deep-well injection and increased earthquake activity (Forrestal, 1975).

An EPA report to Congress on waste disposal practices found that if deep wells are properly placed and monitored, they can pose a

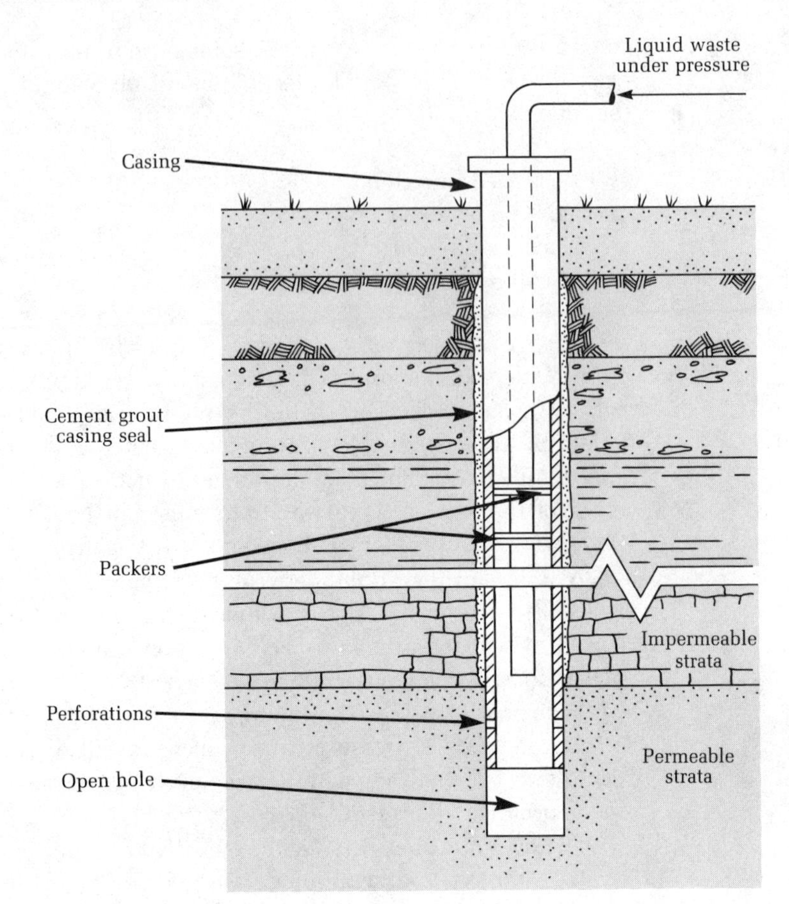

Figure 12.3 Schematic diagram of deep well injection system.

minimal threat to groundwater (Environmental Protection Agency, 1977). Others do not recommend use of deep wells for hazardous wastes because of the danger of groundwater contamination and because of the lack of control of the wastes once they are injected (Ottinger and others, 1973).

Dilution or dispersion. One method of dealing with hazardous wastes is to reduce their concentration by dilution to a point deemed harmless and then to disperse them into the air or water. If the waste in question is biodegradable and relatively non-toxic, then dilution and dispersal may be applicable. If the waste is subject to bioaccumulation or if it is highly toxic, dilution and dispersal may be a very dangerous method.

Chemical landfill. Landfilling is the method of disposal most often used for industrial residues (Environmental Protection Agency, 1976). A potential disadvantage is the possible pollution of groundwater from toxic wastes. If certain specific hazardous wastes are inherently non-leachable or if they have been treated to make them so, then a sanitary landfill is a possible alternative for their disposal. If wastes are leachable, or if they have the possibility of vaporizing, they should not be placed in a sanitary landfill. A chemical landfill is designed to contain those wastes not acceptable for an ordinary sanitary landfill.

One of the most important requirements for a chemical landfill is a secure site. The California Class I (hazardous waste) landfill site criteria state that geological barriers to the horizontal or vertical mixing of waste liquids or gases with usable groundwater must exist; that the leachate should be confined within the site; that sites should not be located over active faulting zones; and that the landfill should not be located in an area of extensive population (Fields and Lindsey, 1975).

In addition to a safe site, other steps are taken to secure the wastes against polluting the environment. Wastes are sorted and separated to prevent mixture of components that might unite to form hazardous gases or an explosion. Percolation barriers are placed on top of the landfill to prevent entrance of water. A leachate drainage and collection system is installed and hazardous leachate is collected and treated. A system of wells to monitor adjacent groundwater is set up (Fields and Lindsey, 1975). Figure 12.4 represents a schematic of a chemical landfill.

Unless geological barriers to water migration are perfect, chemical landfills are usually lined to prevent escape of contaminated fluid. Such liners can be compact native fine-grained soils, clay, asphalt, portland cement, soil sealants (such as penetrating latex or lime), sprayable liquid rubber, or synthetic polymeric membranes (Environmental Protection Agency, 1976). The liner (or combination of liners) used depends on the type of waste to be contained, the allowable level of permeability, the lifetime required, and liner exposure to sunlight and weather. Wastes are often treated by chemical neutralization, or are encapsulated before placing them in the landfill. Bacterial decomposition of the wastes may occur over a long period of time.

The chemical landfill has the advantage of safely disposing of hazardous wastes at a relatively moderate cost. For some hazardous wastes, no technology has been developed to create an alternative to

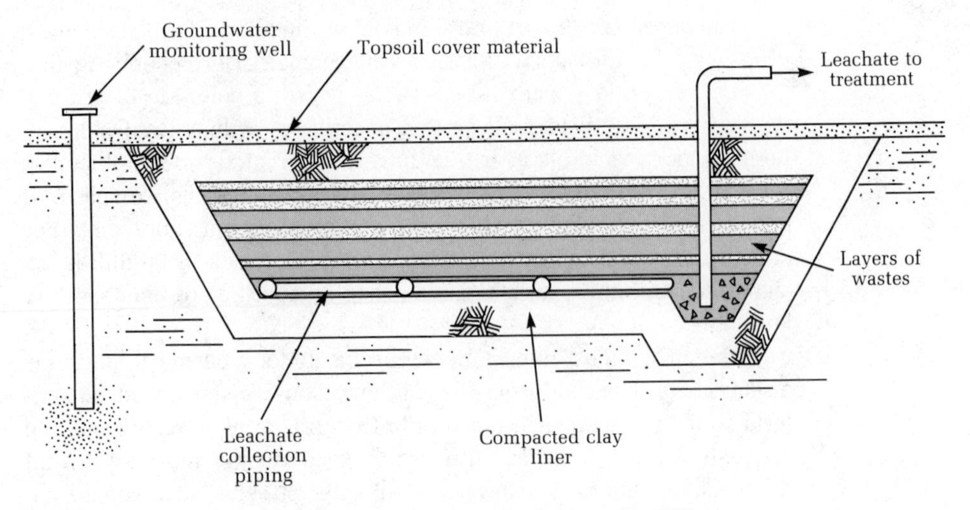

Figure 12.4 Schematic diagram of chemical landfill.

chemical landfilling. Disadvantages are the danger of leakage and the requirement of monitoring the site for many years after it is filled. There likely will be public resistance to having such a landfill nearby.

Ocean dumping. The dumping of hazardous wastes into the deep ocean is much less common now than it was several years ago, although it still occurs. Most of the waste is dumped in bulk from moving barges. The distance can be anywhere from 10 to 125 miles from shore; the speed varies from 3 to 6 knots, and the release rate varies from 4 to 20 tons per minute. Another method of dumping wastes is to encapsulate them in weighted 55-gallon drums, using the ocean bed as a burial place. The drums are released up to 300 miles from shore and should weigh at least 550 pounds to prevent them from floating (Ottinger and others, 1973).

The primary advantage of ocean dumping is its low cost—lowest of all the ultimate disposal methods discussed here. Some wastes are judged to be less harmful to the environment when dumped at sea than when disposed of on land. For those wastes which do not contain elements that are concentratable by marine life, the effects are limited to short time periods (12 hours or less) and to the immediate vicinity of dumping (Ottinger and others, 1973).

The disadvantage of ocean dumping centers primarily on the danger of concentration by marine organisms. The complex barriers to biological uptake that have evolved in terrestrial organisms do not exist in the marine environment. For example, shellfish concentrate hazardous material at a rate unequaled on land (Ember, 1975). Degradation takes place very slowly in the ocean, and a complex species such as a chlorinated pesticide, which would be degraded by soil bacteria on land, is not degraded in the ocean. Research has suggested that there is no free-living population of microorganisms in deep water to carry out degradation (Winsen and Jamseh, 1976).

Energy and material recovery

Energy and material recovery occurs when an economic or political incentive develops to further process a waste stream. If resources are being depleted or raw materials are expensive, those resources in wastes will increase in value. If energy is expensive, energy recovery will occur. If ultimate disposal space is limited, or so regulated as to be expensive, more recycling will take place. In some areas, the above conditions have occurred and recycling is being practiced. Many of the treatment methods outlined in the previous sections can be used as recycling methods, since they separate and concentrate certain chemicals that can be reused or sold. Energy/material recovery techniques discussed here include waste recycling, pyrolysis, and incineration.

Waste recycling. Wastes can be used in the raw state, as for example in a proposed project to revegetate strip mines in Philadelphia using industrial sludge. The sludge contains heavy metals, which can be concentrated in plants; however, the vegetation will be ornamental and not used as food for livestock or humans (Ember, 1975). This project demonstrates that a recovery system that uses raw hazardous wastes must take into account the qualities which make the waste hazardous.

A refinement of waste recycling is the waste clearinghouse. Typically, a waste clearinghouse puts a waste producer in touch with a potential waste user. A clearinghouse gives waste users an economic incentive to discover uses for the wastes (Harness and others, 1977). The producers would have an incentive to adapt their industrial processes to produce a usable waste stream. A study in Philadelphia showed, however, that only 9% of wastes in that city have exchange potential, and of those, only 20% would actually be exchangeable due to economic and practical limitations (Ricci, 1976).

The advantages of waste recycling are obvious. Waste does not have to be disposed of through ultimate disposal, and nonrenewable resources are not wasted. Disadvantages of waste recycling are that most recovery processes are still largely experimental and are prone to expensive failures or inefficient operations. Markets for recovered material are sometimes unstable, and if a contract for receiving the materials is not renewed, dumping costs added to recycling costs make the waste disposal extremely expensive (Wingerter, 1976).

Pyrolysis. Pyrolysis involves the exposure of wastes to high temperature in an airless environment (to prevent combustion). The gaseous products are usually steam, CO, CO_2, H_2, and CH_4. Liquid products are various organics, including methanol, ethanol, acids and tars. The solid product is a form of charcoal, mixed with the inorganic, non-volatile components of the wastes.

The advantage of pyrolysis is its potential for recovery of fuels of economic value. The gases could be burned directly as low-BTU fuel, the liquids could be blended in fuels or processed for chemicals recovery, and the solids, providing they did not contain actively hazardous inorganic components could also be burned for fuel. At the present time, pyrolysis is still in the developmental stage.

Incineration. Incineration of solid wastes has been an accepted method of treatment for some time. Incineration without energy recovery may be considered a form of ultimate disposal, but today most incinerators are designed with some provision for heat recovery, hence incineration is included in this section.

Incineration reduces the volume of materials. If, as is the case with most organics, the toxicity is due to the molecular structure of the material, incineration also serves as a detoxification process. Unlike the landfill disposal, incineration requires only a small land area.

Incineration is more costly than landfill, and the equipment involved is more complicated to operate. If the material to be burned is toxic due to elements it contains (such as heavy metals), these elements will not be destroyed, but will be vaporized or remain in ash, which may create further problems. The burning of any hazardous waste poses the problem of potential air pollution, which must be prevented by use of stack gas pollution-control devices.

If the incineration is conducted at sea, however, fewer impacts are felt. The incinerator ship *Vulcanus* has successfully burned chlorinated hydrocarbons at sea. The gases contain HCl, which is readily absorbed in the ocean water with negligible impact. Also, an old oil rig in the Gulf of Mexico has been proposed as a hazardous waste incineration site (*Orlando Sentinel Star*, 1981).

12-5 Radioactive wastes

Background and definitions

There are many issues surrounding the generation, handling, storage, reprocessing, and disposal of radioactive wastes. Several of these issues evoke strong emotional arguments from opponents of nuclear power. Radioactive waste management is similar in most ways to other major public health issues that have been confronted in the past. "However, no previous set of technical solutions has met with such formidable resistance and has been attacked through the use of such pervasive uncertainties as that which surround ionizing radiation and more particularly, commercial nuclear power and associated waste." (Gloyna, 1980)

The biggest stumbling block appears to be that people want a solution with zero risk, which is impossible. To gain the benefits of nuclear technology, we must be willing to define a level of acceptable risk. Only then can we proceed to engineer solutions to the problems of waste disposal. Recently, the Nuclear Regulatory Commission issued a proposed policy statement dealing with this risk (*Engineering Times*, 1982). They would set the acceptable risk (of an individual contracting cancer caused by living near the nuclear plant) at 0.1% of the risk of contracting cancer from other sources. While this seems reasonable at first, this level of risk has been interpreted to mean possibly 1300 deaths from nuclear reactor accidents for a projected 150 reactors in the United States over the next 30 to 40 years (*Engineering Times*, 1982).

There are two main types of nuclear waste: high-level wastes (HLW) and low-level wastes (LLW). (Much of the following discussion is abstracted from Gloyna, 1980). HLWs are the portion of wastes generated in the reprocessing of spent fuel that contain virtually all of the fission products. The waste is characterized by high levels of penetrating radiation, high heat generation rates, and long radioactive half-lives.

Presently, about 270,000 cubic meters of high-level waste, most resulting from military operations, is stored in steel tanks and bins.

To date, only about 2300 cubic meters of high-level waste has been generated as a result of commercial reprocessing. Since April 1977 no commercial reactor fuel has been reprocessed in the United States, but other countries are reprocessing such fuel.

LLWs contain very slight amounts of radioactivity, require little or no shielding, but have potentially hazardous concentrations or quantities of radionuclides. Present annual production of solid low-level waste in the United States is about 113,200 cubic meters or about 0.45 kg/person/yr. The U.S. Department of Energy produces about 50% of all LLW, but other areas of modern society—including medicine, medical research, university research, commercial power generation, and Navy vessels—also contribute to LLW.

The total exposure of Americans to radiation from all sources averages 182 mrem/yr per person. It comes principally from natural sources such as granite and cosmic rays (102 mrem/yr) and medical diagnostic x-rays (72 mrem/yr). Commercial nuclear power is estimated to contribute less than 0.01 mrem/year (Gloyna, 1980). A millirem (mrem) is one-thousandth of a REM (Radiation Equivalent Man), which is an arbitrary unit of measurement of absorbed radiation energy in human tissue. One REM is equivalent to absorbing 100 ergs of radiation energy for each gram of body tissue. One erg is equal to one ten-millionth of a joule—the amount of energy required to lift a mosquito about 1.0 centimeter.

Disposal alternatives

Generally LLWs are simply buried, but this is not considered optimal because water can leach radioactive materials from the site. Volume reduction is necessary, and this combined with solidification methods is the utlimate key to LLW disposal.

In dealing with high-level waste, reprocessing appears to be necessary. Spent fuel reprocessing permits recovery and reuse of unfissioned uranium and reduces the volume of nuclear waste to as little as 11% of its original volume. A 1000-MW nuclear power plant, using 30 tons of fuel, will produce (after reprocessing) one cubic yard of waste (fission products) per year. As a comparison, a similarly rated coal-fired plant produces about 620,000 tons of waste.

Once reprocessed, high-level wastes should be confined using multiple barriers. The multiple-barrier concept involves a combination of natural and artificial barriers, so that total reliance is not placed on just one barrier (Klingsberg and Duguid, 1982). Solidified waste may be encapsulated in glass, sealed inside steel canisters, and placed in retrievable locations in geologically sound formations. The fully engineered system would logically encompass consideration of the solid waste form, container, overpack, rock formation, and geographic isolation. Isolation from the biosphere must be very good because these concentrated wastes remain highly radioactive for hundreds to thousands of years (Cohen, 1977).

Radioactive wastes generate significant quantities of heat as they decay during the first hundred years or so after removal from a fission reactor. However, the heat rates are predictable and can be handled in the system design (Cohen, 1977). Neither the temperature nor solidification matrix, nor any other single item, need be the dominant factor in a waste disposal system design. The system design must always consider the multibarrier concept.

In conclusion we quote from Gloyna (1980): "Radiation is everywhere in our lives and there never will be zero radiation. . . . The radioactive waste problem is not going to vanish—not as long as we have hungry people. . . . We must continue to utilize nuclear capability to the fullest. . . . We must begin to reprocess our nuclear fuel, and demonstrate to the public that the waste can be placed in suitable containment systems. For this to happen, we need to separate the waste management question, which is technically feasible, from the philosophical debates surrounding no risk."

Recently an event occurred that may mark the beginning of a comprehensive U.S. plan to dispose of its nuclear wastes. In January 1983 a bill was signed into law which permits the selection, development, and operation of at least one permanent geologic repository for high-level wastes. Selection of a site by the Department of Energy is anticipated by 1987, with commencement of operations before the end of the century (Karoff, 1983).

12-6 Problems

12.1 A 100-g sample of sludge is subjected to the EP toxicity test. The final 1.950 liters of liquid contains 0.8 mg/L of cadmium. Assuming that 50% of the original cadmium leached out of the sludge, calculate the weight percent cadmium in the sludge.

12.2 A stream of wastewater containing benzene is subjected to carbon adsorption in which the benzene is separated from the wastewater and concentrated on the carbon. The carbon must eventually be processed to remove the benzene from it, and the benzene then disposed of. Discuss any advantages to the "extra" steps involved using carbon adsorption.

12.3 Define and discuss the pros and cons of chemical landfill and deep well injection.

12.4 Discuss the pros and cons of ocean disposal of hazardous wastes.

12.5 Discuss some of the advantages and disadvantages of burning hazardous wastes (especially chlorinated pesticides) at sea and on land.

12.6 Propose and discuss a process to treat an acidic wastewater stream which contains 400 mg/L dissolved lead, and 300 mg/L COD.

12.7 If the waste sludge (containing significant amounts of chromium, cadmium, and copper hydroxides) from a metal plating operation is tested by the EP toxicity test and found to be nontoxic, do you think this waste should be allowed to be put into an ordinary landfill? Explain.

12.8 In problem 6 above, assume that sodium phosphate (Na_3PO_4) is to be used to precipitate lead (II) phosphate ($Pb_3(PO_4)_2$) as part of the process. Calculate the mass addition rate of sodium phosphate required to produce an effluent stream with less than 0.5 mg/L lead. The wastewater flows at 500 liters per minute.

12.9 Chromium can be removed from metal plating wastewater by first reducing Cr^{+6} to Cr^{+3}, then precipitating $Cr(OH)_3$. One method is to use ferrous sulfate to act as the reducing agent. The reactions are as follows:

$$Cr^{+6} + 3Fe^{+2} \longrightarrow Cr^{+3} + 3Fe^{+3}$$

$$Cr^{+3} + 3Fe^{+3} + 12OH^{-} \longrightarrow Cr(OH)_3 + 3Fe(OH)_3$$

1. Calculate the addition rate of $FeSO_4$ required to remove 99% of the chromium from a 100 L/min stream containing 200 mg/L of Cr^{+6}.
2. Calculate the total amount of sludge produced in the process if NaOH is used in the precipitation step (assume 30% solids content of sludge).
3. Calculate the total amount of sludge produced if $Ca(OH)_2$ is used in the precipitation step (assume 30% solids content of sludge).

References

Atkins, P. R. 1972. The pesticide manufacturing industry—Current waste treatment and disposal practices. Washington, DC: Environmental Protection Agency, Program No. 12020 FYE.

Chemical Week. 1982. Bugs tame hazardous spills and dumpsites. May 5, 1982.

CDM News. 1982. Superfund: The first step. *CDM News* (published by Camp Dresser & McKee, Inc) 15(1).

Cohen, B. L. 1977. The disposal of radioactive wastes from fission reactors. *Scientific American* 236(6).

Cooper, C. D. 1978. Environmental planning for the treatment and disposal of hazardous wastes in the east central Florida area. Orlando: University of Central Florida.

Ember, L. R. 1975. Ocean dumping: Philadelphia's story. *Environmental Science and Technology* 9, October 1975, 916.

Engineering Times. 1982. NRC proposes guidelines for "acceptable risk" from nuclear power plants. *Engineering Times* 4(4).

Fields, T., Jr. and Lindsey, A. W. 1975. Landfill disposal of hazardous wastes: A review of literature and known approaches. Washington, DC: Environmental Protection Agency.

Forrestal, L. 1975. Deep mystery. *Environment* 17, November 1975, 25–31.

Gloyna, E. F. 1980. Radiactive waste—A perspective look. Distinguished Lecture Series, College of Engineering, University of Central Florida.

Groer, A. 1982. EPA lists 418 toxic dumps in nation—25 in Florida. *Orlando Sentinel*, Dec. 21, 1982.

Harness, R. L. and others. 1977. Waste clearinghouse promotes recycling efforts. *Civil Engineering-ASCE* 14(3), 71–73.

Karoff, P. 1983. National plan for burial of nuclear wastes ends years of U.S. indecision. *Engineering Times*, February 1983.

Klingsberg, C. and Duguid, J. 1982. Isolating radioactive wastes. *American Scientist* 70, March–April 1982.

Munnecke, D., Day, H. R. and Trask, H. W. 1976. Review of pesticide disposal research. Washington, DC: Environmental Protection Agency, Publication SW-527.

Orlando Sentinel Star. 1981. Gulf rig may burn noxious wastes. December 28, 1981.

Ottinger, R. S. and others. 1973. *Recommended methods of reduction, neutralization, recovery, or disposal of hazardous waste*, 16 volumes. Cincinnati: National Environmental Research Center.

Ricci, L. J. 1976. Chemical waste swapping: Promising but no panacea. *Chemical Engineering* 83(14), 44.

U.S. Environmental Protection Agency. 1975. Hazardous wastes. Publication SW-138.

U.S. Environmental Protection Agency. 1976. *Residual management by land disposal*, Proceedings of the Hazardous Waste Research Symposium. Cincinnati: Environmental Protection Agency.

U.S. Environmental Protection Agency. 1977. Report to Congress: Waste disposal practices and their effects on groundwater.

U.S. Environmental Protection Agency. 1980. Everybody's problem: Hazardous waste. Publication SW-826.

Wilhelm, A. R. and Ely, R. B. 1976. A two step process for toxic wastewater. *Chemical Engineering*, February 16, 1976.

Wingerter, E. J. 1976. Resource recovery. *Solid Waste Management* 19, October 1976, 30.

Winsen, C. O. and Jamseh, H. W. 1976. Decomposition of solid organic materials in the deep sea. *Environmental Science and Technology* 10, September 1976, 880.

13

Environmental Impact Assessment

13-1 Introduction

In some respects, the topics that have been presented in this text might be considered dissimilar: air pollution, water pollution, solid wastes, energy resources. Yet, in other respects, these topics are unified and thus quite similar to each other. One unifying theme is that of the materials balance approach, which can be applied to the analysis of typical problems in each area.

Another important unifying factor of this text is that all human activities have some sort of environmental impact. In discussing each of the major topics of this book, we have concentrated on the environmental impacts and engineering solutions to those impacts. Certainly it should be recognized that large-scale projects will have impacts on the environment in many functional areas. Construction and operation of a new airport, for instance, will likely result in air, water, land-use, noise, biological, cultural, and socioeconomic impacts in the community.

The field of environmental impact assessment has evolved to help us see the larger picture in this type of situation. Through methodical and open-minded analysis of a proposed action, we try to forecast the size and significance of all the project's impacts, to judge its benefits and its costs, to propose alternatives (in advance of the project's implementation), to reduce its impacts, and ultimately to make a decision whether or not to proceed with the project. The previous sentence summarizes, in a nutshell, the "why" of environmental impact assessment. In this chapter, we shall briefly explore the who, how, and what of this important field.

13-2 Background

The following quote is related to the assessment of European mining projects:

"... the strongest argument of the detractors is that the fields are devastated by mining operations ... Also, they argue that the woods and groves are cut down, for there is need of an endless amount of wood for timbers, machines, and the smelting of metals. And when the woods and groves are felled, then are exterminated the beasts and birds, very many of which furnish a pleasant and agreeable food for man. Further, when the ores are washed, the water which has been used poisons the brooks and streams, and either destroys the fish or drives them away ... Thus, it is said, it is clear to all that there is greater detriment from mining than the value of the metals which the mining produces." (Georgius Agricola, 1556)

As indicated by the date of the quote, the concept of environmental impact assessment is not new to the world. Probably the first American text in this field was *Man and Nature; or Physical Geography as Modified by Human Action*, written by George Marsh in 1871. Coincidentally, the first of our national parks, Yellowstone, was dedicated in 1872.

Around the turn of the century, the first real efforts at federal land preservation were initiated by President Theodore Roosevelt and his advisor, Gifford Pinchot. Pinchot developed the idea of multiple uses for forest lands, and under the Roosevelt administration millions of acres were set aside for National Forests. It was not until the 1930s, under President Franklin Roosevelt, that another major federal conservation effort was begun, partly in response to the needs to provide jobs during the depression.

In the 1950s and 1960s, several federal laws were passed to deal with individual environmental issues such as water pollution and air pollution. However, none of these laws required prior comprehensive environmental planning for large projects with potentially serious environmental consequences. Toward the latter part of the 1960s, it became clear that such a law was needed.

On January 1, 1970, Public Law 91–190, the National Environmental Policy Act (NEPA), was signed into law by President Richard Nixon. To quote from the law itself, its purposes were "to declare a national policy which will encourage productive and enjoyable harmony between man and his environment; to promote efforts which will prevent or eliminate damage to the environment and biosphere and stimulate the health and welfare of man; [and] to enrich the understanding of the ecological systems and natural resources important to the Nation."

13-3 How the impact assessment system works

Prior to NEPA, management decisions were based primarily on technical and economic factors. After NEPA, it was required that societal and environmental concerns be considered on an equal basis. The mechanism originally specified by NEPA was to require a detailed environmental impact statement (EIS) on all "major federal actions significantly affecting the quality of the human environment." Since January 1, 1970, there have been numerous publications, agency regulations, and court rulings to interpret and clarify just what must be contained in an EIS, who must prepare one, when must it be prepared, and how the assessment must be conducted.

There are dozens of different local, state, or federal agencies that may be involved (directly or indirectly) in the planning and permitting process for a large project such as a new nuclear power plant, a new phosphate rock mine, or a new oil pipeline. There have been scores of methodologies ("how-to" documents) published. There have been hundreds of state and federal court rulings on the thousands of environmental documents that have been written. Therefore, any attempt to unravel and explain all the intricacies of "the system" in a few paragraphs will not be totally successful. However, we must make the attempt.

The terminology

Before we undertake a discussion about the agencies and the processes they perform, it is wise to define a few terms commonly used in this field.

The first is *environmental inventory*. The environmental inventory is "a complete description of the environment as it exists in an area where a particular proposed action is being considered" (Canter, 1977). It includes descriptive parameters from various areas of the physical environment (air quality, water quality), the biological environment (the major types of plants and animals), and the societal environment (socioeconomics, culture, local history).

The next major term is *environmental impact assessment* (EIA). An EIA is the essence of the goals of NEPA; it is the "systematic, interdisciplinary approach" required by NEPA to attempt to predict and evaluate the consequences of the project on the environment. According to Canter (1977), an EIA must do three main things:

1. It must predict the anticipated changes in the key parameters identified by the environmental inventory.
2. It must evaluate the magnitude of the predicted changes.
3. It must judge the importance of the predicted changes.

A third term that should be defined is *environmental impact statement* (EIS). Even though we have previously used this term, let us now try to pin down what it is and what information it must contain. The EIS contains a brief but sufficient description of the proposed project and its reasonable alternatives, as well as a detailed summary of all the pertinent information contained in the environmental inventory and the environmental impact assessment. It is a document prepared by a team of well-qualified people that presents their findings as well as their reasoning in reaching their conclusions. It is written in a certain format as specified by the law and by guidelines issued by various authorities.

Based on NEPA, every EIS must contain information on the following five items:

1. "The environmental impact of the proposed action."
2. "Any adverse environmental effects which cannot be avoided."
3. "Alternatives to the proposed action."
4. "The . . . local short-term use of man's environment and the . . . long-term productivity,"
5. "Any irreversible and irretrievable commitments of resources [due to] the proposed action."

In addition, agency guidelines and court cases have added some further requirements for the contents of EISs. Some of these include:

6. A full range of responsible opinions.
7. A section describing the proposed project.

8. A section describing the existing environment.
9. A section devoted to discussion of the comments (objections) of reviewers.
10. Inclusion of all reasonable alternatives, even those not within the existing authority of the agency preparing the EIS.

Since governmental agencies are so involved with environmental impact statements, it is useful to discuss some of these agencies.

The agencies

The Council on Environmental Quality (CEQ) was created by NEPA to be the President's advisor on environmental matters. Although the agency is small (total staff of less than 100 people), part of its job is to ensure that all federal agencies comply with the intent of NEPA. The CEQ periodically issues guidelines and regulations for other federal agencies specifying how they should operate with respect to their environmental responsibilities. As one example, in June 1978 the CEQ issued revised guidelines on the preparation of EISs to reduce paperwork and time delays in the NEPA process. The theme of a portion of the regulations was to "be analytic, not encyclopedic." In one case, an EIS was over 9600 pages in 17 volumes (Council on Environmental Quality, 1978). The new regulations to all federal agencies suggested that most EISs be less than 150 pages (or 300 for complex issues).

The CEQ helps to troubleshoot problem projects and may occasionally review a controversial EIS. Also, it publishes an annual report to Congress on the state of environmental quality in the nation. These reports are excellent sources of statistical data and trends of environmental quality in the United States.

The Environmental Protection Agency (EPA) was established at the recommendation of the CEQ to be responsible for environmental regulation in the United States. It has thousands of employees and is organized into ten operating regions, each having jurisdiction in a different geographical area. The EPA is responsible for doing research and development, for writing environmental standards, for enforcing compliance with environmental regulations, and for preparing and reviewing EISs. This last responsibility is no small task. In recent years, there have been about 1200 EISs to be reviewed each year (Privitera, 1982). In addition, the EPA administers a considerable amount of money to help improve environmental quality in various places throughout the country.

Many states have passed their own (similar) versions of NEPA and have established agencies to meet the responsibilities outlined above. In addition to various state agencies, with sometimes overlapping responsibilities, there are local (city or county) agencies that usually are involved in the detailed planning and permitting process. Sometimes it can be frustrating for the engineer working on a project to have to deal with so many agencies. Not only must today's engineers have a broad understanding of the engineering disciplines covered in this text as well as the social sciences (sociology, psychology, and such), but they must also be able to communicate with people not technically trained, and do so in language that is clear and simple.

The process

The process that culminates in a formal EIS goes something like this:

1. A project is proposed.
2. A preliminary environmental inventory of the area is made, along with an assessment of the scope of the project and its environmental impacts.
3. The assessment is reviewed to decide whether an EIS is required.
4. If an EIS is warranted, then a draft EIS (DEIS) is prepared (includes detailed inventory and EIA).
5. The DEIS is circulated to all interested parties for comment. Public hearings are held.
6. A final EIS (FEIS) is prepared, which may be significantly revised according to the comments received on the DEIS.
7. A go or no-go decision is made on the project.

The above process may take years depending on the complexity of the project, the magnitude and importance of the environmental impacts, and the extent of legal challenges to the findings of the DEIS. However, the process ensures that a considerable amount of planning goes into the proposed project to begin with. Furthermore, the public is assured of having a voice in the decision-making process.

Not all projects require an EIS. If after an environmental assessment (step 4 above) the project is found to have only small and acceptable impacts, then a document known as a Finding of No Significant Impact (FONSI) is issued. Whereas EISs typically cost $100,000 or more to prepare and are typically hundreds of pages long, FONSIs are usually much cheaper and shorter. However, a FONSI still includes an environmental inventory and an environmental impact assessment. When they are applicable, FONSIs serve a useful purpose in the system.

In recent years, the number of EISs written has declined from the early 1970s. At the same time, the numbers of EIAs and FONSIs have increased. This trend may be due to fewer large-scale projects, or may be due to (as some critics have claimed) a decreased emphasis on the importance of the environment. However, it may be more likely that environmental awareness is becoming more widespread. Therefore, proper environmental planning is occurring earlier in the process, resulting in better projects with fewer significant impacts. It may just be that some of the original goals of NEPA are being achieved.

13-4 Methodologies

A methodology may be defined as a systematic, reproducible, interdisciplinary approach to making an environmental impact assessment. The reasons for using a certain methodology include the following:

1. To ensure that all important items are considered and none are left out.
2. To evaluate dissimilar environmental quality parameters on a more or less common basis.
3. To document the team's efforts and reasoning for review by the public.
4. To allow for quantitative comparison of alternatives.

Since January 1, 1970, there have been hundreds of methodologies proposed. Canter (1977) presents brief descriptions of many of these methodologies. Modern techniques can be classified into three main types: matrix methods, checklists, and overlays. Because of space limitations, we shall briefly discuss only one of the matrix methods here.

One well-known matrix method is that of the U.S. Geological Survey (Leopold and others, 1971). The full Leopold matrix lists 100 actions on the horizontal axis and 88 environmental characteristics on the vertical axis. Although 8800 interactions are possible on this matrix, usually less than 100 will be found for any particular project. It is important to note that the matrix can be expanded or contracted as necessary.

In using this methodology, the team of experts must first determine which project actions interact with which environmental parameters. At each interaction, a slash (/) is placed in the appropriate box. These impacts are then quantified as to magnitude and importance by the project team.

A number from 1 to 10 indicating the magnitude of the impact is placed in the upper left-hand corner of the box with 1 being the smallest and 10 being the greatest impact. In the lower right-hand corner, another number (from 1 to 10) is placed to indicate the relative importance of the impact. It should be noted that impacts can be of large magnitude but of little importance or vice versa. An example of an interaction with a low magnitude but high importance might be the operation of pesticide storage tanks near a river and its impact on aquatic life. The magnitude is low because a spill is unlikely and normal discharges are zero. But the importance is high because of the sensitivity of the aquatic system to this toxic chemical and the overall value of the aquatic system to the local ecosystem.

After the matrix is filled out, the text of the EIS should offer an explanation of all significant impacts—those columns or rows with many boxes checked, or those individual boxes with the larger numbers. Figure 13.1 reproduces an example matrix.

A brief explanation of this figure is necessary. This matrix was prepared for evaluation of a 1969 permit application for a phosphate mining operation in the Los Padres National Forest in Ventura County, California. As can be seen from Figure 13.1, the actions with the most impacts were "surface excavation" and "emplacement of tailings." The environmental characteristics of most importance were "scenic views and vistas," "wilderness qualities," and "rare and unique species."

The reason this last characteristic was considered so important (importance factors of 10) was that a California condor sanctuary was located 15 miles east of the proposed mining site. The magnitudes of the possible impacts were considered moderate, but any effects on these extremely rare birds were thought to be very important. The qualities "scenic views and vistas" and "wilderness qualities" were impacted by a number of actions and were considered of moderate importance because of the use of the area as a recreational refuge by many people. In a national forest, these qualities are important. The reason the magnitude of the impacts were all low to moderate was because the mining plan called for an "annual excavation of 4 to 5 acres with reclamation following closely in the mined-out area" (Leopold and others, 1971).

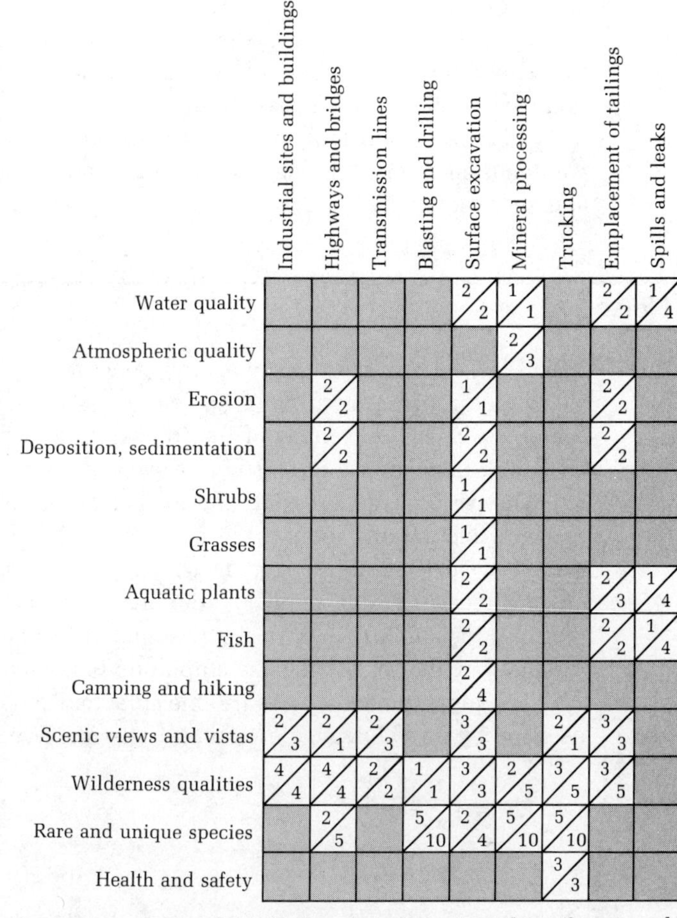

	Industrial sites and buildings	Highways and bridges	Transmission lines	Blasting and drilling	Surface excavation	Mineral processing	Trucking	Emplacement of tailings	Spills and leaks
Water quality					2/2	1/1		2/2	1/4
Atmospheric quality						2/3			
Erosion		2/2			1/1			2/2	
Deposition, sedimentation		2/2			2/2			2/2	
Shrubs					1/1				
Grasses					1/1				
Aquatic plants					2/2			2/3	1/4
Fish					2/2			2/2	1/4
Camping and hiking					2/4				
Scenic views and vistas	2/3	2/1	2/3		3/3		2/1	3/3	
Wilderness qualities	4/4	4/4	2/2	1/1	3/3	2/5	3/5	3/5	
Rare and unique species		2/5		5/10	2/4	5/10	5/10		
Health and safety							3/3		

Figure 13.1 A reduced matrix for a phosphate mining lease (from Leopold and others, 1971).

There are several sub-areas of the total environment for which impacts must be assessed in an EIS. These are the air, water, noise, biological, cultural-historical, and socioeconomic environments. Discussions of the air, water, and biological-ecological environments have been presented in earlier chapters of this text.

In general, the procedures for impact assessment for each of the above-mentioned areas are similar. First, an inventory or description is made of the existing environment with respect to the particular factor under consideration. Next, applicable local, state, and federal laws are researched to ensure legal compliance of the project. Next, the impacts of the project and its alternatives are predicted. Any measures that can be taken to reduce or mitigate the effects of the project are identified. Finally, the work is documented and presented clearly and fairly for public review and comment. In the next few sections, we shall briefly discuss those areas of the human environment for which environmental impact assessment is required but which were not mentioned in earlier chapters: noise, cultural-historical, and socioeconomic.

13-5 Noise

Sound is a form of mechanical energy, generated by a vibrating solid object, and transmitted from its source to a hearer's ear by a periodic compression and expansion of the transmitting medium (such as air or water). In air at sea level, sound travels at approximately 340 m/s.

The periodic compression and expansion of air results in cyclic pressure fluctuations about the mean atmospheric pressure. If a sound wave is visualized as a trigonometric sine function, two parameters are of importance: the frequency (the inverse of the period of oscillation) and the amplitude. With a pure sound, the frequency is related to the pitch or the tone while the amplitude is related to the loudness.

The instantaneous pressure due to sound is a function of time as given by the following equation (Mestre and Wooten, 1980):

$$P(t) = P_A + p(t) \tag{13.1}$$

where

$P(t)$ = instantaneous pressure

P_A = atmospheric pressure

$p(t)$ = pressure disturbance due to sound

The average pressure level of this type of cyclic function is often defined by a root-mean-square (rms) type of averaging. From basic physics, the relation between the rms and peak values of a sinusoidal function is

$$P = p_{rms} = \frac{p_{max}}{\sqrt{2}} \tag{13.2}$$

Thus,

$$P_T = P_A + P \tag{13.3}$$

where

P_T = average total pressure
P = rms sound pressure

From now on, the term sound pressure refers to the rms value.

Sound pressures are very small compared with atmospheric pressure. According to Mestre and Wooten (1980), for two people about 1 meter apart, normal conversation results in sound pressures received by the hearer of about one-millionth of one bar (1 bar = 100 kPa; 1.013 bar = 1 atm). Sound pressures from various sources can be received simultaneously. When rms sound pressures are added, the method is to first square the pressure values, then add them, then take the square root as shown in the following example.

Example Problem 13.1

Two sounds individually result in sound pressures at the same receiver of 2.0 and 3.0 μbar. What is the sound pressure received when both sounds are combined?

Solution

$$P_{comb} = \sqrt{(2.0)^2 + (3.0)^2}$$
$$= 3.6\ \mu\text{bar}$$

Noise has been defined commonly as unwanted or unpleasant sound (Canter, 1977). This definition implies that noise is disturbing to or has adverse effects on humans, domestic animals, or wildlife. Noise is almost always a mixture of a multitude of sounds, composed of many frequencies at different loudness levels (Mestre and Wooten, 1980). People's sensitivities differ, and the distinction between noise and sound is subjective and sometimes a difficult one to make (as in the matter of popular music). Important factors in how annoying the noise is perceived to be are how the noise varies with time (steady or intermittent), its duration (short or long), its location (downtown or suburbs), and the time of day it occurs. Because noise perception is rather subjective, we next briefly discuss human hearing.

Human hearing response

Although human hearing is not as good as that of many animals, the human ear is truly remarkable. We can hear sounds with frequencies from about 15 hertz (Hz) to about 20,000 Hz (hertz is the same as cycles per second). Furthermore, the human ear can hear sounds over a tremendous range of sound pressures—from about 0.0002 μbar to about 10,000 μbar (Canter, 1977).

Hearing response is highly nonlinear with respect to both pressures and frequencies. Because of the nonlinearity of human hearing response, a weighted logarithmic scale has been developed to measure sound pressures. Noise or sound is reported as "sound pressure level" defined as the logarithmic ratio of sound pressures. The reference pressure is the threshold of human hearing, 0.0002 μbar. The defining equation is

$$SPL = 10 \log_{10}(P^2/P_o{}^2) \qquad (13.4)$$

where

SPL = sound pressure level, decibels

P = sound pressure, μbar

P_o = reference pressure, 0.0002 μbar

The units of SPL are decibels (dB). The scale weighted to correct for human hearing response to different frequencies is called the A-weighted decibel (dBA) scale. Sound meters have an electronic weighting network to automatically simulate human hearing response when they measure sound pressure levels in units of dBA. Table 13.1 lists

TABLE 13.1 Some common sounds, sound pressures, and SPLs.

Sound	*Pressure, μbar*	*SPL, dBA*
Threshold of hearing	0.0002	0
Breathing	0.00063	10
Quiet bedroom at night	0.0063	30
Library	0.020	40
Normal conversation	0.20	60
Automobile at 20 feet	1.00	74
Busy intersection	6.3	90
Power lawn mower; garbage truck	20.0	100
Loud motorcycle at 20 feet	63.0	110
Peak level from rock band	200.0	120
Jet aircraft at 200 feet	200.0	120
Jet aircraft at 20 feet	2,000.0	140

some common sounds and their corresponding sound pressures and SPLs.

It is important to note how to add noise levels using dBA units. Because of the definition of SPL, we would be wrong to simply add the dBA values together. Rather, we first must convert dBA to sound pressures (μbar) using equation (13.4). Then, because sound pressures represent rms averages, we must add their squares to get the square of the combined sound pressure. Last, we convert the square of the combined sound pressure to dBA.

Example Problem 13.2

Two individual sounds are 80 dBA and 70 dBA, respectively. What is the SPL of the two sounds combined? Recalculate the answer for two sounds of 80 dBA each.

Solution For the first case:

$$\mathrm{SPL}_1 = 80 \text{ dBA}$$

$$\left(\frac{P_1}{P_o}\right)^2 = \text{inv } \log_{10}\left(\frac{80}{10}\right) = 1.0 \times 10^8$$

$$\mathrm{SPL}_2 = 70 \text{ dBA}$$

$$\left(\frac{P_2}{P_o}\right)^2 = 1.0 \times 10^7$$

$$\left(\frac{P_3}{P_o}\right)^2 = 1.0 \times 10^8 + 1.0 \times 10^7 = 1.1 \times 10^8$$

$$\mathrm{SPL}_3 = 10 \log_{10}\left(\frac{P_3}{P_o}\right)^2 = 80.4 \text{ dBA}$$

For the second case:

$$\mathrm{SPL}_1 = \mathrm{SPL}_2 = 80 \text{ dBA}$$

$$\left(\frac{P_3}{P_o}\right)^2 = 1.0 \times 10^8 + 1.0 \times 10^8 = 2.0 \times 10^8$$

$$\mathrm{SPL}_3 = 10 \log_{10}\left(\frac{P_3}{P_o}\right)^2 = 83.0 \text{ dBA}$$

Effects of noise

It has been reported that prolonged exposure to sound of 80–90 dBA can cause hearing damage, that 120 dBA is the human pain threshold, and that 150 dBA can rupture the eardrum (Miller, 1982). Excessive noise can cause physical damage to the inner ear, can contribute to headaches and high blood pressure, and can cause psychological distress in individuals. Risk criteria for hearing damage have been established by the Occupational Safety and Health Administration. These criteria set maximum allowable exposure times to noise at various levels. Longer exposure times to noise at the listed levels may result in partial or complete permanent hearing loss. These criteria are listed in Table 13.2.

The noise levels in Table 13.2 are for the workplace and are generally much higher than would be tolerated in the community. Noise in the community has adverse effects including the following: (1) causes temporary or permanent partial loss of hearing, (2) interferes with conversation or work, (3) interferes with sleep, and (4) causes general annoyance.

Various localities have different sensitivities to noise (Mestre and Wooten, 1980). Facilities such as schools or hospitals or wildlife sanctuaries are very sensitive. Residential communities, restaurants, and offices are moderately sensitive. Industrial plants, truck stops, and automobile racetracks are insensitive. The response of residential communities to peak noises received from nearby sources depends on the levels of noise. Mestre and Wooten (1980) indicated a rough correlation in which response varies from acceptance (50–60 dBA) through letters of complaint (70–80 dBA) to vigorous organized community protests including legal action (80–90 dBA).

TABLE 13.2 Federal risk criteria for hearing loss.

Noise level, dBA (slow response)	*Maximum allowable exposure, hours per day*
90	8
92	6
95	4
97	3
100	2
102	1.5
105	1
110	0.5
115	0.25 or less

Noise impact assessment

The first step in making a noise impact assessment is to inventory the existing noise environment. Both actual measurements and modeling techniques are used. Researching existing noise laws is important because ordinances vary widely from state to state. Next, the expected noise levels due to the proposed project must be predicted, both for the construction phase and the operation phase. Finally, any mitigating measures for noise reduction should be identified for possible implementation.

As an example of a mitigating measure, an increased buffer zone could be acquired between the project and the neighborhood. Sound levels from a point source dissipate quickly with distance as shown by the following equation:

$$\mathrm{SPL}_2 = \mathrm{SPL}_1 - 20 \log_{10} \frac{D_2}{D_1} \tag{13.5}$$

where

SPL_2 = sound pressure level received at distance D_2 from the source

SPL_1 = sound pressure level received at distance D_1 from the source

Other measures include using quieter equipment, erecting sound barriers, or allowing only certain types of users of the project such as restricting access to airports to only the quieter planes. It has been stated, for example, that aircraft like the L-1011 or DC-10 are about one-fourth as loud as the Concorde (Council on Environmental Quality, 1978).

13-6 Cultural-historical impacts

The cultural-historical environment includes buildings and areas of architectural, religious, archeological, and historical importance. By nature of the resources involved, all existing or potential historic or prehistoric sites must be considered. Certainly it is recognized that as human population increases, more pressures are brought to bear on old neighborhoods or undeveloped areas. Not every potential site can always be preserved. However, as remnants of past eras are destroyed to make room for present civilizations, people realize more and more the importance of preserving some of their national heritage.

If for no other reason, cultural resources are recognized under NEPA because they are essentially nonrenewable. However, federal, state, and local laws recognized the importance of these resources years ago. An early federal law was the Antiquity Act of 1906, which provided for the protection of all historic and prehistoric ruins on federal land (Canter, 1977). The Historic Sites Act of 1935 established a policy of preservation of historic buildings, sites, and objects. The National Historic Preservation Act of 1966 provided for the National Register of Historic Places to register sites, buildings, and other features that are significant in American history, architecture, culture, or archeology (Canter, 1977). This law also provided for funds for the restoration and preservation of important sites.

The first step in impact assessment on the cultural-historical environment is a description or inventory of the resources in and around the project area. Identification of established sites may be made by contacting the National Register or local and state historic societies. Generally, heavy reliance must be placed on local groups because there are many cultural resources that have not been placed on the National Register. If the project encompasses or passes through a large land area (such as a new pipeline project), local universities can be contacted to discuss the possibilities of archeological sites in the project area. In a methodology proposed by Stover (1973), a major area called "human well-being" includes the sub-areas "community goals and objectives" in which local cultural desires and values are considered as resources.

The next step involves the prediction and quantification of impacts. Impacts on cultural resources include damage, disruption, or destruction (Canter, 1977). They can result directly from the construction or the operation phase or indirectly from theft or vandalism. In the EIS, an attempt should be made to estimate the magnitude and importance of these impacts. This is easier said than done because it involves value judgment with little mathematical basis. However, in many cases, a good inventory done early enough in the planning stages of a project can help reduce impacts to a minimum.

A good example is the environmental assessment made for an area in New Jersey through which a pipeline was to be built (Bordman and others, 1977). A complete archeological-historical inventory resulted in identification of several old cemeteries, some possible prehistoric sites, and several buildings of historical importance.

Laying the pipeline as originally planned would have destroyed two prehistoric sites and had negative impacts on some historical sites. However, by careful consideration of alternate routes and some minor rerouting on the major alternatives, an alignment was recommended that had almost no negative impacts on these resources.

13-7 Socioeconomic impacts

The socioeconomic environment is, of course, a major part of the human environment; however, in the early years of EIA, it was overlooked in favor of concentrating on the physical-chemical and biological environment. This was probably because the socioeconomic environment is large and diverse with many complex interrelations. One suggestion for organization in the approach to inventorying and assessing it is to group and analyze data in one of two classifications: (1) regional economic growth (or decline) and (2) community social processes (Corwin and others, 1975). In Table 13.3 are listed some socioeconomic factors in each of these major classes.

As with the other parts of the total human environment, the first step is an inventory and description. This step can be a monumental task. Often a prescreening step is useful. That is, based on preliminary judgments by the project team, certain key parameters can be picked out. Then, the main efforts of the inventory can be concentrated on these key parameters.

TABLE 13.3 Examples of socioeconomic factors.

Factors related to regional economic growth
Production of goods and services (economic base)
Population characteristics and trends
Provision of major access, water and sewer, or other utilities
Land-use patterns and regional land-use projections and plans
Major educational (college) and recreational amenities
Employment and unemployment trends
Factors related to community social processes
Primary and secondary schools
Income levels and trends
Family housing patterns
Local shopping and recreational facilities
Transportation systems
Community identities and ethnic groups
Health and social services
Local zoning ordinances
Police and fire protection

The next step is quantitatively to predict, or at least qualitatively describe any changes in the socioeconomic parameters described in the inventory. This is done for the project and its alternative, including the "do-nothing" alternative. These predictions are often carried out far into the future when considering major projects that are expected to operate for 30 to 50 years. Of course, such projections are often outdated only a few years from the time they are made. However, in making the projections, all alternatives are compared using the same basis and methods, so the process is still useful.

A major area of concern in socioeconomic impact analysis is that of secondary impacts. For example, consider a small town in which a steel mill provides the entire economic base. Other jobs are present (such as in the barbershop, drugstore, and food market), but these are secondary and depend on the steel workers. Changes in the secondary sector are usually small and result in no change in the steel mill employment. However, a change at the steel mill (such as a new line or a shutdown of an old line) may result in large changes in the non-basic sector in addition to the direct (basic) changes. These changes in the non-basic sector are called secondary impacts.

If the ratio of non-basic to basic employment is known, secondary impacts can be estimated quantitatively (Corwin and others, 1975). For example, say the ratio for the above small town is 0.6; that is, for every ten jobs in the steel mill, there are six outside the mill. Thus, if the mill adds 100 jobs, then 60 more jobs will be created in the town. Furthermore, planners generally have data on the amounts of community services (such as schools, housing, sewer and water services, and police) needed for the new families that will be added to the community.

One of the most far-reaching types of projects, especially with regard to long-term secondary socioeconomic impacts, is providing a major new access to an area. New freeways, for instance, or even new interchanges on existing freeways have resulted in profound impacts on areas. The usual results include commercial development along the highway, new basic industries, large population growth (from both direct and secondary impacts), increases in land values (and property taxes), and increased demands for community services.

13-8 Externalities and environment

Can we account for all our actions and impacts on ourselves and others? This book has helped in an understanding of the impacts from the disposal of residuals (mass and energy) from a modern society with high-rate consumption and production activities. Also, solutions to problems with unacceptable impacts have been developed.

We found it useful to examine environmental pollution and its control using a materials-balance and an energy-balance approach. Kneese and others (1970) examined this approach for the general economy. The impacts to our society and smaller systems of our society consist of the chemicals and energy necessary to sustain life. The processes of systems necessarily cause external effects on other systems. These inefficiencies are called externalities and can be quantified in terms of costs to others. Examples of externalities are: air pollution from energy conversion, air pollution from automobiles, water pollution from wastewater treatment, solid waste from food processing, solid waste from domestic consumption, stormwater discharges of metals and other toxins, disposal of hazardous waste from industries, and noise from automobiles (Kneese and others, 1970).

Those responsible for generating the external pollution or residual effects are sometimes required to pay the price of clean-up. Other times, the public must pay for the clean-up. Thus, the external costs can be internalized (paid for by the user-pollutor) or paid for by the government (paid for by taxes or other methods). It matters who pays and who benefits. Thus, there are other issues of equity, national wealth distribution, quality of life, and economic efficiency. Kneese and others (1971) did a rough estimate of the increase in costs in the early 1970s to achieve "substantial" reductions in environmental pollution. These totals, shown in Table 13.4, were predicted to result in about a 20% increase in the gross national product. Thus, pollution control can become an important factor in the economic position of a country.

Once pollution control is achieved, benefits in terms of dollars and aesthetics result. In fact, government agencies must ensure that the benefits resulting from environmental controls exceed the costs before federal expenditures are made. These types of calculations are frequently called benefit-to-cost ratios. Benefits to the public were

TABLE 13.4 Cost increases to achieve pollution control (billions of $ 1975).

Water (no separation of sewers)	$26
Air	10
Solids	4
Others	15
TOTAL:	$55

estimated from the clean-up of stormwater for a lake in Orlando, Florida (Harper and others, 1982). The benefit-to-cost ratio was about 7:1, thus, the project was justified on a benefit-to-cost analysis. In 1913, the Mellon Institute (O'Conner, 1913) published benefits of atmospheric pollution control for Pittsburgh alone. In 1913, these benefits were about $20 per person. The benefits were calculated based on the (1) cost of imperfect combustion, (2) individual laundry and cleaning, (3) household costs such as painting, cleaning, and corrosion, (4) damaged goods, maintenance, and lighting, and (5) building repairs and maintenance. These costs did not include annual health costs resulting from air pollution. Such costs were estimated at about 2 billion dollars in 1963 (Lave and Seskin, 1970). If all benefits are quantified, in general $1 spent on air pollution control will prevent $16 worth of air pollution damage (Stern, 1968). Of course, this benefit-to-cost ratio is an oversimplification because benefits and costs vary with the degree of control.

Engineers can design and operate residual management facilities and, thus, control pollution. The basic concepts of materials and energy balances are important and can be integrated with economic concerns.

13-9 Problems

13.1 A woman's house on the rural outskirts of a growing city has a 100-foot dirt drive that she wishes to pave. The backyard overlooks a small lake. Prepare a Leopold-type matrix for the construction and operation of a concrete driveway for this house. Include three actions: clearing and grading, paving, and spills and oil leaks. Include five environmental quality parameters: water quality, water recharge, air quality, employment, and cultural. Briefly explain your reasoning in filling out the matrix.

13.2 You are the lead development engineer for a new theme park to be created in central Florida. You are in the initial planning stages for the park which will feature a glass-bottomed boat ride through a tropical garden, so that the tourists can view both tropical fish and tropical birds. Organize this problem in your mind; make an organized checklist of all the environmental interactions that must be considered in building and operating this attraction.

13.3 What sound pressure level (in dBA) would one expect from a loud motorcycle at 10 feet?

13.4 What SPL (in dBA) would result from combining two sounds with sound pressures of 6.0 μbar and 5.0 μbar, respectively?

13.5 The SPL received by a person 50 feet from a city bus is about 85 dBA. Calculate the total SPL received by a person who is 30 feet away from a departing bus and 60 feet away from an approaching bus.

13.6 Do some research in your community and make a list of its major cultural-historical resources. (Suggestion: Contact a local chamber of commerce or a local historical society.)

13.7 Describe some of the socioeconomic impacts you might expect from a new airport in your town.

References

Agricola, Georgius. 1556. *De Re Metallica.* 1950 (transl. H. C. Hoover and L. H. Hoover). New York: Dover.

Bordman, S. L., Dresnack, R., Golub, E., Khera R., Olenik, T. and Salek, F. 1977. Environmental assessment of the Raritan Confluence Force Main. Newark, NJ: Center for Technology Assessment, New Jersey Institute of Technology.

Canter, L. W. 1977. *Environmental Impact Assessment.* New York: McGraw-Hill.

Corwin, R., Heffernan, P. H., Johnston, R. A., Remy, M., Roberts, J. A. and Tyler, D. B. 1975. *Environmental Impact Assessment.* San Francisco: Freeman Cooper.

Council on Environmental Quality, 1977. *Environmental Quality—1977* (Eighth Annual Report). Washington, DC.

Council on Environmental Quality, 1978. *Environmental Quality—1978* (Ninth Annual Report). Washington, DC.

Harper, H. H., Wanielista, M. P., Yousef, Y. A. and DiDomenico, D. 1982. *Lake Restoration.* Tallahassee: Florida Department of Environmental Regulation.

Kneese, A. V., Ayres, R. U. and d'Arge, R. C. 1970. *Economics and the Environment—A Materials Balance Approach.* Washington, DC: Resources for the Future.

Kneese, A. V., Rolfe, S. E. and Harned, J. W. 1971. *Managing the Environment.* New York: Praeger.

Lave, L. B. and Seskin, E. P. 1970. Air pollution and human health. *Science* 169, 723–733.

Leopold, L. B., Clark, F. E., Hanshaw, B. B. and Balsley, J. R. 1971. *A Procedure for Evaluating Environmental Impact.* Washington, DC: U.S. Geological Survey Circular 645.

Mestre, V. E. and Wooten, D. C. 1980. Noise impact analysis. In *Environmental Impact Analysis Handbook,* ed. J. G. Rau and D. C. Wooten. New York: McGraw-Hill.

Miller, G. T., Jr. 1982. *Living in the Environment*, 3rd ed. Belmont, CA: Wadsworth.

O'Conner, J. J. 1913. The economic cost of the smoke nuisance to Pittsburgh. Pittsburgh, PA: Mellon Institute, Smoke Investigation Bulletin No. 4.

Privitera, M. L. 1983. NEPA and federal decision making. In *Sixth Annual Environmental Review Conference, October 21–22, 1982*. Washington, DC: Environmental Protection Agency, Publication EPA 904/9-83-110.

Stern, A. C. 1968. *Air Pollution*, 2nd ed. New York: Academic Press.

Stover, L. V. 1973. *Environmental Impact Assessment: A Procedure*. Washington, DC: STV, Inc.

Appendix A—Notation

A	cross-sectional area
A_s	surface area of tank
ac	acre
atm	atmosphere
BOD	biochemical oxygen demand
BOD_5	BOD exerted after five days
BOD_u	ultimate BOD
BTU	British Thermal Unit
C	Hazen-Williams friction coefficient
°C	Celsius degree
C_A	concentration for material "A"
c_p	specific heat
C_s	saturation concentration
cal	calorie
cfm	cubic feet per minute
cfs	cubic feet per second
cm	centimeter
COD	chemical oxygen demand

D	dissolved oxygen deficit (saturation dissolved oxygen concentration − actual concentration)
DO	dissolved oxygen
d	particle diameter
E	activation energy
F	Volumetric flow rate
°F	Fahrenheit degrees
fpm	feet per minute
fps	feet per second
ft	foot
g	gravitational constant
g	gram
gal	gallon
gpd	gallon/day
GWT	groundwater table
H	water depth
H	enthalpy
HC	a general symbol for hydrocarbons
ha	hectare
HP	horsepower
HP-hr	horsepower-hour
hr	hour
I	infiltration
I_A	initial abstraction
in.	inch
J	joule
K_1	deoxygenation reaction rate constant
K_2	reaeration rate constant
K_d	BOD first-order decay constant
K_a	reaction rate constant in DO model
K	Kelvin
k	reaction rate constant
k_o	frequency factor in Arrhenius equation
kg	kilogram
km	kilometer

kW	kilowatt
kWh	kilowatt-hour
L	concentration of BOD
L_t	BOD remaining at time t
L_u	BOD ultimate
L	liters
lb	pound
$\dot{m}$	mass flow rate
m	mass
m	meter
M	molarity (mole per litre)
mg	million gallons
mgd	million gallons/day
mg/L	milligrams/liter
min	minute
mL	milliliter
mm	millimeter
mph	miles per hour
N	newton
N	coliform organism numbers
NH_4-N	ammonia nitrogen
NO_3-N	nitrate nitrogen
NO_x	nitrogen oxides
p	wetted perimeter
ppb	parts per billion
ppm	parts per million
pH	hydrogen-ion index ($pH = -\log H^+$)
ppm	parts per million
psf	pound per square foot
psi	pound per square inch
Q	Water flow rate
Q	Heat flow rate
R	hydraulic radius
R	rainfall excess
R	recycle
r	reaction rate

S	slope of energy grade line
S	cross-sectional area
S	substrate concentration
s	second
SO_x	sulfur oxides
SVI	sludge volume index
t	time
T	absolute temperature (kelvin)
V	volume
v	linear velocity
W	watt
X	reactor biomass concentration-mixed liquid suspended solids
yd	yard
μ	fluid viscosity
Δp	pressure drop
ρ	fluid density
ρ_s	particle density
ν	linear velocity
τ	detention time or hydraulic residence time
λ	heat of phase change
θ_h	hydraulic retention time
θ_c	biological cell retention time

Appendix B—Metric Units with Customary Equivalents and Physical Constants

Length

Metric units

micrometer, μm	1000 μm = 1 mm
millimeter, mm	10 mm = 1 cm
centimeter, cm	100 cm = 1 m
meter, m	1000 m = 1 km
kilometer, km	

Equivalents

m	× 39.37	= in.	×	0.0254	=	m
m	× 3.28	= ft	×	0.3049	=	m
km	× 0.62	= mi	×	1.6129	=	km
mm	× 0.039	= in.	×	25.4	=	mm
cm	× 0.394	= in.	×	2.54	=	cm

5280 ft = 1 mi = 1760 yd

Area

Metric units

square millimeter, mm^2	10^2 mm^2 = cm^2
square centimeter, cm^2	10^4 cm^2 = m^2
square meter, m^2	10^6 m^2 = km^2
hectare, ha	10^2 ha = km^2

$mm^2 \times 0.00155 = in.^2 \times 645.16 = mm^2$
$cm^2 \times 0.155 = in.^2 \times 6.45 = cm^2$
$m^2 \times 10.764 = ft^2 \times 0.093 = m^2$
$km^2 \times 0.384 = mi^2 \times 2.604 = km^2$
$km^2 \times 247.10 = ac \times 0.004 = km^2$
$ha \times 2.471 = ac \times 0.405 = ha$
$ha \times 0.00386 = mi^2 \times 259 = ha$

43,560 ft^2 = 1 ac
4,840 yd^2 = 1 ac
144 $in.^2$ = 1 ft^2
640 ac = 1 mi^2

Volume

Metric units

Cubic centimeter, cm^3	10^6 cm^3 = m^3
cubic meter, m^3	10^3 L = m^3
liter, L	10^3 cm^3 = L

Equivalents

$cm^3 \times 0.061 = in.^3 \times 16.393 = cm^3$
$m^3 \times 35.314 = ft^3 \times 0.028 = m^3$
$L \times 1.057 = qt \times 0.946 = L$
$L \times 0.264 = gal \times 3.785 = L$
$L \times 0.81(10^{-6}) = ac\text{-}ft \times 1.235 \times 10^6 = L$

7.48 gal = 1 ft^3
1728 in^3 = 1 ft^3
28.32 L = 1 ft^3
8.34 lb = 1 gal (of water)
62.43 lb = 1 ft^3 (of water)
1 L = 1 kg (of water)

Force

Newton (N) × 0.22481 = lb (weight)
× 7.24 = lb (force)

Pressure

2.307 ft of H_2O = 1 lb/in.^{2}
2.036 in. of Hg = 1 lb/in.^{2}
14.70 psia = 1 atm
29.92 in. of Hg = 1 atm
33.93 ft of H_2O = 1 atm
76.0 cm of Hg = 1 atm
4.882 kg/m^2 = 1 lb/ft^2

Energy

3412 BTU = 1 kWh
1000 kWh = 1 MWh
1000 Wh = 1 kWh
778 ft-lb = 1 BTU
1055 joules = 1 BTU
3600 kJ = 1 kWh
1000 N-m = 1 kJ
2865 kJ = 1 HP-hr
4.185 kJ = 1 kcal

Power

2545 BTU/hr = 1 HP
3412 BTU/hr = 1 kW
550 Ft-lb/s = 1 HP
0.746 kW = 1 HP
1000 kW = 1 MW
1000 W = 1 kW
1000 N-m/s = 1 kW
1341 HP = 1 MW

Flow Discharge

m^3/s	$\times$ 15.850	= gpm	$\times$ 0.063	= m^3/s
m^3/s	$\times$ 2.12×10^3	= ft^3/min	$\times$ 0.472×10^{-3}	= m^3/s
m^3/s	$\times$ 35.314	= ft^3/s	$\times$ 0.0283	= m^3/s
L/s	$\times$ 15.850	= gpm	$\times$ 0.063	= L/s
m^3/day	$\times$ 0.264×10^{-3}	= mgd	$\times$ 3.788×10^3	= m^3/day
L/day	$\times$ 0.264	= gpd	$\times$ 3.788	= L/day
m^3/s	$\times$ 22.83	= mgd	$\times$ 0.0438	= m^3/s

Density

kg/cm^3	$\times$ 0.0624	= lb/ft^3	$\times$ 16.026	= kg/cm^3
lb/gal	$\times$ 1.2×10^5	= mg/L	$\times$ 8.33×10^{-6}	= lb/gal
kg/m^3	$\times$ 0.065	= lb/ft^3	$\times$ 15.38	= kg/m^3

Mass

Metric units

milligram, mg	1,000 mg = g
gram, g	1,000 g = kg
kilogram, kg	1,000 kg = 1 tonne (t)

Equivalents

mg	$\times$ 0.01543	= g	$\times$ 64.809	= mg
g	$\times$ 0.0022	= lb	$\times$ 457.1	= g
g	$\times$ 15.43	= grain	$\times$ 0.065	= g
kg	$\times$ 2.205	= lb	$\times$ 0.454	= kg
kg	$\times$ 0.0011	= ton	$\times$ 907.20	= kg
t	$\times$ 1.1023	= ton	$\times$ 0.907	= t

Using 2000 lb ton (Short ton)

2000 lb = 1 ton
454 g = 1 lb
7000 gr = 1 lb
2240 lb = 1 long ton

Linear Velocity

m/s × 3.280 = ft/s × 0.305 = m/s
m/s × 2.230 = mph × 0.448 = m/s

44.70 cm/s = 1 mph
1.467 ft/s = 1 mph

Appendix C—Chemical Data

TABLE C.1 Atomic numbers, weights, and symbols of some chemical elements.

Element	*Symbol*	*Atomic number*	*Atomic weight*
Aluminum	Al	13	26.98
Arsenic	As	33	74.91
Barium	Ba	56	137.36
Boron	B	5	10.82
Bromine	Br	35	79.92
Cadmium	Cd	48	112.41
Calcium	Ca	20	40.08
Carbon	C	6	12.01
Chlorine	Cl	17	35.46
Chromium	Cr	24	52.01
Cobalt	Co	27	58.94
Copper	Cu	29	63.54
Fluorine	F	9	19.00
Gold	Au	79	197.2
Hydrogen	H	1	1.008
Iodine	I	53	126.92
Iron	Fe	26	55.85
Lead	Pb	82	207.21
Magnesium	Mg	12	24.32
Manganese	Mn	25	54.93
Mercury	Hg	80	200.61
Nickel	Ni	28	58.69
Nitrogen	N	7	14.01
Oxygen	O	8	16.00
Phosphorus	P	15	30.98
Platinum	Pt	78	195.23
Potassium	K	19	39.10
Selenium	Se	34	78.96
Silicon	Si	14	28.09
Silver	Ag	47	107.88
Sodium	Na	11	23.00
Strontium	Sr	38	87.63
Sulfur	S	16	32.07
Tin	Sn	50	118.70
Zinc	Zn	30	65.38

TABLE C.2 Formulas of chemicals and ions commonly used in environmental engineering.

Name	*Formula*	*Molecular Weight*
Activated carbon	C	12.0
Aluminum sulfate (alum)	$Al_2(SO_4)_3 \cdot 14H_2O$	600
Aluminum hydroxide	$Al(OH)_3$	78.0
Ammonia	NH_3	17.0
Ammonium fluosillicate	$(NH_4)_2SiF_6$	178
Ammonium sulfate	$(NH_4)_2SO_4$	132
Calcium carbonate	$CaCO_3$	100
Calcium fluoride	CaF_2	78.1
Calcium hydroxide	$Ca(OH)_2$	74.1
Calcium hypochlorite	$Ca(ClO)_2 \cdot 2H_2O$	179
Calcium oxide (lime)	CaO	56.1
Carbon dioxide	CO_2	44.0
Chlorine	Cl_2	71.0
Chlorine dioxide	ClO_2	67.0
Copper sulfate	$CuSO_4$	160
Ferric chloride	$FeCl_3$	162
Ferric hydroxide	$Fe(OH)_3$	107
Ferric sulfate	$Fe_2(SO_4)_3$	400
Ferrous sulfate (copperas)	$FeSO_4 \cdot 7H_2O$	278
Fluosilicic acid	H_2SiF_6	144
Hydrochloric acid	HCl	36.5
Magnesium hydroxide	$Mg(OH)_2$	58.3
Oxygen	O_2	32.0
Potassium permanganate	$KMnO_4$	158
Sodium aluminate	$NaAlO_2$	82.0
Sodium carbonate	Na_2CO_3	106
Sodium chloride	NaCl	58.4
Sodium fluoride	NaF	42.0
Sodium hydroxide	NaOH	40.0
Sodium hypochlorite	NaClO	74.4
Sodium silicate	Na_4SiO_4	184
Sodium fluosilicate	Na_2SiF_6	188
Sodium thiosulfate	$Na_2S_2O_3$	158
Sulfur dioxide	SO_2	64.1
Sulfuric acid	H_2SO_4	98.1
Ammonium.	NH_4^+	18.0
Hydroxyl	OH^-	17.0
Bicarbonate	HCO_3^-	61.0
Carbonate	CO_3^{-2}	60.0
Orthophosphate	PO_4^{-3}	95.0
Orthophosphate, mono-hydrogen	HPO_4^{-2}	96.0
Orthophosphate, di-hydrogen	$H_2PO_4^-$	97.0
Bisulfate	HSO_4^-	97.0
Sulfate	SO_4^{-2}	96.0
Bisulfite	HSO_3^-	81.0
Sulfite	SO_3^{-2}	80.0
Nitrite	NO_2^-	46.0
Nitrate	NO_3^-	62.0
Hypochlorite	OCl^-	51.5

TABLE C.3 Solubility constants of some common compounds.

Compound	K_{sp}
Bromides	
$PbBr_2$	4.6×10^{-6}
Hg_2Br_2	1.3×10^{-22}
$AgBr$	5.0×10^{-13}
Carbonates	
$BaCO_3$	1.6×10^{-9}
$CdCO_3$	5.2×10^{-12}
$CaCO_3$	4.7×10^{-9}
$CuCO_3$	2.5×10^{-10}
$FeCO_3$	2.1×10^{-11}
$PbCO_3$	1.5×10^{-15}
$MgCO_3$	1×10^{-15}
$MnCO_3$	8.8×10^{-11}
Hg_2CO_3	9.0×10^{-17}
$NiCO_3$	1.4×10^{-7}
Ag_2CO_3	8.2×10^{-12}
$SrCO_3$	7×10^{-10}
$ZnCO_3$	2×10^{-10}
Fluorides	
BaF_2	2.4×10^{-5}
CaF_2	3.9×10^{-11}
PbF_2	4×10^{-8}
MgF_2	8×10^{-8}
SrF_2	7.9×10^{-10}
Hydroxides	
$Al(OH)_3$	5×10^{-33}
$Ba(OH)_2$	5.0×10^{-3}
$Cd(OH)_2$	2.0×10^{-14}
$Ca(OH)_2$	1.3×10^{-6}
$Cr(OH)_3$	6.7×10^{-31}
$Co(OH)_2$	2.5×10^{-16}
$Co(OH)_3$	2.5×10^{-43}
$Cu(OH)_2$	1.6×10^{-19}
$Fe(OH)_2$	1.8×10^{-15}
$Fe(OH)_3$	6×10^{-38}
$Pb(OH)_2$	4.2×10^{-15}
$Mg(OH)_2$	8.9×10^{-12}
$Mn(OH)_2$	2×10^{-13}
$Hg(OH)_2(HgO)$	3×10^{-26}
$Ni(OH)_2$	1.6×10^{-16}
$AgOH(Ag_2O)$	2.0×10^{-8}
$Sr(OH)_2$	3.2×10^{-4}
$Sn(OH)_2$	3×10^{-27}
$Zn(OH)_2$	4.5×10^{-17}
Iodides	
PbI_2	8.3×10^{-9}
Hg_2I_2	4.5×10^{-29}
AgI	8.5×10^{-17}
Chlorides	
$PbCl_2$	1.6×10^{-5}
Hg_2Cl_2	1.1×10^{-18}
$AgCl$	1.7×10^{-10}
Chromates	
$BaCrO_4$	8.5×10^{-11}
$PbCrO_4$	2×10^{-16}
Hg_2CrO_4	2×10^{-9}
Ag_2CrO_4	1.9×10^{-12}
$SrCrO_4$	3.6×10^{-5}
Phosphates	
$Ba_3(PO_4)_2$	6×10^{-39}
$Ca_3(PO_4)_2$	1.3×10^{-32}
$Pb_3(PO_4)_2$	1×10^{-54}
Ag_3PO_4	1.8×10^{-18}
$Sr_3(PO_4)_2$	1×10^{-31}
Sulfates	
$BaSO_4$	1.5×10^{-9}
$CaSO_4$	2.4×10^{-5}
$PbSO_4$	1.3×10^{-8}
Ag_2SO_4	1.2×10^{-5}
$SrSO_4$	7.6×10^{-7}
Sulfides	
Bi_2S_3	1.6×10^{-72}
CdS	1.0×10^{-28}
CoS	5×10^{-22}
CuS	8×10^{-37}
FeS	4×10^{-19}
PbS	7×10^{-29}
MnS	7×10^{-16}
HgS	1.6×10^{-54}
NiS	3×10^{-21}
Ag_2S	5.5×10^{-51}
SnS	1×10^{-26}
ZnS	2.5×10^{-22}
Miscellaneous	
$NaHCO_3$	1.2×10^{-3}
$KClO_4$	8.9×10^{-3}
$K_2[PtCl_6]$	1.4×10^{-6}
$AgC_2H_3O_2$	2.3×10^{-3}
$AgCN$	1.6×10^{-14}
$AgCNS$	1.0×10^{-12}

TABLE C.4 Ionization constants of some common compounds.

Monoprotic acids		
acetic	$HC_2H_3O_2 \rightleftharpoons H^+ + C_2H_3O_2^-$	1.80×10^{-5}
benzoic	$HC_7H_5O_2 \rightleftharpoons H^+ + C_7H_5O_2^-$	6.0×10^{-5}
chlorous	$HClO_2 \rightleftharpoons H^+ + ClO_2^-$	1.1×10^{-2}
cyanic	$HOCN \rightleftharpoons H^+ + OCN^-$	1.2×10^{-4}
formic	$HCHO_2 \rightleftharpoons H^+ + CHO_2^-$	1.8×10^{-4}
hydrazoic	$HN_3 \rightleftharpoons H^+ + N_3^-$	1.9×10^{-5}
hydrocyanic	$HCN \rightleftharpoons H^+ + CN^-$	4.0×10^{-10}
hydrofluoric	$HF \rightleftharpoons H^+ + F^-$	6.7×10^{-4}
hypobromous	$HOBr \rightleftharpoons H^+ + OBr^-$	2.1×10^{-9}
hypochlorous	$HOCl \rightleftharpoons H^+ + OCl^-$	3.2×10^{-8}
nitrous	$HNO_2 \rightleftharpoons H^+ + NO_2^-$	4.5×10^{-4}
Polyprotic acids		
arsenic	$H_3AsO_4 \rightleftharpoons H^+ + H_2AsO_4^-$	$K_1 = 2.5 \times 10^{-4}$
	$H_2AsO_4^- \rightleftharpoons H^+ + HAsO_4^{-2}$	$K_2 = 5.6 \times 10^{-8}$
	$HAsO_4^{-2} \rightleftharpoons H^+ + AsO_4^{-3}$	$K_3 = 3 \times 10^{-13}$
carbonic	$CO_2 + H_2O \rightleftharpoons H^+ + HCO_3^-$	$K_1 = 4.2 \times 10^{-7}$
	$HCO_3^- \rightleftharpoons H^+ + CO_3^{-2}$	$K_2 = 4.8 \times 10^{-11}$
hydrosulfuric	$H_2S \rightleftharpoons H^+ + HS^-$	$K_1 = 1.1 \times 10^{-7}$
	$HS^- \rightleftharpoons H^+ + S^{-2}$	$K_2 = 1.0 \times 10^{-14}$
oxalic	$H_2C_2O_4 \rightleftharpoons H^+ + HC_2O_4^-$	$K_1 = 5.9 \times 10^{-2}$
	$HC_2O_4^- \rightleftharpoons H^+ + C_2O_4^{-2}$	$K_2 = 6.4 \times 10^{-5}$
phosphoric	$H_3PO_4 \rightleftharpoons H^+ + H_2PO_4^-$	$K_1 = 7.5 \times 10^{-3}$
	$H_2PO_4^- \rightleftharpoons H^+ + HPO_4^{-2}$	$K_2 = 6.2 \times 10^{-8}$
	$HPO_4^{-2} \rightleftharpoons H^+ + PO_4^{-3}$	$K_3 = 1 \times 10^{-12}$
phosphorous	$H_3PO_3 \rightleftharpoons H^+ + H_2PO_3^-$	$K_1 = 1.6 \times 10^{-2}$
(diprotic)	$H_2PO_3^- \rightleftharpoons H^+ + PO_3^{-2}$	$K_2 = 7 \times 10^{-7}$
sulfuric	$H_2SO_4 \rightarrow H^+ + HSO_4^-$	strong
	$HSO_4^- \rightleftharpoons H^+ + SO_4^{-2}$	$K_2 = 1.3 \times 10^{-2}$
sulfurous	$SO_2 + H_2O \rightleftharpoons H^+ + HSO_3^-$	$K_1 = 1.3 \times 10^{-2}$
	$HSO_3^- \rightleftharpoons H^+ + SO_3^{-2}$	$K_2 = 5.6 \times 10^{-8}$
Bases		
ammonia	$NH_3 + H_2O \rightleftharpoons NH_4^+ + OH^-$	1.8×10^{-5}
aniline	$C_6H_5NH_2 + H_2O \rightleftharpoons C_6H_5NH_3^+ + OH^-$	4.6×10^{-10}
dimethylamine	$(CH_3)_2NH + H_2O \rightleftharpoons (CH_3)_2NH_2^+ + OH^-$	7.4×10^{-4}
hydrazine	$N_2H_4 + H_2O \rightleftharpoons N_2H_5^+ + OH^-$	9.8×10^{-7}
methylamine	$CH_3NH_2 + H_2O \rightleftharpoons CH_3NH_3^+ + OH^-$	5.0×10^{-4}
pyridine	$C_5H_5N + H_2O \rightleftharpoons C_5H_5NH^+ + OH^-$	1.5×10^{-9}
trimethylamine	$(CH_3)_3N + H_2O \rightleftharpoons (CH_3)_3NH^+ + OH^-$	7.4×10^{-5}

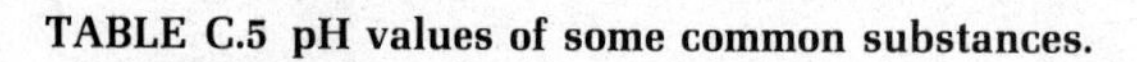
TABLE C.5 pH values of some common substances.

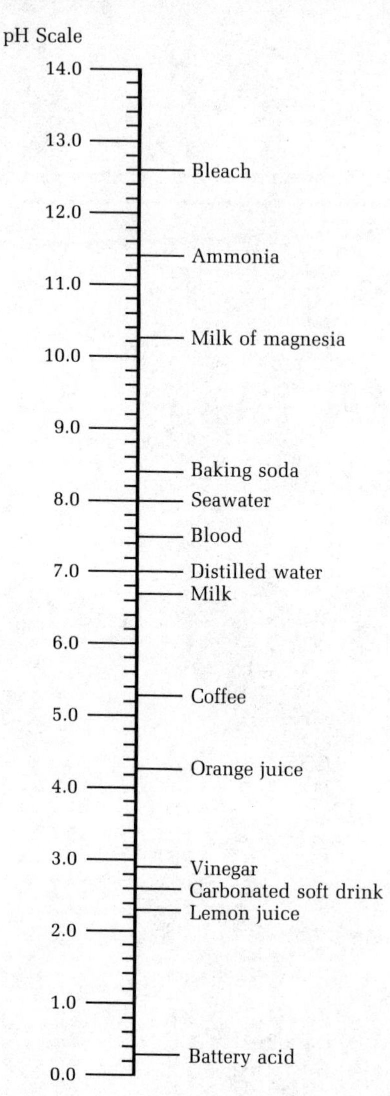

Appendix D— Fluid Characteristics

TABLE D.1 Viscosity and density of water

Temperature, C	*Density* (g/cm^3)	*Absolute viscosity (centipoises)*[a]	*Kinematic viscosity (centistokes)*[b]	*Temperature, F*
0	0.99987	1.7921	1.7923	32.0
2	0.99997	1.6740	1.6741	35.6
4	1.00000	1.5676	1.5676	39.2
6	0.99997	1.4726	1.4726	42.8
8	0.99988	1.3872	1.3874	46.4
10	0.99973	1.3097	1.3101	50.0
12	0.99952	1.2390	1.2396	53.6
14	0.99927	1.1748	1.1756	57.2
16	0.99897	1.1156	1.1168	60.8
18	0.99862	1.0603	1.0618	64.4
20	0.99823	1.0087	1.0105	68.0
22	0.99780	0.9608	0.9629	71.6
24	0.99733	0.9161	0.9186	75.2
26	0.99681	0.8746	0.8774	78.8
28	0.99626	0.8363	0.8394	82.4
30	0.99568	0.8004	0.8039	86.0

[a] 1 centipoise = 10^{-2} g/cm-s. To convert to lbf-s/ft^2 multiply centipoise by 2.088×10^{-5}

[b] 1 centistoke = 10^{-2} cm^2/s. To convert to ft^2/s multiply centistoke by 1.075×10^{-5}

TABLE D.2 Vapor pressure of water and surface tension of water in contact with air

Temperature, C	0	5	10	15	20	25	30
Vapor pressure (p_w), mm Hg	4.58	6.54	9.21	12.8	17.5	23.8	31.8
Surface tension (σ), dyne/cm	75.6	74.9	74.2	73.5	72.8	72.0	71.2

TABLE D.3 Hazen-Williams formula flow-chart.

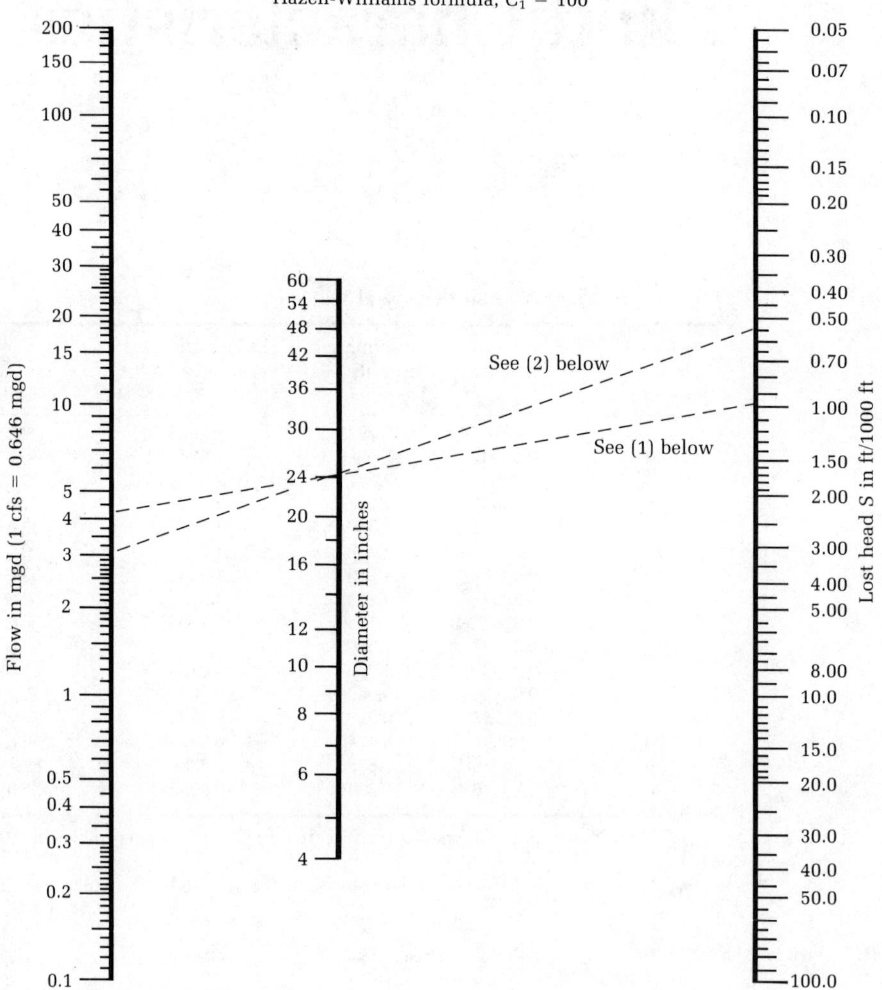

Use of Chart

(1) Given $D = 24$ in., $S = 1.0$ ft/1000 ft, $C_1 = 120$; find flow Q.
Chart gives $Q_{100} = 4.2$ mgd.
For $C_1 = 120$, $Q = (120/100)(4.2) = 5.0$ mgd.

(2) Given $Q = 3.6$ mgd, $D = 24$ in., $C_1 = 120$; find lost head.
Change Q_{120} to Q_{100}: $Q_{100} = (100/120)(3.6) = 3.0$ mgd.
Chart gives $S = 0.55$ ft/1000 ft.

Appendix E—Answers to Odd-numbered Problems*

1.3 700 kg S/week, 1543 lb S/week
1.5 0.32 ppm

2.3 4.5 cm
2.11 $+1$ g/m^2-yr, others balance

3.1 1800 kg/day
3.3 Stream C = 55 L/s; Stream D has 628 mg/L; Stream E = 5 L/s and has 59,715 mg/L; Stream A′ = 150 L/s and has 2210 mg/L
3.5 (a) 40.4% (b) 995 kW
3.7 (a) 0.50 hr (b) 213 hr (c) 3.43 m^3
3.9 81%
3.11 SO_2 = 80 tons/day; bottom ash = 333 tons/day; fly ash collected = 653 tons/day; fly ash emitted = 13 tons/day
3.13 37%, 9.22×10^8 BTU/hr, 3.98×10^9 BTU/day
3.15 95.9% and 575 kg/hr
3.17 3.87×10^8 lb/hr, 1722 CFS, 8.55 ft
3.19 1100 kg of 5% ethanol and 900 kg of 25% ethanol
3.21 6897 gal of high-octane, 3103 gal of regular
3.23 Bottom stream = 3.0 kg-mol/hr with 0.082 mole fraction benzene

Stream (1) = 454.2 kg/hr, (2) = 181.65 kg/hr, (3) = 272.55 kg/hr

* Some problems require assumptions. If so, no answer is provided.

3.25 Materials-balance table is:

Stream	*Coal*	*Water*	*Total*
1	20.31	20.31	40.62
2	0.31	15.31	15.62
3	20.0	5.0	25.0

4.1 23.1 M, 46.2 N, 3.70×10^9 meq/day
4.3 56.7 kg
4.5 11.6 mg/L
4.7 5 min (zero-order), 6.93 min (first-order), 10 min (second-order)
4.9 First order, 1100 s, 0.00062 s^{-1}, 1600 s
4.11 pKsp: 8.73 (10°C), 8.29 (25°C), 7.89 (40°C)
4.13 3380 kg
4.15 31,800 kg O_2; 53,000 kg N_2
4.17 154.3 kg/day

5.2 41.7 mg/L
5.3 4.57 mg/L O_2
5.5 2.5 mg/L-day; 4.5 mg/L-day; 7.0 mg/L-day
5.7 0.277 day^{-1}; 284 mg/L
5.9 0.43 hr^{-1}
5.11 300 mg/L; 0.094 day^{-1}
5.13 7.7 mg/L dissolved oxygen

6.1 7.2×10^7 m^3/yr; 7.56×10^7 m^3/yr
6.3 12,500 m^3; 12,500 m^2
6.5 1.37 cm/hr
6.7 Insufficient volume and depth
6.9 1650×10^3 m^3

7.1 (a) 1.18×10^4 gal (b) 6.94×10^6 gal (c) PFTR
7.3 (a) 200 mg/L (b) 100 mg/L (c) 258 mg/L (d) 158 mg/L
7.5 (a) 330 mg/L (b) 180 mg/L (c) 0 (d) 50 mg/L
(e) 150 mg/L (f) 0
7.7 More effective at pH = 7
7.9 6.26 mg/L Cl_2
7.11 3.34 m/s, 66.8 m^3/s
7.13 4.71 m

8.1 3.3 days
8.3 (a) 4.8×10^5L (b) 164 kg/day
8.5 69.5%
8.7 0.150 cm/s; 129 $\frac{m^3}{m^2\text{-day}}$

9.1 $80.4 billion
9.3 2.42×10^9 kg S; 4.84×10^9 kg SO_2
9.5 (b) 1.44×10^5 kg/hr (c) 3.72×10^{10} kJ
9.7 19.1%; 32.6%
9.9 7.4%
9.11 Maximum efficiency of nuclear plant = 49.2%; maximum efficiency of fossil fuel plant = 63.0%
9.13 161 unit trains/year

10.1 188 ppm
10.3 1.2 $\mu g/m^3$
10.5 0.50 ppm
10.7 98.25%
10.9 NO_x = 4095 lb/hr; SO_x = 8190 lb/hr; particulates = 205 lb/hr
10.11 CO = 340 kg/yr; NO_x = 100 kg/yr; VOC = 41 kg/yr
10.13 CO = 13,200 kg/day; particulates = 1150 kg/day
10.15 No

11.1 (a) 3.79×10^5 lb/day (b) 8.71×10^5 lb/day
11.7 166 mg/L CN^-
11.9 98.7%

12.1 0.00312%
12.9 (a) 173.4 g/min (b) 537 g/min (c) 1055 g/min

13.3 116 dBA
13.5 90.4 dBA

Index